Antelope Conservation

Conservation Science and Practice Series

Published in association with the Zoological Society of London

Wiley-Blackwell and the Zoological Society of London are proud to present our *Conservation Science and Practice* series. Each book in the series reviews a key issue in conservation today from a multidisciplinary viewpoint.

Each book proposal will be assessed by independent academic referees, as well as our Series Editorial Panel. Members of the Panel include:

Previously published

Elephants and Savanna Woodland Ecosystems: A Study from Chobe National Park, Botswana
Edited by Christina Skarpe, Johan T. du Toit, and Stein R. Moe
ISBN: 978-0-470-67176-4 Hardcover; May 2014

Biodiversity Monitoring and Conservation: Bridging the Gap between Global Commitment and Local Action
Edited by Ben Collen, Nathalie Pettorelli, Jonathan E. M. Baillie, and Sarah M. Durant
ISBN: 978-1-4443-3291-9 Hardcover;
ISBN: 978-1-4443-3292-6 Paperback; April 2013

Biodiversity Conservation and Poverty Alleviation: Exploring the Evidence for a Link
Edited by Dilys Roe, Joanna Elliott, Chris Sandbrook, and Matt Walpole
ISBN: 978-0-470-67478-9 Paperback;
ISBN: 978-0-470-67479-6 Hardcover; December 2012

Applied Population and Community Ecology: The Case of Feral Pigs in Australia
Edited by Jim Hone
ISBN: 978-0-470-65864-2 Hardcover; July 2012

Tropical Forest Conservation and Industry Partnership: An Experience from the Congo Basin
Edited by Connie J. Clark and John R. Poulsen
ISBN: 978-0-4706-7373-7 Hardcover; March 2012

Reintroduction Biology: Integrating Science and Management
Edited by John G. Ewen, Doug. P. Armstrong, Kevin A. Parker, and Philip J. Seddon
ISBN: 978-1-4051-8674-2 Paperback;
ISBN: 978-1-4443-6156-8 Hardcover; January 2012

Trade-offs in Conservation: Deciding What to Save
Edited by Nigel Leader-Williams, William M. Adams, and Robert J. Smith
ISBN: 978-1-4051-9383-2 Paperback; ISBN: 978-1-4051-9384-9 Hardcover; September 2010

Urban Biodiversity and Design
Edited by Norbert Müller, Peter Werner, and John G. Kelcey
ISBN: 978-1-4443-3267-4 Paperback;
ISBN: 978-1-4443-3266-7 Hardcover; April 2010

Wild Rangelands: Conserving Wildlife While Maintaining Livestock in Semi-Arid Ecosystems
Edited by Johan T. du Toit, Richard Kock, and James C. Deutsch
Edited by Glyn Davies and David Brown
ISBN: 978-1-4051-6779-6 Paperback; December 2007

Reintroduction of Top-Order Predators
Edited by Matt W. Hayward and Michael J. Somers
ISBN: 978-1-4051-7680-4 Paperback;
ISBN: 978-1-4051-9273-6 Hardcover; April 2009

Recreational Hunting, Conservation, and Rural Livelihoods: Science and Practice
Edited by Barney Dickson, Jonathan Hutton, and Bill Adams
ISBN: 978-1-4051-6785-7 Paperback;
ISBN: 978-1-4051-9142-5 Hardcover; March 2009

Participatory Research in Conservation and Rural Livelihoods: Doing Science Together
Edited by Louise Fortmann
ISBN: 978-1-4051-7679-8 Paperback; October 2008

Bushmeat and Livelihoods: Wildlife Management and Poverty Reduction
Edited by Glyn Davies and David Brown
ISBN: 978-1-4051-6779-6 Paperback; December 2007

Managing and Designing Landscapes for Conservation: Moving from Perspectives to Principles
Edited by David Lindenmayer and Richard Hobbs
ISBN: 978-1-4051-5914-2 Paperback; December 2007

Protected Areas: Are They Safeguarding Biodiversity?
Edited by Lucas Joppa, Jonathan Baillie and John Robinson
ISBN: 9781118338155 Paperback;
ISBN: 9781118338162 Hardcover; March 2016

Conservation Science and Practice Series

Antelope Conservation: From Diagnosis to Action

Edited by

Jakob Bro-Jørgensen
David P. Mallon

WILEY Blackwell

ZSL
LIVING CONSERVATION

Registered Office
John Wiley & Sons, Ltd, The Atrium, Southern Gate, Chichester, West Sussex, PO19 8SQ, UK

Editorial Offices
9600 Garsington Road, Oxford, OX4 2DQ, UK
The Atrium, Southern Gate, Chichester, West Sussex, PO19 8SQ, UK
111 River Street, Hoboken, NJ 07030-5774, USA

For details of our global editorial offices, for customer services and for information about how to apply for permission to reuse the copyright material in this book please see our website at www.wiley.com/wiley-blackwell.

Library of Congress Cataloging-in-Publication data applied for

ISBN Hardback: 9781118409640
ISBN Paperback: 9781118409633

A catalogue record for this book is available from the British Library.

Wiley also publishes its books in a variety of electronic formats. Some content that appears in print may not be available in electronic books.

Set in 10.5/12.5pt Minion by SPi Global, Pondicherry, India

1 2016

Contents

Contributors

Rajan Amin Conservation Programmes, Zoological Society of London, United Kingdom

Anagaw Atickem Centre for Ecological and Evolutionary Synthesis (CEES), Department of Biosciences, University of Oslo, Norway; Department of Zoological Science, Addis Ababa University, Addis Ababa, Ethiopia

Andrew E. Bowkett Field Conservation and Research Department, Whitley Wildlife Conservation Trust, Paignton Zoo, United Kingdom

Jakob Bro-Jørgensen Mammalian Behaviour and Evolution Group, Department of Evolution, Ecology and Behaviour, Institute of Integrative Biology, University of Liverpool, United Kingdom

Nils Bunnefeld School of Natural Sciences, Biological and Environmental Sciences, University of Stirling, Scotland, United Kingdom

Philippe Chardonnet International Game Foundation, Paris, France

Jacobus du Toit Du Toit Wildlife Services, Pretoria, Republic of South Africa

Sarah M. Durant Institute of Zoology, Zoological Society of London, United Kingdom

John E. Fa Division of Biology and Conservation Ecology, School of Science and the Environment, Manchester Metropolitan University, United Kingdom; Center for International Forestry Research (CIFOR), CIFOR Headquarters, Bogor, Indonesia

Sam Ferreira Conservation Services, South African National Parks, Republic of South Africa

Adam T. Ford Department of Integrative Biology, University of Guelph, Canada

John M. Fryxell Department of Integrative Biology, University of Guelph, Canada

Angela Gaylard Conservation Services, South African National Parks, Republic of South Africa

Tania Gilbert Marwell Wildlife, Colden Common, Winchester, United Kingdom

Christina Grant Conservation Services, South African National Parks; Centre for African Conservation Ecology, Nelson Mandela Metropolitan University, Port Elizabeth, Republic of South Africa

Marna Herbst Conservation Services, South African National Parks, Republic of South Africa

Markus Hofmeyr Conservation Services, South African National Parks, Republic of South Africa

Stephen Holness Conservation Services, South African National Parks, Republic of South Africa; Centre for African Conservation Ecology, Nelson Mandela Metropolitan University, Port Elizabeth, Republic of South Africa

J. Grant C. Hopcraft Institute of Biodiversity, Animal Health and Comparative Medicine, University of Glasgow, Scotland, United Kingdom

Michael H. Knight Conservation Services, South African National Parks, Republic of South Africa; Centre for African Conservation Ecology, Nelson Mandela Metropolitan University, Port Elizabeth, Republic of South Africa

Richard Kock Department of Pathology & Infectious Diseases, Royal Veterinary College, Hatfield, United Kingdom

Leif E. Loe Norwegian University of Life Sciences, Ås, Norway

Eline D. Lorenzen Natural History Museum of Denmark, University of Copenhagen, Denmark

E. J. Milner-Gulland Department of Zoology, University of Oxford, United Kingdom

John Newby Sahara Conservation Fund, L'Isle, Switzerland

Peter Novellie Conservation Services, South African National Parks, Republic of South Africa; Sustainability Research Unit, Nelson Mandela Metropolitan University, George, Republic of South Africa

Nathalie Pettorelli Institute of Zoology, Zoological Society of London, United Kingdom

Mark R. Stanley Price WildCRU, The Recanati-Kaplan Centre, Oxford University, United Kingdom

Herbert H. T. Prins Resource Ecology Group, Wageningen University, Wageningen, The Netherlands

Claire Risley Institute of Biological, Environmental and Rural Sciences (IBERS), Aberystwyth University, Wales, United Kingdom

Eli K. Rueness Centre for Ecological and Evolutionary Synthesis (CEES), Department of Biosciences, University of Oslo, Norway

Anthony R. E. Sinclair Biodiversity Research Centre, University of British Columbia, Canada

Navinder J. Singh Department of Wildlife, Fish and Environmental Studies, Faculty of Forest Sciences, Swedish University of Agricultural Sciences, Sweden

Nils C. Stenseth Centre for Ecological and Evolutionary Synthesis (CEES), Department of Biosciences, University of Oslo, Norway; Department of Zoological Science, Addis Ababa University, Addis Ababa, Ethiopia

Tim Wacher Conservation Programmes, Zoological Society of London, United Kingdom

Tim Woodfine Marwell Wildlife, Colden Common, Winchester, United Kingdom

Preface and Acknowledgements

From hot deserts to snowy plains, from lowland rainforests to mountain peaks, antelopes form a critical part of natural ecosystems throughout Africa and Asia. They are distributed in every country of mainland Africa and across the Arabian Peninsula, Middle East to Central Asia, extending west of the Caspian Sea in Russia, eastwards to Mongolia and northern China, and southwards to the Indian subcontinent and all the way to the Indochinese peninsula. They occur at elevations from 150 m below sea level at Lac Assal in Djibouti to more than 5,000 m above on the Qinghai-Tibet Plateau. Their habitats encompass hyperarid deserts, high-altitude cold desert, semi-deserts, steppes, open bush, dry scrub, deciduous woodland, humid rainforest, wetlands, swamps, rocky outcrops as well as mountains.

However, this fascinating and diverse group of animals is under increasingly severe threat of further species extinctions and tends to slip under the conservation radar compared to many other aspects of biodiversity. The addax and dama gazelle are currently very close to becoming extinct in the wild, and a few other species have been reduced to perilously low numbers. At the other end of scale, wildebeest and Mongolian gazelle still number over one million, and the annual migration of the common wildebeest remains one of the world's most impressive wildlife spectacles. But for most species, the vast herds of the recent past have disappeared, and almost everywhere antelope numbers are declining and their ranges shrinking and fragmenting in the face of over-harvesting, conversion of land to agriculture and over-grazing.

Although important work is being done by passionate conservationists and researchers, as a community antelope specialists appear less vocal compared to many other interest groups - and conservation concerns relating to this magnificent facet of the wild are often overlooked even though they are in many ways central to the current conservation debate. To galvanize the community of antelope researchers, we brought together leading experts in antelope conservation at a highly stimulating two-day symposium at Zoological Society of London (ZSL) in November 2011. The meeting included researchers and conservationists as well as students and lay people from across the globe. The forum enabled engaging discussions and exchange of views on a range of topics, from common best practices in conservation to specific concerns pertaining to antelopes. This book is inspired by the meeting and reflects its key themes, even if the priority of getting a balanced book addressing general conservation issues means that the contributing authors do not correspond exactly with the speaker list at the meeting.

We would like to extend our deepest gratitude to the Zoological Society of London who made the meeting possible by providing an atmospheric venue and financial support. The organizational team, especially Linda DaVolls and Joy Hayward, were a pleasure to work with. Likewise, we are indebted to Wiley-Blackwell whose patience and support during the preparation of this book have been critical in our efforts to ensure that the coverage of the topic became as comprehensive as possible; Ward Cooper, Kelvin Matthews and Kavitha Chandrasekar and her team here deserve a special mention. All chapters have undergone rigorous scientific peer review by experts in the various topics covered, and we are indebted to all the reviewers, who shall remain anonymous, for their often extremely thorough and insightful reviews. We are also deeply grateful to all the photographers who kindly allowed their exquisite photos to be reproduced on the colour plates. Finally, we would like to thank all the speakers at the 2011 meeting and all the contributors to this volume for their thought-provoking contributions.

Some authors have special acknowledgements. The study in Chapter 6 was supported by the European Commission under the HUNT project of the 7th Framework Programme for Research and Technological Development, and EJMG further acknowledges the support of a Royal Society Wolfson Research Merit award. Rajan Amin, Andrew E. Bowkett, and Tim Wacher thank their colleagues at the Kenya Wildlife Service in relation to case study 1 in Chapter 11, and, for presenting this study at the ZSL symposium on their behalf, Olivia Needham. Mark R. Stanley Price acknowledges that collating accurate information on antelope reintroductions for Chapter 12 required the assistance of many people and with pleasure thanks the following who helped: Teresa Abaigar Ancín, Karolina Brandlová, Donald Bunge, Hamish Currie, Ehab Eid, Tania Gilbert, Brahim Haddane, Haut Commissariat aux Eaux et Forêts et à la Lutte Contre la Désertification (Morocco), Louisa van Huyssteen, Zafar-ul Islam, Mansoor Al Jahdhami, David Mallon, Charles Musyoki, Widade Oubrou, Marie Petretto, Husam el Alqamy, Paul Reillo, Mick Reilly, David Saltz, Greg Simkins, Jake Veasey, Tim Wacher, and Tim Woodfine. John Newby, Tim Wacher, Sarah M. Durant, Nathalie Pettorelli and Tania Gilbert wish to thank Mr Vincent Turmine for preparation of the map of the Sahelo-Saharan zone, Mike Hoffmann for assistance with the distribution maps, and Amina Fellous for help with the bibliography for Chapter 13. For Chapter 15, E. J. Milner-Gulland and Navinder J. Singh acknowledge the support from the Leverhulme Trust and a Royal Society Wolfson Research Merit Award to EJMG. They are grateful to all their colleagues who have worked on the ecology and conservation of the saiga antelope over many years, 'making significant, but sometimes underappreciated, contributions to the understanding of saiga biology and conservation'. They acknowledge in particular the work of their collaborators and friends Yuri Arylov, Amankul Bekenov, Amgalan Buyana, Elena Bykova, Alexander Esipov, Yuri Grachev, Anna Lushchekina, and Valery Neronov. Jakob Bro-Jørgensen is grateful to E. J. Milner-Gulland for inspiring comments.

Jakob Bro-Jørgensen
David P. Mallon

Foreword

The outlook for conservation of antelopes and other large mammals in this century depends on the willingness of mankind to provide safe havens for wildlife. Maintaining and increasing existing protected areas is essential. The words of the Triennial Report of the IUCN Antelope Specialist Group (ASG) in 1988–1990 hold just as true today, a quarter of a century later: '*Preservation of any substantial areas of wildlife habitat outside of existing parks and reserves depends on making use of antelope and other wildlife on a sustainable basis more rewarding than other land uses. If their value can be demonstrated in time, it would still be possible to set aside buffer zones surrounding parks that would be accepted as inviolate sanctuaries necessary for renewing stocks of species to be hunted for food or trophies in designated hunting areas.... Nearly all problems of wildlife conservation stem from the continuing uncontrolled growth and impoverishment of the human population.*' In Africa – a continent of paramount importance for antelope conservation – the human population has now passed one billion and continues to increase at 2–3% a year. Huge amounts of natural habitat have already been transformed to meet human needs. Even parks and game reserves set aside for wildlife are under pressure, and development between and around them is turning them into islands. Hemmed in by settlement, wildlife in these protected areas will be unable to extend their range in response to climate change.

Faced with this advanced stage of habitat attrition, the present book is a valuable source of information on central issues in antelope conservation, and the varied contributions provoke reflections on how best to safeguard biodiversity for the future. Reflecting on my own experience as Chair of the ASG between 1978 and 2005, three priorities appear particularly important to slow or even stop the elimination of wildlife habitat in the antelope range states: making wildlife utilization more valuable than competing forms of land use, supporting the growing number of Transboundary Conservation Areas (TBCAs) jointly managed by neighboring countries, and combatting climate change.

The willingness of landowners to save natural habitat and tolerate wildlife ultimately depends on whether the benefits outweigh the costs. Most antelopes live in developing countries occupied by people living at subsistence level. How can communal and private landowners be persuaded to share their property with antelopes and other large herbivores that compete for forage and space with their livestock and crops? One option to make wildlife a more valuable resource is giving landowners increasing ownership. In 1975, Zimbabwe passed a Parks and Wildlife Act that conferred custodianship of wildlife on their land to landholders, followed in 1982 by the CAMPFIRE program that allowed rural communities to benefit economically from their wildlife. The success of these programs led neighboring countries to follow suit. By the end of the 1980s, most of South Africa's larger antelope species were more numerous than at any time since the 1800s (Estes & East, 2009). Neighboring landowners formed large wildlife conservancies and hunting blocks by taking down dividing fences, both in southern and eastern Africa.

Moreover, a guaranteed share of income from wildlife-based tourism should benefit the communities bordering parks and game. Communal and private landowners may also be persuaded to protect wildlife by subsidy, ideally implemented by an effective new international agency. It will be merely one of the expensive but essential commitments by mankind to safeguard natural habitat otherwise sure to be transformed by the ever-growing human footprint. World Heritage Sites and Man and Biosphere Reserves should head the list.

Incorporating vast landscapes encompassing both protected and unprotected areas, TBCAs could play a key role in restoring fragmented ecosystems of migratory ungulates. In southern Africa, TBCAs (here also known as 'Peace Parks') are intended not only for the conservation and sustainable use of biological and cultural resources; they also have the objective of facilitating and promoting regional peace and cooperation and socioeconomic development. Since the Global Transboundary Conservation Network was formed at the 2003 World Parks Congress, over 200 TBCAs have been catalogued in the UNEP/WCMC database.

With such initiatives, there is hope that we can also restore mutually beneficial, harmonic associations between wild and domestic ungulates on open rangelands including buffer zones bordering parks, reserves and game-controlled areas (Marshall, 1990; Homewood *et al.*, 2001; Reid, 2012). Livestock and wildlife have shared the savannas for thousands of years; why give up hope that they can do so again? Transhumance worked and can work again if livestock numbers can be controlled to prevent habitat degradation. Instead of counting wealth in the number of livestock however, pastoralists stand to gain far greater wealth by satisfying the demand for meat in cities, where much poached bushmeat is currently marketed. Combining wildlife and livestock production on private property, as is common in southern Africa, can be very profitable and could work for pastoralists on public land and within some protected areas, the Ngorongoro Conservation Area offering a possible model (Estes, 2014). Where projections of increasingly arid regional climates come true, farmers may indeed find their land better suited to pastoralism than agriculture.

But all these measures may be too little, too late. Suddenly we find that global warming is proceeding faster than predicted. President Obama of the United States, the world's leading economic power and second-largest emitter of greenhouse gases, has sounded the alarm, seconded by Pope Francis: '*Human activity is disrupting the climate in many ways faster than we previously thought. The science is stark. It is sharpening. It proves that this once distant threat is now very much in the present. It is happening here, it is happening now. Already disrupting our agriculture and ecosystems, our water and food supplies, our energy, our infrastructure, human health, human safety, now, today*' (Obama, 2015). Maintaining biodiversity needs to be closely linked with efforts to combat global climate change. Conservation of antelopes and the other large charismatic mammals of Africa's and Asia's rangelands would go a long way toward conserving the planet's biodiversity, given their role as umbrella and keystone species. The Bovidae, the last and greatest mammalian radiation, can show the way.

Richard D. Estes

References

Estes, R. D. (2014): *The gnu's world: Serengeti wildebeest ecology and life history*. Berkeley: University of California Press.

Estes, R. D. & East, R. (2009): Status of the wildebeest (*Connochaetes taurinus*) in the wild, 1967–2005. Working Paper 37. Bronx, NY: Wildlife Conservation Society.

Homewood, K., Lambin, E. F., Coast, E., Kariuki, A., Kikula, I., Kivelia, J., Said, M., Serneels, S. & Thompson M. (2001): Long-term changes in Serengeti-Mara wildebeest and land cover: pastoralism population or policies? *Proceedings of the National Academy of Sciences* **98**: 12544–12549.

Marshall, F. B. (1990): Origins of specialized pastoral production in East Africa. *American Anthropologist* **92**(4): 873–894.

Obama, B. (2015): Remarks by the president at the GLACIER Conference – Anchorage, AK.https://www.whitehouse.gov/the-press-office/2015/09/01/remarks-president-glacier-conference-anchorage-ak (as at 20 February 2016).

Reid, R.S. (2012): *Savannas of our Birth: People, Wildlife, and Change in East Africa*. Berkeley: University of California Press.

1

Our Antelope Heritage – Why the Fuss?

Jakob Bro-Jørgensen

Mammalian Behaviour and Evolution Group, Department of Evolution, Ecology and Behaviour, Institute of Integrative Biology, University of Liverpool, United Kingdom

Introduction

Why a book dedicated to antelope conservation? Our planet has witnessed a decrease of more than 50% in its vertebrate populations since 1970, and this drastic decline has hit antelopes particularly hard, according to the Living Planet Index (BBC, 2008; McLellan, 2014; see also Craigie *et al.*, 2010). Many will agree that antelopes constitute an outstanding aspect of the world's biodiversity and that the prospect of losing this heritage is a concern in its own right. A savanna bereft of flickering herds of gazelles (Figure 1) or a rainforest where duikers no longer lurk in the understorey may be likened to bodies that have lost their souls. But leaving subjective sentiments aside, antelopes are also of fundamental importance for the functioning of many ecosystems across Africa and Asia. They have important roles as architects of habitats, as dispersers of seed, as the prey base for endangered carnivores and indeed in nutrient cycling in general (Sinclair & Arcese, 1995; Sinclair *et al.*, 2008; Gallagher, 2013). Maintaining healthy antelope populations is therefore vital for the management of many ecosystems, and the motivation for this book comes from an urgent concern not only at the species level but also relating to wider repercussions at the ecosystem level.

Antelopes moreover provide a well-suited model to obtain insights into the operation of threat processes affecting wildlife populations more generally. Because they share the same basic biology, yet display a striking variation in habitats and threats, this species-rich group presents an extraordinary opportunity to pinpoint how human impact on wildlife populations depends on the interaction between threats and specific species traits. Many of the issues facing antelopes are central to

Antelope Conservation: From Diagnosis to Action, First Edition. Edited by Jakob Bro-Jørgensen and David P. Mallon.

Figure 1 **Thomson gazelles and impalas in Maasai Mara National Reserve, Kenya (© Jakob Bro-Jørgensen).**

the current conservation debate, including the sustainable use of wildlife (for meat and trophies), protection of migratory as well as highly habitat-specific species in a world of climate change and habitat fragmentation, and the coexistence of wildlife with people and their livestock without conflict. Typically, antelope conservation takes place in developing countries with growing human populations and severely under-resourced wildlife authorities, which brings the issue of how to integrate conservation and development to the forefront. Valuable long-term data sets are present for several antelope species, placing them in a strong position to provide some general lessons for conservation biology, especially in relation to the particular challenge of preserving large mammals (MacDonald *et al.*, 2013).

However, following a surge in pioneering field studies of many antelope species in the 1960s and 1970s, the reality is that antelope research seems to have lost its general appeal, and the attention from the general public is modest compared to that received by many of their mammalian relatives, such as carnivores and primates, which are widely seen as more charismatic. This book is intended to reinvigorate the interest in antelope research and give a deeper understanding of the threat drivers facing antelopes today, thereby providing a basis for reflection on common best practices in conservation. As a background, this introductory chapter will first take an evolutionary perspective to understanding the ecological importance of global antelope biodiversity and then outline the current conservation status of this world heritage.

Antelopes – an evolutionary success story … so far

A green world presents a tremendous opportunity for the evolution of efficient plant-eaters, and here antelopes have been an extraordinary success story. A major evolutionary breakthrough took place in the Eocene some 50 Myrs BP when the compartmentalized ruminant stomach evolved (Fernández & Vrba, 2005). This enabled a more efficient breakdown of fibrous plant material by chewing cud and using microbial symbionts to digest cellulose. The antelopes are members of the ruminant family Bovidae, characterized by permanent horns consisting of a bone core covered by a sheath of keratin. The first known bovid fossil, *Eotragus,* dates back to the early Miocene some 20 Myrs BP (Gentry, 2000; Fernandez & Vrba, 2005), and since then, an astonishing adaptive radiation has taken place as bovid species have evolved to occupy a wide range of ecological niches. The majority of these species are antelopes: 88 extant species are represented by 14 species in Asia and 75 species in their main stronghold in Africa, with only the dorcas gazelle (*Gazella dorcas*) found on both continents. Antelopes vary in size from the 1.5 kg of a royal antelope (*Neotragus pygmaeus*) (Plate 3) to nearly a ton in a full-grown giant eland bull (*Tragelaphus derbianus*) (cover).

So what distinguishes antelopes? Treating antelopes as a group is questionable from a strict evolutionary perspective because it violates the ideal of keeping together all species descending from a given distinctive ancestor. The group is created by cutting off two distinct monophyletic branches from the bovid tree: (i) the wild oxen *Bovini,* characterized by their heavier build and water-dependence, and (ii) the wild goats and sheep *Caprinae,* characterized by their extreme adaptation to rocky habitats (Figure 2). However, antelopes are not defined only by what they are not (i.e., as a bovid that is neither an oxen nor a goat). They can be succinctly described as horned ruminants lightly built for swift movement in habitats with predominantly even ground. This has resulted in a characteristic graceful and elegant morphology, often adorned with spectacular ornaments and weapons due to strong sexual selection in the more social species (Stoner *et al.*, 2003; Bro-Jørgensen, 2007).

The broad array of ecological adaptations in antelopes is apparent when considering the variety between the 12 tribes (Plates 1, 2, & 3). The spiral-horned antelopes of Africa **Tragelaphini** (elands, kudus, nyalas and allies), together with their Asian relatives **Pseudorygini** (saola *Pseudoryx nghetinhensis*) and **Boselaphini** (nilgai *Boselaphus tragocamelus,* four-horned antelope *Tetracerus quadricornis*), represent a highly diverse ancient line from within which the wild oxen descended. Except for the browsing saola, they are mixed feeders; that is, feeding on both browse and grass. They vary more than tenfold in size and are found from dense forests (bongo *Tragelaphus eurycerus,* saola) to semi-deserts (common eland *Tragelaphus oryx*), and from swamps

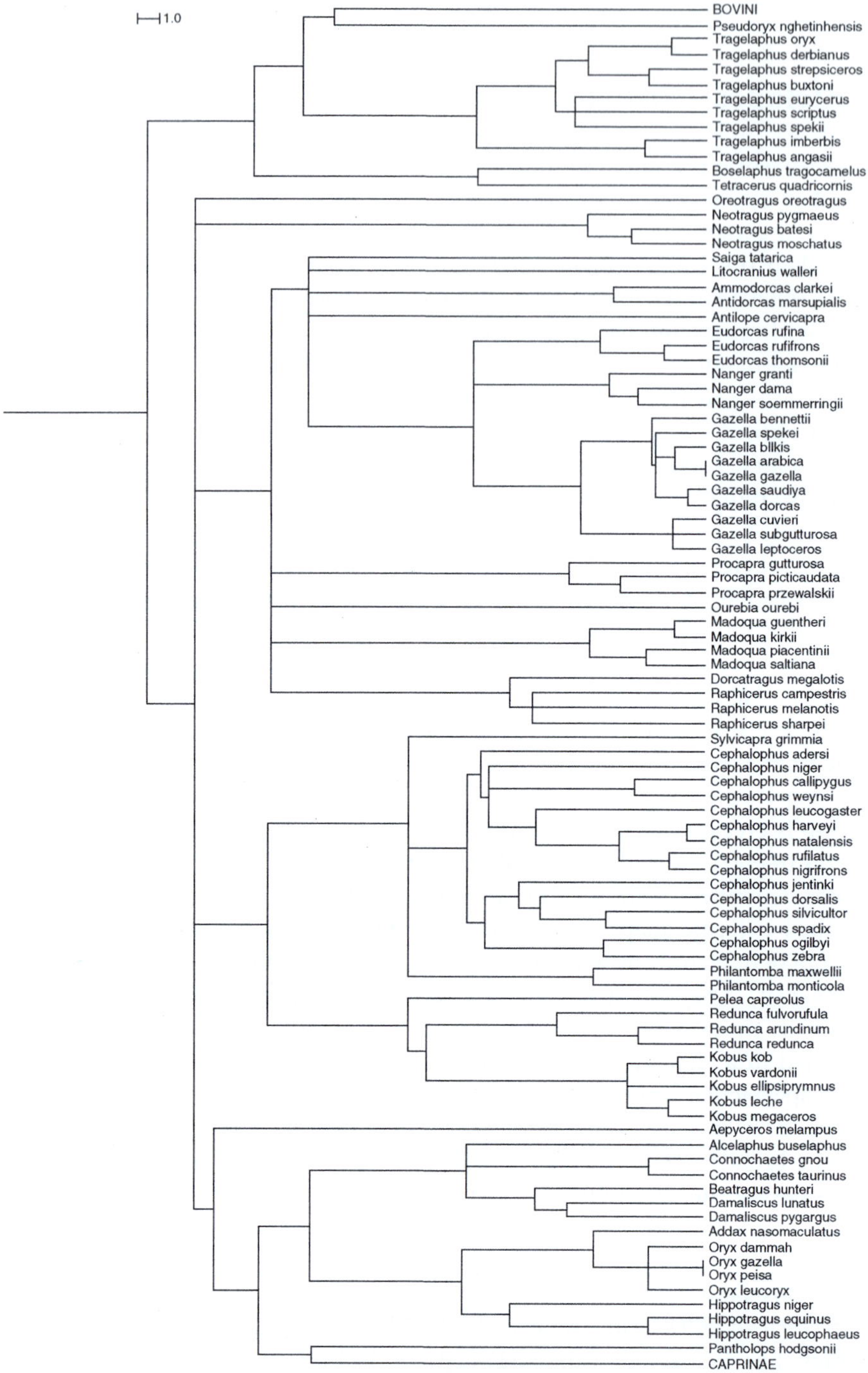

Figure 2 **The evolution within Bovidae since the divergence from deer 32 million years ago (for common names, see the Appendix). Bar indicates one million years. Based on Fernández & Vrba 2005; drawn in Dendroscope, Huson & Scornavacca 2012.**

(sitatunga *Tragelaphus spekii*) to mountains (mountain nyala *Tragelaphus buxtoni*). Other mixed feeders include the arid-adapted gazelles **Antilopini** which span from hot to rather cold regions, and the horse antelopes **Hippotragini,** which predominantly graze and occur from relatively moist savannas (roan *Hippotragus equinus* and sable antelope *Hippotragus niger*) to semi-deserts (oryxes) and deserts (addax *Addax nasomaculatus*). Both the latter tribes have representatives in Africa as well as Asia. Also mixed-feeders, the African impala (*Aepyceros melampus*) and rhebok (*Pelea capreolus*) are the only living representatives of the tribes **Aepycerotini** and **Peleini** respectively. The grazing tribes include the reduncines **Reduncini** (lechwes, reedbucks and allies), adapted to relatively moist savannas and wetlands, and the alcelaphines **Alcelaphini** (wildebeests and allies), adapted to drier savannas; both are exclusively African. The Tibetan antelope (*Pantholops hodgsonii*), the only representative of the caprine-related **Pantholopini**, also feeds on grass, as well as herbs, on the often snowy steppes of the Tibetan Plateau. Smaller antelopes include the duikers **Cephalophini,** which are adapted to the ecology of African forests, where they feed on high-quality browse and fruits, and the dwarf antelopes **Neotragini** which are ecologically diverse, mainly browsers and frugivores, but some also feeding on grass (notably the oribi *Ourebia ourebi*), and inhabiting a wide range of habitats spanning from forests (royal antelope) and thickets (suni *Neotragus moschatus*, dik-diks), to rocky outcrops (klipspringer *Oreotragus oreotragus*, beira *Dorcatragus megalotis*) and fairly open savannas (oribi); several neotragines are actually likely to be more closely related to gazelles than to the genus *Neotragus*. In contrast to the gregarious species of the open land, the smaller species in dense habitats are usually solitary or found in groups of minimal size (Jarman, 1974; Brashares *et al.*, 2000).

Antelopes as an integral part of the structure and function of ecosystems

In an evolutionary and ecological sense, antelopes have thus been an immensely successful group, occupying a remarkable range of habitats. Moreover, within each habitat, a proliferation of species often occupies distinct niches in terms of their diet and antipredator behaviour. For example, 16 species coexist alongside each other in the Serengeti-Mara ecosystem. Throughout Africa and Asia, antelopes often dominate the community of larger herbivores in undisturbed wilderness areas. Their numerical abundance – at least historically – combined with their long period of coevolution with plants and predators means that they are intrinsically linked to the function of the ecosystems they inhabit. Some of their ecological roles are fairly obvious whereas other important links are more subtle and indirect and some dynamics undoubtedly still await discovery.

Antelopes have a major impact on both the structure and function of the plant community. In some cases, the loss of antelope populations may even cause wilderness areas to switch from one biome to another. For example, the grazing pressure from the great migration of wildebeest in Serengeti-Mara is crucial for maintaining the open landscape to which the wider savanna community is adapted. In the absence of wildebeest, thickets proliferate, and the whole system could eventually reach an ecological tipping point where the habitat becomes unfavourable for today's rich community of grazers and gravitates towards an alternative, more wooded state (Sinclair *et al.*, 2007). Antelopes may also have important effects on the vegetation that are less conspicuous. For example, impala distribute themselves in a 'landscape of fear' as they avoid areas of thick cover due to high predation risk from leopards and hunting dogs (Ford *et al.*, 2014). As a consequence, impala browsing pressure on acacia is highest in open habitats, and this gives acacia species protected by thorns a competitive advantage in such areas. In this way, browsing by impala has been shown to shape the spatial structure of the woody community of African savannas (Ford *et al.*, 2014).

Antelopes can also have a profound effect on the vegetation by acting as seed dispersers. Frugivores in forest habitats, such as the duikers, are highly important in this regard (Jordano, 2013). They act as vectors of seeds, and seed germination may even depend on being passed through the gut of an antelope consumer. In tropical forests, many of the most carbon-rich hardwood trees rely on animals such as forest antelopes for their dispersal, and loss of seed-dispersers through bushmeat hunting has been linked to a reduction in hardwoods (Brodie & Gibbs, 2009). Because hardwoods are particularly important in sequestering CO_2, this could compromise the role of the forest as a carbon sink, which in turn reduces its potential to mitigate the adverse effects of climate change.

Antelopes are of crucial importance also as a prey base for larger predators: without thriving antelope populations, efforts to preserve carnivores will often make little sense. From a management perspective, it is important to recognize the intricate relationships between predators and their prey. Predator species show marked differences in their prey preference profiles. For instance, lions (*Panthera leo*) prefer large, relatively slow prey species that are not suitable prey for smaller predators (Sinclair *et al.*, 2003). In turn, cheetahs (*Acinonyx jubatus*) prefer smaller, but fast prey species that are less preferred by lions (Hayward *et al.*, 2006b), while leopards (*Panthera pardus*), ambush predators, also prefer smaller, but slower prey (Hayward *et al.*, 2006a). Such relationships are the result of long-term coevolutionary processes (Bro-Jørgensen, 2013), and it is unreasonable to expect that different prey species can readily substitute for each other. A decline in the population size of one species can have knock-on effects on others, and to maintain natural ecosystem dynamics the full breadth of species diversity within both predator and prey communities requires conservation.

Threats facing antelopes today

As a key component of natural ecosystems, antelopes are an integral part of global life support systems. In areas of poverty, they can directly benefit human livelihoods as sources of food for subsistence or sale and through other income-generating activities such as ecotourism and trophy hunting. The physiological efficiency and high productivity of bovids is shown by the fact that the taxon includes the ancestors of the most important domesticated livestock: that is, cattle, sheep and goats. Yet, the evolutionary potential of antelopes and the ecosystem services they provide are usually grossly undervalued in the formal economy, and human development therefore takes place without the relevant costs from squandering areas of wilderness being integrated into land use planning.

Consequentially, human activities are rapidly decimating many of the remaining antelope populations: 31% (27/88) of the extant antelope species assessed by the IUCN Red List are now formally categorized as threatened (including 64% [9/14] of the Asian species) and a further 9% (8/88) as near-threatened (IUCN 2015). The extinction in the wild of the scimitar-horned oryx (*Oryx dammah*) in year 2000, and the global extinction of the bluebuck (*Hippotragus leucophaeus*) in 1800, and probably also the kouprey (*Bos sauveli*) in recent years, clearly point to the serious danger that further bovid extinctions are imminent. Particular hot spots of highly threatened species include the desert regions of North Africa, the horn of Africa, the West African rainforests and the Asian steppes. Taxonomically, species with high threat status are dispersed throughout the phylogeny. Conservation concerns are not limited to red-listed species: the population trend is decreasing for 64% (54/84) of all the species assessed, stable for 33% (28/84) and increasing for only 2% (2/84) (i.e., the springbok *Antidorcas marsupialis* and black wildebeest *Connochaetes gnou* in Southern Africa). As many as 76% (67/88) of all species are threatened by exploitation through hunting and trapping primarily for meat, but also for horns (used predominantly as trophies and in traditional medicine), hides and – specifically in the Tibetan antelope – underfur ('shahtoosh') used for shawls. Various human land-use changes affect 69% (61/88) of species, practically all of which are simultaneously affected by exploitation; specifically, 45% (40/88) are affected by livestock farming and ranching, and 48% (42/88) are affected by encroaching human settlements. In addition, 13% (11/88) of species are threatened by war or other civil unrest; half of these are in the Horn of Africa and also the Sudano-Sahelian savannas belt is severely affected. Currently, 18% (16/88) of species are referred to as affected by climate change, but our knowledge in this area is still limited, and the figure may rise as more information becomes available (Akçakaya *et al.*, 2014; Payne & Bro-Jørgensen, 2016).

Outline of this book

In summary:

- Antelopes are a high conservation priority of significant ecological importance
- Multiple threats face this ecologically diverse set of species
- Conservation generally takes place in developing economies with growing human populations so social sustainability of any conservation action is a priority

Given these conditions, which approaches can most effectively secure antelope populations into the future? The chapters in this book seek a deeper understanding of the key threat processes facing antelopes today and critically evaluate the various options for action. Whereas a broad consensus emerges on several issues, a diversity of opinion also manifests itself on certain points, reflecting the varied experience of the authors. To begin with, Chapter 2 provides an overview of ecosystem functioning and conservation challenges pertaining to savannas, a habitat of vital importance for antelope biodiversity. Chapter 3 goes on to present a conceptual framework for understanding what regulates antelope populations in natural ecosystems and uses this insight to explore the potential impact of climate change alongside other threat drivers. Following on from this, Chapter 4 focuses specifically on interspecific interactions over resources and provides a critical review of the current evidence that competition and facilitation significantly affect antelope population performance. Chapter 5 reviews the role of disease in antelope ecology, both as part of natural systems and as a threat associated with human activities.

In Chapter 6, attention turns to human exploitation of antelope populations with a review of the conservation impact of subsistence hunting of antelopes for meat, emphasising forest systems. Next, Chapter 7 examines the potential of trophy hunting to contribute to antelope conservation. Considering a broader set of management interventions, Chapter 8 takes its outset in the South African context and discusses the usefulness of a range of options to promote antelope conservation. Chapter 9 in turn outlines ways in which molecular techniques can be applied to inform antelope conservation; and Chapter 10 focuses specifically on the application of landscape genetics as a tool in conservation. Chapter 11 introduces another novel conservation technique, the use of camera-trapping in population monitoring. Chapter 12 provides a review of the use of reintroduction in antelope conservation, and Chapters 13 and 14, by concentrating on the critical conservation status of Sahelo-Saharan desert antelopes, stresses the urgent need for action to preserve the most threatened antelopes. Rounding off, Chapter 15 reflects, based on experience from saiga (*Saiga tatarica*) conservation, on the factors that can create opportunities and present obstacles when it comes to safeguarding antelope populations in practice. Finally in Chapter 16, key challenges facing antelope conservation over the next century are summarized in a synthesis.

References

Akçakaya, H. R., Butchart, S. H. M., Watson, J. E. M. & Pearson R. G. (2014): Preventing species extinctions resulting from climate change. *Nature Climate Change* **4**: 1048–1049.

BBC (2008): Wildlife populations 'plummeting'. http://news.bbc.co.uk/1/hi/uk/7403989.stm (as at 2 July 2015).

Brashares, J. S., Garland Jr, T. & Arcese, P. (2000): Phylogenetic analysis of coadaptation in behaviour, diet, and body size in the African antelope. *Behavioral Ecology* **11**: 452–463.

Bro-Jørgensen, J. (2007): The intensity of sexual selection predicts weapon size in male bovids. *Evolution* **61**: 1316–1326.

Bro-Jørgensen, J. (2013): Evolution of sprint speed in African savannah herbivores in relation to predation. *Evolution* **67**: 3371–3376.

Brodie, J. F. & Gibbs, H. K. (2009): Bushmeat hunting as climate threat. *Science* **326**: 364–365.

Craigie, I. D., Baillie, J. E. M., Balmford, A., Carbone, C., Collen, B., Green R. E. & Hutton J. M. (2010): Large mammal population declines in Africa's protected areas. *Biological Conservation* **143**: 2221–2228.

Fernández, M. H. & Vrba, E. S. (2005). A complete estimate of the phylogenetic relationships in Ruminantia: a dated species-level supertree of the extant ruminants. *Biol. Rev. Camb. Philos. Soc.* **80**: 269–302.

Ford, A. T., Goheen, J. R, Otieno, T. O., Bidner, L., Isbell, L. A., Palmer, T. M., Ward, D., Woodroffe, R. & Pringle, R. M. (2014): Large carnivores make savanna tree communities less thorny. *Science* **346**: 346–349.

Gallagher, R. S. (Ed.): (2013): *Seeds: The Ecology of Regeneration in Plant Communities* (3rd edn). New York: CABI Publishing.

Gentry, A. W. (2000): The ruminant radiation. In *Antelopes, deer, and relatives*: 11–25. Vrba E. S. & Schaller G. B. (Eds). New Haven: Yale University Press.

Hayward, M. W., Henschel, P., O'Brien, J., Balme, G., Kerley, G. I. H. (2006a): Prey preferences of the leopard (*Panthera pardus*). *Journal of Zoology* **270**: 298–313.

Hayward, M. W., Hofmeyr, M., O'Brien, J. & Kerley, G. I. H. (2006b): Prey preferences of the cheetah (*Acinonyx jubatus*) (Felidae: Carnivora): morphological limitations or the need to capture rapidly consumable prey before kleptoparasites arrive? *J. Zool.* **270**: 615–627.

Huson, D. H. & Scornavacca C. (2012): Dendroscope 3- an interactive viewer for rooted phylogenetic trees and networks. *Systematic Biology* **61**: 1061–1067.

IUCN (2015): The IUCN Red List of Threatened Species. Version 2015.1. http://www.iucnredlist.org (as at 21 June 2015).

Jarman, P. (1974): The social organisation of antelope in relation to their ecology. *Behaviour* **48**: 215–267.

Jordano, P. (2013): Fruits and frugivory. In *Seeds: the ecology of regeneration in plant communities* (3rd edn): 18–61. Gallager, R. S. (Ed.) New York: CABI Publishing.

MacDonald, D. W., Boitani, L., Dinerstein, E., Frtiz, H. & Wrangham, R. (2013): Conserving large mammals: are they a special case? In *Key topics in conservation biology*. **2**: 277–312. MacDonald, D. W., Willis, K. J. (Eds). Oxford: Wiley-Blackwell.

McLellan, R. (Ed.) (2014): *Living planet report 2014*. WWF International, Gland.

Payne, B. L. & Bro-Jørgensen J. (2016): Disproportionate climate-induced range loss forecast for the most threatened African antelopes. *Current Biology* **26**: (in press).

Sinclair, A. R. E. & Arcese, P. (Eds) (1995): *Serengeti II*. Chicago: Chicago University Press.

Sinclair, A. R. E, Mduma, S. A. R., Hopcraft, J. G. C., Fryxell, J. M., Hilborn, R., & Thirgood S. (2007): Long-term ecosystem dynamics in the Serengeti: lessons for conservation. *Conservation Biology* **21**: 580–590.

Sinclair, A. R. E., Mduma, S. A. R. & Brashares, J. S. (2003): Patterns of predation in a diverse predator-prey system. *Nature* **425**: 288–290.

Sinclair, A. R. E., Packer, C., Mduma, S. A. R. & Fryxell, J. M., (Eds) (2008): *Serengeti III*. Chicago: University of Chicago Press.

Stoner, C. J., Caro, T. M. & Graham, C. M. (2003): Ecological and behavioral correlates of coloration in artiodactyls: systematic analyses of conventional hypotheses. *Behavioral Ecology* **14**: 823–840.

2

Conservation Challenges Facing African Savanna Ecosystems

Adam T. Ford[1], John M. Fryxell[1], and Anthony R. E. Sinclair[2]

[1]*Department of Integrative Biology, University of Guelph, Canada*
[2]*Biodiversity Research Centre, University of British Columbia, Canada*

People evolved in African savannas and rapidly spread across the rest of the globe, assuming a defining role in the acquisition of space, energy, and other resources. Despite an epoch of co-existence with wildlife, the tides are shifting towards an increasingly troubled future for antelope in these ancient landscapes. In this chapter, we review the biophysical characteristics of savannas, some of the key interactions that define how savannas function, and how people are changing this functionality. We end by highlighting conservation efforts that can help maintain ecosystem function in an increasingly human-dominated world. Within this context, we focus on issues relating to large herbivores in general and in particular the antelopes that often dominate the ecology of many savannas.

Key characteristics of savanna systems

At the global scale, savannas occur between grasslands and forests, typically where rainfall ranges between 50 and 130 cm per year and where average temperatures rarely fall below 20°C (Sankaran *et al.*, 2005; Lehmann *et al.*, 2014). At the landscape scale, savannas are defined by pronounced heterogeneity in vegetation cover, manifesting itself as a discontinuous overstory of woody cover, interspersed with herbaceous understorey and bare ground. Indeed, it is the co-domination of trees and grasses that most clearly distinguishes a savanna

Antelope Conservation: From Diagnosis to Action, First Edition. Edited by Jakob Bro-Jørgensen and David P. Mallon.

ecosystem from that of forest or grasslands. Over time, the relative composition of tree and grass cover changes in a shifting mosaic largely determined by the interaction of bottom-up factors – rainfall, fire, soil nutrients – and top-down factors – herbivory and, indirectly, via predation and pathogens.

The temporal pattern of rainfall underlies ecological interactions in savannas. Savanna ecosystems typically experience a period of four or more months of very low rainfall, coupled with one or more periods of high rainfall. Marked seasonality results in dramatic pulses in vegetation growth in both understorey and overstory plants. In these low rainfall environments, soil moisture is often the major limiting factor for plant growth (Breman & de Wit, 1983; McNaughton, 1985). As such, when sufficient rainfall occurs, a carpet of nutritious new grass-shoots rapidly emerge, either from the seed bank in the case of annuals or from root or crown reserves in the case of perennial grasses.

As with rainfall, both the frequency and intensity of fire affects the composition and spatial patterning of tree-grass cover in savannas. In landscapes where fires are frequent, fires tend to be less severe, and cover tends to be dominated by grasses. Within days (Green *et al.*, 2015) to years (Sensenig, Demment *et al.*, 2010) after a fire, nutrient release increases the quality of forage, which then attracts smaller and medium-sized ungulates. Fire also reduces the accumulation of plant biomass and increases visibility, thereby enhancing the ability of medium and larger-sized antelope to detect and avoid their predators. Thus, recently burned areas may offer both forage and antipredator benefits to antelope.

The biomass potential of vegetation communities in African savannas is also underlain by soil fertility. Concentrations of key nutrients for plant growth depend on geological materials and the degree of weathering, erosion, and leaching that has taken place (Hopcraft *et al.*, 2010). Over time, soils may become weathered and consequently low in crucial minerals. In other places, volcanism, such as that associated with the African rift, is responsible for deposition of nutrient-rich ash that can be highly productive. The distribution of nutrient-rich and poor soils can be highly patchy relative to the movements of large mammalian herbivores (Goheen & Palmer, 2010), further adding to the spatial heterogeneity of savanna vegetation. At the patch scale, 'hotspots' of nutrients provide access to high-quality forage, attract grazers, and keep grasses in a state of high productivity (Anderson *et al.*, 2010). These hotspots can be derived through natural variation in nutrient availability or through human sources, such as abandoned cattle corrals (Augustine, 2003; Augustine *et al.*, 2003; Augustine, 2004).

One of the most conspicuous top-down forces in African savannas is herbivory by the species-rich and abundant populations of ungulates. For example, the biomass of large herbivores in Laikipia, Kenya, is 1.74 t/km^2 among 9 species (Georgiadis *et al.*, 2007), and 0.94 t/km^2 among 31 species in the Serengeti-Mara ecosystem (Frank *et al.*, 1998). Conversely, in temperate systems, such as Yellowstone National Park in the United States, there are

0.37 t/km^2 of grazer among eight species (Frank *et al.*, 1998). The composition of antelope communities varies widely among different vegetation types and land uses in Africa. Landscapes with higher tree cover tend to support a great abundance of browsers, while grassier areas are dominated by grazers. Within these broad diet types, there exists a pronounced diversity of feeding styles, body sizes, and digestive strategies that confer differential use of forage resources (Jarman, 1974; Demment & Van Soest, 1985). For example, impala (*Aepyceros melampus*) tend to browse more in the dry season when grasses are rank, but will switch to grass in the wet season (Augustine, 2010). Dik-dik (*Madoqua* spp.) eat a diversity of browse species year-round and rarely consume grasses (Manser & Brotherton, 1995). Buffalo (*Syncerus caffer*), zebra (*Equus quagga*), and wildebeest (*Connochaetes taurinus*) compete with smaller antelope like Thomson's gazelle (*Eudorcas thomsonii*) for the same plant species, but each consumes different parts of grasses to maximize intake of biomass (for larger species) or nutrient-dense forage (for smaller species) (Jarman, 1974; Demment & Van Soest, 1985; Cerling *et al.*, 2003).

Because of the pronounced environmental heterogeneity borne out over large (landscape) and fine (patch) scales, many antelopes are highly mobile, both within their home ranges (Augustine, 2010) and via seasonal migrations (Fryxell & Sinclair, 1988; Holdo *et al.*, 2009). The abundance, diversity, and mobility of African antelope enable them to effectively exploit forage that can vary widely in quality, quantity, and palatability (Fryxell *et al.*, 2005). Together, both small and large herbivores alter shrub cover (Prins & Vanderjeugd, 1993; Augustine & McNaughton, 2004; Sankaran *et al.*, 2005; Bond, 2008; Goheen *et al.*, 2013; Pringle *et al.*, 2014) and the distribution of grass biomass (du Toit & Olff, 2014). Consequently, factors influencing the movement and abundance of antelope can profoundly affect the vegetation structure of African savannas.

In addition to resources (e.g., water, forage), the distribution and abundance of African antelope is also driven by pathogens and predators. For example, rinderpest caused widespread declines in antelope in the Serengeti prior to the 1950s. With the loss of herbivory, grass biomass accumulated, prompting a series of intense fires that carried into the canopy and reduced the amount of woody cover (Holdo *et al.*, 2009). On smaller and closed systems, such as the game reserves of South Africa, tuberculosis can spread among antelope and spill over into the livestock herds on neighbouring range lands (Michel *et al.*, 2006). Indeed, species diversity in reserves increases risk of infection for impala, as does decreasing reserve size (Ezenwa, 2004). Thus, the characteristic mobility and diversity of antelope assemblages mediates the extent to which pathogens shape savanna ecosystem function.

Predation can powerfully shape both the abundance and distribution of antelope. Smaller species (<100 kg) are more vulnerable to predators than larger species, which are more vulnerable to starvation (Sinclair *et al.*, 2003). This allometric relationship arises because larger species are better able to defend themselves, and because there is a greater diversity, dietary overlap, and abundance of smaller predators than larger ones (Sinclair *et al.*, 2003). Spatial variation in predation is also linked to the patchy habitat structure of savannas. Lions (*Panthera leo*), for example, prefer to hunt in areas with greater cover (Hopcraft *et al.*, 2005), as do wild dogs and leopards (*P. pardus*) (Ford *et al.*, 2014). As a result, landscapes characterized by greater cover may have stronger top-down effects than more open areas (Hopcraft *et al.*, 2010). Likewise, homogenization of this structure can heighten top-down regulation of predators on antelope (Beale *et al.*, 2013). In addition, the behaviour (Underwood, 1982) and movement patterns (Thaker *et al.*, 2011) of antelope are shaped by the perceived risk of predation relative to foraging gains, with many species acting more vigilant and more cryptic, or altogether avoiding areas with dense cover. Temporal variation in predator activity is seemingly influenced by both prey vulnerability and thermoregulatory demands. Savanna carnivores are typically more active at night, during crepuscular periods, and on full (or new) moons (Packer *et al.*, 2011; Cozzi *et al.*, 2012). As result, antelope like impala (Ford *et al.*, 2014) and dik-dik (Ford & Goheen, in press), shift between habitats with dense cover in the day to more open habitats at night.

People and the function of savannas: salient conservation issues for antelopes

Of all terrestrial biomes, African savannas may be the ecosystem with the longest history of human occupancy, and it is difficult to separate how these landscapes would have evolved in the absence of human activity. However, there is little doubt that human population growth here as elsewhere has been implicated in much of the loss of biodiversity that has occurred over the past 200 years (Vitousek *et al.*, 1997; McKee *et al.*, 2004). In 1800, the world supported 1 billion humans. That number doubled by the middle of the 20th century and is currently slightly higher than 7 billion. This growth occurred despite profound changes in human demographic rates across the globe, commonly termed the 'demographic transition' (Bongaarts, 2009). As countries develop, initial drops in mortality rate due to improved health care drive rapid growth. This typically slows down as fertility rates more slowly equilibrate with mortality rates, but human populations continue to grow at a decelerating rate throughout that socio-economic transition period and indeed for several decades more, as the population converges on a new stable configuration. These patterns are well in place across much of the globe, but even optimistic forecasts call for a human population of 9 billion by the time humans approach zero population growth. In other words, we can anticipate further demands on all natural resources simply to meet the minimal demands of a larger population of humans, let alone cope with rising economic expectations and consumption patterns as development proceeds. This will further accentuate the already pronounced demand by humans (Vitousek *et al.*, 1997) for freshwater supplies, arable land, and wood fibre for fuel and home construction as well as implying increased loading of nitrogen and siltation in watercourses.

Next, we consider how global shifts in climate, human population growth, and the changing structure of socio-economic systems in Africa are changing the structure and function of savanna ecosystems, and along with it, the distribution and viability of antelope populations. We then review how, along with these large-scale changes, savannas are being transformed by local changes in livestock utilization, disease, fire management, invasive species, abundance of predators, consumption by people, and habitat loss.

Climate change. Climate change is altering rainfall and temperature patterns throughout the savannas of eastern and southern Africa. Climate models suggest that while the total amount of rainfall is expected to increase slightly, variation in rainfall intensity is predicted to increase dramatically (Hulme *et al.*, 2001). This means that droughts are drier, or last longer, and that rainy seasons are wetter. Such changes may impact antelope in a number of ways; for example, extended droughts may reduce the availability of forage, increase fire frequency, and increase the prevalence of grasses vs. trees (Bond, 2008). Furthermore, temperature increases (Hulme *et al.*, 2001) may reduce the hunting effort of some carnivores, like African wild dogs (*Lycaon pictus*) (Woodroffe, 2011), potentially reducing overall predation on antelope prey like dik-dik and impala. Surface water availability may also be reduced because of climate change (De Wit & Stankiewicz, 2006), and should have a greater impact on larger, water-dependent antelope (e.g., topi, *Damaliscus lunatus*) than smaller species (e.g., dik-dik) (Georgiadis *et al.*, 2007; Augustine, 2010). Many large herbivores require water on a daily basis, thereby limiting spatial distribution at certain times of year to areas adjacent to permanent water sources (Redfern *et al.*, 2003; Hopcraft *et al.*, 2012).

Variation in weather caused by the El Niño/Southern Oscillation (ENSO) can further exacerbate changes in savanna ecology. Years with strongly positive ENSO scores typically experience less rainfall in the wet season, but more rainfall in the dry season, than occurs on average. Some antelope, such as Serengeti wildebeest, have highly synchronized breeding, timed to coincide with the latter part of the wet season. As a consequence, the survival of young is highly dependent on food conditions over the dry season; hence positive ENSO years lead to higher food availability and improved offspring recruitment in wildebeest (Sinclair *et al.*, 2013). However, other species of Serengeti herbivores breed throughout the year and so experience little benefit from positive ENSO events. Top carnivores, such as lions, benefit in turn from improved numbers of young prey, which form the bulk of their diet, so leading to improved cub recruitment (Mosser *et al.*, 2009; Sinclair *et al.*, 2013). Because ENSO-like events are predicted to increase in frequency with climate change (Timmermann *et al.*, 1999), uncertainty in top-down and bottom-up forcing for antelope will likely increase.

Human population growth. The growth of the human population is having an adverse impact on antelope populations by fomenting conflict over scarce resources, altering land-use to more intensive production methods, and redistributing people to more agriculturally marginal areas that are often adjacent to protected areas. Indeed, human populations at the margins of such protected areas are growing even faster than those in other comparable rural landscapes (Wittemyer *et al.*, 2008). This might reflect the economic advantages of living close to protected areas, which receive external funding not available to communities in other rural areas. Therefore, the human footprint around protected areas is likely to increase in the near future. Of particular concern is competition for scarce water supplies in semi-arid environments and declines (e.g., for charcoal production) or increases (e.g., because of over-grazing) in woody vegetation to meet the needs of a growing human population (Wittemyer *et al.*, 2008). For example, models suggest profound changes to the Serengeti ecosystem may accompany increased usage of permanent water supplies, such as the Mara River (Gereta *et al.*, 2002). Such speculation is not unreasonable, given the rapid population growth that has been well documented around the Mara Reserve and Serengeti National Park (Ogutu *et al.*, 2009; Ogutu *et al.*, 2011; Bhola *et al.*, 2012). These same economic forces will also increase pressure for access to scarce grasslands within protected areas to feed livestock as well as suitable sites for raising cereal crops, such as wheat.

Economic development. Since World War 2, economic linkages between European-based colonial administrations and African countries have been supplemented by stronger ties between Africa and the emerging economies of the Middle East and Asia. These new economic partnerships are transforming patterns of development in savanna ecosystems (Cotula, 2009). For example, Kenya and China announced a USD $5 billion deal involving Chinese investment in Kenya transportation and energy; foreign investment in Ethiopia's biofuel sector is concentrated in areas where livestock grazing predominates land use, one of the few land-uses that are compatible with wild populations of antelope (Vermeulen & Cotula, 2010). Throughout 2004–2009, over 2.5 million ha were approved as foreign land allocation in Ethiopia, Ghana, Madagascar, Mali, and Sudan (Cotula, 2009). This systematic shifting of land-use and economic valuation of uncultivated land may further erode the habitat available for antelope in areas outside of protected areas.

Livestock. In addition to these global drivers of change, local variation in human activity can have pronounced effects on antelope in African savannas. Where people and antelope co-exist, livestock and ranching are the primary means of livelihood. In many areas, wildlife and livestock readily co-exist – and have for millennia. Indeed, wild grazers facilitate forage availability for cattle in some seasons (Odadi *et al.*, 2011), while traditional ranching practices facilitate safety for impala (Ford *et al.*, 2014). In other areas, high stocking densities deplete forage reserves for wild herbivores (Georgiadis *et al.*, 2007; Georgiadis *et al.*, 2007; Kinnaird & O'Brien, 2012), and cause shrub encroachment into grassy areas (Roques *et al.*, 2001; Sankaran *et al.*, 2005; Ward, 2005). Shrub encroachment has been implicated in the declines of species preferring more open habitats, such as the critically endangered hirola (*Beatragus hunteri*) (Dahiye, 2009).

Pastoralism and more recently, commercial ranching, re-create some of the processes driven by wild herbivores, though their impacts on vegetation are not entirely interchangeable. Unlike many antelope populations, livestock often are supplemented with artificial mineral, water, and forage resources. Livestock are also defended by herders from most predation, and their movement is determined largely by access to forage and water, rather than risk avoidance. Unless there is catastrophic drought, these adjuncts (forage resources, safety) allow livestock to acquire forage that wild antelope are reluctant to use and, as such, may occur at biomass densities much greater than that of wild herbivores (Ottichilo *et al.*, 2000; Georgiadis *et al.*, 2007; Kinnaird & O'Brien, 2012). Indeed, livestock can alter the distribution of woody cover, especially at high stocking densities, leading to shrub encroachment and a reduction in palatable grasses (Joubert *et al.*, 2008; Eldridge *et al.*, 2011). Understanding the dynamic coupling of people, their livestock, and savanna ecology is a critical gap in our knowledge of antelope conservation in human-occupied-areas.

While there is some indication that cattle and zebra may facilitate one another (Odadi *et al.*, 2011), it is less clear how most species of antelope and livestock interact. In some of the pastoralist and commercial ranching properties of central Kenya, abandoned cattle corrals provide access to nutrient-enriched grasses (Augustine, 2004) that also facilitate predator-avoidance for impala (Ford *et al.*, 2014). This suggests that cattle ranching may facilitate large, grazing antelope that exploit both the forage and safety benefits accruing from the open spaces. However, predators may preferentially hunt in these abandoned cattle corrals, potentially contributing to apparent competition between abundant grazers (e.g., zebra) and rarer antelope (e.g., roan (*Hippotragus equinus*), sable (*H. niger*), tsessebe (*Damaliscus lunatus lunatus*)) that aggregate together (Owen-Smith & Mills, 2006).

The type of livestock also bears on the question of how people and wildlife interact. For example, in parts of East Africa, camel ranching is becoming increasingly more common as a means to buffer against catastrophic droughts that more greatly threaten cattle husbandry. Moreover, camels and goats consume browse, whereas cattle and sheep forage on grasses. Thus, species composition of livestock herds determines the extent to which wild and domesticated herbivores compete or facilitate one another. While there is a tendency for antelope and other wild herbivores to occur at lower – and declining – densities as livestock density increases (Georgiadis *et al.*, 2007a; Georgiadis *et al.*, 2007b), pastoralism is only one of many anthropogenic factors that are changing the function of savanna ecosystems and the abundance of antelope.

Disease. Livestock can also serve as a reservoir for disease transmission to wild antelope (Prins & Weyerhaeuser, 1987; Mbise *et al.*, 1991; Prins & Vanderjeugd, 1993; Holdo *et al.*, 2009). Diseases such as malignant catarrhal fever, rinderpest, and brucellosis are shared across wild and domesticated ungulates. Rinderpest transmission from cattle caused massive declines in many wild ungulates in the early 20th century, such as the Serengeti wildebeest (Sinclair *et al.*, 2007). A more recent outbreak has had similar negative demographic effects, albeit on a smaller scale (Kock *et al.*, 1999). Fear of transmission of malignant catarrhal fever to cattle is often claimed for the reason that Massai pastoralists keep their cattle herds segregated from wildebeest during the calving season. The extent to which livestock ranching and antelope coexist in the future will depend in part on stocking densities as well as other activities occurring in human-occupied landscapes, such as fire management, predator persecution, and hunting.

Fire. Suppression of fire is contributing to shrub encroachment in African savannas (Roques *et al.*, 2001; Wigley *et al.*, 2010; Hoffmann *et al.*, 2012). Long-standing policies enacted under colonial administrations actively sought to suppress all wild fires, with limited 'controlled' burns occurring in wet-seasons (Laris & Wardell, 2006; Rohde & Hoffman, 2012). The loss of fire from many landscapes in Africa is changing the dynamic, spatial heterogeneity of tree-grass co-domination. For example, the dense understories of grass that form when fire is suppressed may prevent seedling establishment for woody plants (Riginos & Young, 2007; Riginos & Grace, 2008). When fires eventually occur in these areas, the accumulation of fuel burns intensely, potentially killing grass roots and changing landscapes to an alternative stable state (Laris & Wardell, 2006; Joubert *et al.*, 2008). In addition, the accumulation of cover in fire-absent landscapes may reduce habitat availability for obligate grazers (e.g., hirola, topi), and increase vulnerability to predation for medium-sized species like impala and Grant's gazelle (*Nanger granti*).

Invasive species. Non-native plants are reducing forage availability for many African antelope. In South Africa, *Chromolaena odorata* and other alien species are established in over 10 M ha. In Kenya, the *Opuntia* spp. cactus also appears to be spreading, caused in part, by seed ingestion and dispersal by antelope (Strum *et al.*, 2015). The extent to which these invasions alter savanna ecology remains to be seen. For example, Macdonald & Frame (1988) suggest that some species will not spread under natural regimes of fire frequency, and are restricted to heavily disturbed areas near roads. Beale *et al.* (2013) report that the costs of eradication can be markedly lower if removal and monitoring programs are implemented before invasive plants can spread more broadly. Besides weedy plants, introduced diseases and pathogens can have a pronounced impact on wildlife (see above).

Predators. Large-carnivores, especially lions, often are persecuted in human-occupied areas (Woodroffe & Frank, 2005; Kinnaird & O'Brien, 2012). The loss of some larger predators may release small carnivores from predation and this 'meso-predator release' may change the trophic dynamics of food webs for antelope. For example, following the loss of leopards and lions in Ghana, the abundance of olive baboons increased, with a knock-on effect of heightened predation on small antelope such as oribi (*Ourebia ourebi*) (Brashares *et al.*, 2010). Larger antelope in Ghana were spared from increased predation by baboons (*Papio anubis*), but did not benefit from the absence of large predators because of human hunting. Some carnivores, such as African wild dogs, are also particularly susceptible to pathogen exposure in ranching areas (Woodroffe & Ginsberg, 1999; Woodroffe *et al.*, 2012). The fluctuating abundance of African wild dogs has, in turn, changed the abundance of their preferred prey – dik-dik – in central Kenya, but not the abundance of less-commonly consumed prey (Ford *et al.*, in press). As such, loss of carnivores may impact competition-facilitation dynamics of antelope, depending on the body size of the antelope as well as the extent to which other forces (e.g., hunting) are also limiting populations.

Hunting. One of the most direct impacts of humans on the functioning of tropical ecosystems is via hunting and snares (Redford, 1992), and tropical savannas are no exception. In Ghana, wildlife rangers have recorded the numbers of wildlife observed while walking 10–15 km transects for decades. Time-series analysis shows long-term declines of over an order of magnitude in all reserves (Brashares *et al.*, 2001). However, in larger reserves, such as Mole National Park, the abundance of wildlife sharply increased with distance from the park border, where the hunting is most intense. Not surprisingly, a wide number of species have disappeared entirely, with the loss most pronounced in small reserves (Brashares *et al.*, 2001). In Serengeti, it has been estimated that massive declines in buffalo and elephants (*Loxodonta africana*) were largely caused by poaching (Hilborn *et al.*, 2006; Sinclair *et al.*, 2007). Budget limitations to fund patrols and other warden activities led to sharp increase in poaching in Serengeti, but this situation was reversed, fortunately, when operating funds were made available from the central Tanzania Government and from conservation agencies (Hilborn *et al.*, 2006). Evaluations of the efficiency of ranger patrols in Ghana have similarly identified human effort as a key ingredient to enforcement success (Jachmann, 2008).

Most hunting in African savannas occurs to meet the subsistence demand of impoverished rural communities. Comparisons of household economic survey data across a number of countries in sub-Saharan Africa clearly demonstrate that the importance of wildlife as a food source declines with increasing income and opportunity cost (Knapp, 2007; Brashares *et al.*, 2011). The farther villages are from nature reserves, the more expensive wildlife meat becomes, and the less it tends to be used by poorer households (Brashares *et al.*, 2011). On the other hand, wealthy households can afford to acquire bush meat in faraway urban centers, generating a worrisome demand for antelope as a 'luxury' food commodity. Such patterns suggest that reducing the demand for wildlife meat may require substitution of alternative food sources (e.g., poultry) and wider opportunity for sustainable livelihoods that do not involve poaching.

Poaching for commercial wildlife products, such as hides and horns, is also well organized by criminal elements across the African savannas (Ayling, 2013). As human populations grow, it is inevitable that such pressures will increase accordingly. As a luxury good, prices are sharply linked with supply so long as there are consumers worldwide that generate demand. As a result, while increased enforcement can discourage illegal activity, this can be counterbalanced by skyrocketing prices for scarce commodities like ivory. Nonetheless, economic incentives (such as fines or tariffs on goods at market or improved economic opportunities) are less likely to be effective than increased probability of detection by wildlife rangers in the bush (Milner-Gulland & Leader-Williams, 1992). Moreover, in areas where civil and military conflict have occurred, formal protections for wildlife are eroded, and wildlife are often hunted excessively to feed belligerent forces (De Merode *et al.*, 2007; Beyers *et al.*, 2011).

Habitat loss and fragmentation. Land cover is changing rapidly in sub-Saharan Africa, with desertification occurring at the northern and more arid fringes of savannas, urbanization occurring along transportation networks, and more intensive agriculture replacing dispersed ranching and pastoralism (Little, 1996; Newmark, 2008; Western *et al.*, 2009; Vuorio *et al.*, 2014). These forces are changing both cultures and wildlife in savannas – as people intensify land use, wildlife are displaced, human-human and human-wildlife conflicts may increase, and direct mortalities and persecution follow (Hoare & Du Toit, 1999; Reid *et al.*, 2004; Graham *et al.*, 2009). Habitat loss is contributing to the decline of antelope both within and outside of protected areas (Ogutu *et al.*, 2011). Changes in land use in areas adjacent to protected areas is blocking the long-distance movements required of many antelope to access spatially variable resources in different seasons (Caro *et al.*, 2009; Jones *et al.*, 2012). One of the most poignant examples arose when a proposed road-upgrade project had the potential to cut off the iconic wildebeest migration of the Serengeti-Mara Ecosystem Tanzania and Kenya (Dobson *et al.*, 2010; Holdo *et al.*, 2011). The expansion or creation of new road, rail, and pipeline schemes are being proposed throughout many African countries (Beale *et al.*, 2013; Caro *et al.*, 2014), with potential to further fragment savanna landscapes and isolate populations of antelope. Civil conflicts, which already contribute to direct mortality of wildlife (De Merode *et al.*, 2007), can also alter habitats. For example, internally displaced people were settled on pristine grasslands in Laikipia following the post-election violence of 2008 in Kenya. These settlements shifted land-use from traditional grazing to farming, heightening conflicts with local pastoralists, and reducing the availability of habitat for antelope further. The dynamics of socio-economic systems in Africa are thus interwoven with the functionality of savanna ecosystems and the conservation of antelope therein.

Moving forward: a prospectus for the conservation of African antelope in the 21st century

Persistence of viable populations of African antelope depends on our ability to identify the most pressing conservation issues, and act effectively. At least three, non-exclusive and broadly defined actions can help to resolve these critical conservation issues: protection, co-existence, and restoration.

Protected areas are often seen as the flagship or archetypal conservation tool, and African savannas are famous for the proliferation of protected areas in the last 50 years. However, strict reliance on protected areas as the sole means with which to conserve antelope is not feasible in the 21st century (Sinclair, 2015). Human pressures within parks (Setsaas *et al.*, 2007; Craigie *et al.*, 2010; Metzger *et al.*, 2010; Watson *et al.*, 2014; Watson *et al.*, 2015) and on their boundaries (Woodroffe & Ginsberg, 1998) strongly affects wildlife abundance. Efforts to reinforce such boundaries through exclusion fencing may create additional problems for local food webs within the park (Owen-Smith & Mills, 2006; Creel *et al.*, 2013; Packer *et al.*, 2013; Packer *et al.*, 2013; Durant *et al.*, 2015). Additionally, the economic benefits of tourism often are used to justify the establishment and enforcement of protected areas, but the seemingly benign and non-consumptive tourism industry associated with protected areas can stress wildlife (Hayward & Hayward, 2009). More critically, perhaps, is how the benefits of tourism in protected areas accrue for local communities. For example, do all members of a local community have opportunities to interact with tourists, or are some locals excluded from these benefits?

In the absence of reciprocity between the park and local community, there could be considerable resentment by locals, and heightened use of parks for grazing, charcoal burning, thatch collection, or hunting (Reid *et al.*, 2004; Beale *et al.*, 2013; Humavindu & Stage, 2015). An alternative approach integrates local communities within the management of protected areas (Mbaiwa & Stronza, 2011), though these programs need adequate financial management and revenue to achieve conservation success (Humavindu & Stage, 2015).

In addition to protected areas, co-existence and multiple-use landscapes can further augment antelope conservation (Gardner *et al.*, 2007). For example, in Laikipia, Kenya, a patchwork of land-uses and landownership meets through the Laikipia Wildlife Forum to develop a coherent strategy to reduce human-wildlife conflict (Gadd, 2005). Through this forum and its partners, some human-wildlife conflicts have been abated (Woodroffe *et al.*, 2005), including geo-fencing to prevent elephants from damaging farms (Wall *et al.*, 2014) and aversive conditioning for lions (Hazzah & Dolrenry, 2007), while commercial ranches continue to operate with intact communities of large antelope and their predators (Augustine *et al.*, 2011). At the regional scale, multiple-use areas can be integrated with core protected areas to facilitate the migratory movements of wildlife (Beale *et al.*, 2013); critical to this approach is the need to identify compatible land uses, ensure sound management of these areas, and identify priority areas to target incentives for conservation (Caro *et al.*, 2009).

Finally, restoration can reshape modern landscapes and trophic dynamics to historical levels, providing the ecological structure needed to conserve antelope. In some of the fenced reserves of South Africa, managers identified that artificial waterholes inflated the abundance of zebra, supporting a dense population of lions, which then drove down the population of sable, tsessebe, and roan (reviewed in Beale *et al.*, 2013). Removal of water holes, increasing fence permeability, and restoring fires have increased the structural heterogeneity of these areas and improved population connectivity. In Eastern Kenya, local communities are actively seeking restoration of elephants to reduce woody cover which in turn, will improve range conditions for the critically endangered hirola (Ali *et al.*, in review). Prescribed burning is becoming more commonplace in African parks (Brockett *et al.*, 2001; Archibald *et al.*, 2005), and such efforts could restore the characteristic spatial heterogeneity of many areas. One challenge will be to ensure that fire management and wildlife conservation objectives are compatible and that an adaptive management or an experimental framework is used to measure the efficacy of prescribed burns (Parr & Chown, 2003).

Given the rapidly changing nature of local and global drivers of change in African savannas, support for systematic, regional conservation planning in Africa (e.g., Beier *et al.*, 2011) is an urgent priority if the iconic populations of Africa's antelope are to persist through the 21st century.

References

Anderson, T. M., Hopcraft, J. G. C., Eby, S., Ritchie, M., Grace, J. B. & Olff, H. (2010): Landscape-scale analyses suggest both nutrient and antipredator advantages to Serengeti herbivore hotspots. *Ecology* **91**(5): 1519–1529.

Archibald, S., Bond, W., Stock, W. & Fairbanks, D. (2005): Shaping the landscape: fire-grazer interactions in an African savanna. *Ecological Applications* **15**(1): 96–109.

Augustine, D. J. (2003): Long-Term, Livestock-mediated redistribution of nitrogen and phosphorus in an East African savanna. *Journal of Applied Ecology* **40**(1): 137–149.

Augustine, D. J. (2004): Influence of cattle management on habitat selection by impala on Central Kenyan rangeland. *Journal of Wildlife Management* **68**(4): 916–923.

Augustine, D. J. (2010): Response of native ungulates to drought in semi-arid Kenyan rangeland. *African Journal of Ecology* **48**(4): 1009–1020.

Augustine, D. J. & Mcnaughton, S. J. (2004): Regulation of shrub dynamics by native browsing ungulates on East African rangeland. *Journal of Applied Ecology* **41**(1): 45–58.

Augustine, D. J., Mcnaughton, S. J. & Frank, D. A. (2003): Feedbacks between soil nutrients and large herbivores in a managed savanna ecosystem. *Ecological Applications* **13**(5): 1325–1337.

Augustine, D. J., Veblen, K. E., Goheen, J. R., Riginos, C. & Young, T. P. (2011): Pathways for positive cattle-wildlife interactions in semiarid rangelands. *Smithsonian Contributions to Zoology* **632**: 55–71.

Ayling, J. (2013): What Sustains wildlife crime? Rhino horn trading and the resilience of criminal networks. *Journal of International Wildlife Law & Policy* **16**(1): 57–80.

Beale, C. M., Van Rensberg, S., Bond, W. J., Coughenour, M., Fynn, R., Gaylard, A., Grant, R., Harris, B., Jones, T., Mduma, S., Owen-Smith, N. & Sinclair, A. R. E. (2013): Ten lessons for the conservation of African savannah ecosystems. *Biological Conservation* **167**: 224–232.

Beier, P., Spencer, W., Baldwin, R. F. & Mcrae, B. H. (2011): Toward best practices for developing regional connectivity maps. *Conservation Biology* **25**(5): 879–892.

Beyers, R. L., Hart, J. A., Sinclair, A. R., Grossmann, F., Klinkenberg, B. & Dino, S. (2011): Resource wars and conflict ivory: the impact of civil conflict on elephants in the Democratic Republic of Congo – the case of the Okapi Reserve. *PloS one* **6**(11): e27129.

Bhola, N., Ogutu, J. O., Said, M. Y., Piepho, H.-P. & Olff, H. (2012): The distribution of large herbivore hotspots in relation to environmental and anthropogenic correlates in the Mara Region of Kenya. *Journal of Animal Ecology* **81**(6): 1268–1287.

Bond, W. J. (2008): What limits trees in C-4 grasslands and savannas? *Annual Review of Ecology Evolution and Systematics* **39**: 641–659.

Bongaarts, J. (2009): Human population growth and the demographic transition. *Philosophical Transactions of the Royal Society B: Biological Sciences* **364**(1532): 2985–2990.

Brashares, J. S., Arcese, P. & Sam, M. K. (2001): Human demography and reserve size predict wildlife extinction in West Africa. *Proceedings of the Royal Society of London B: Biological Sciences* **268**(1484): 2473–2478.

Brashares, J. S., Golden, C. D., Weinbaum, K. Z., Barrett, C. B. & Okello, G. V. (2011): Economic and geographic drivers of wildlife consumption in rural Africa. *Proceedings of the National Academy of Sciences* **108**(34): 13931–13936.

Brashares, J. S., Prugh, L. R., Stoner, C. J. & Epps, C. W. (2010): Ecological and conservation implications of mesopredator release. In Trophic cascades: predators, prey, and the changing dynamics of nature. Terborgh J. & Estes, J. A. (Eds.), 221–240. Washington, DC: Island Press, 2010.

Breman, H. & De Wit, C. (1983): Rangeland productivity and exploitation in the Sahel. *Science* **221**: 1341–1347.

Brockett, B., Biggs, H. & Van Wilgen, B. (2001): A patch mosaic burning system for conservation areas in southern African savannas. *International Journal of Wildland Fire* **10**(2): 169–183.

Caro, T., Dobson, A., Marshall, A. J. & Peres, C.A. (2014): Compromise solutions between conservation and road building in the tropics. *Current Biology* **24**(16): R722–R725.

Caro, T., Jones, T. & Davenport, T. R. B. 2009. Realities of documenting wildlife corridors in tropical countries. *Biological Conservation* **142**(11): 2807–2811.

Cerling, T. E., Harris, J. M. & Passey, B. H. 2003. Diets of East African bovidae based on stable isotope analysis. *Journal of Mammalogy* **84**(2): 456–470.

Cotula, L. (2009): Land grab or development opportunity?: Agricultural investment and international land deals in Africa. iied. http://www.ifad.org/pub/land/land_grab.pdf (as at 8 February 2016).

Cozzi, G., Broekhuis, F., Mcnutt, J. W., Turnbull, L. A., Macdonald, D. W. & Schmid, B. (2012): Fear of the dark or dinner by moonlight? Reduced temporal partitioning among Africa's large carnivores. *Ecology* **93**(12): 2590–2599.

Craigie, I. D., Baillie, J. E. M., Balmford, A., Carbone, C., Collen, B., Green, R. E. & Hutton, J. M. (2010): Large mammal population declines in Africa's protected areas. *Biological Conservation* **143**(9): 2221–2228.

Creel, S., Becker, M. S., Durant, S. M., M'soka, J., Matandiko, W., Dickman, A. J., Christianson, D., Droge, E., Mweetwa, T., Pettorelli, N., Rosenblatt, E., Schuette, P., Woodroffe, R., Bashir, S., Beudels-Jamar, R. C., Blake, S., Borner, M., Breitenmoser, C., Broekhuis, F., Cozzi, G., Davenport, T. R. B., Deutsch, J., Dollar, L., Dolrenry, S., Douglas-Hamilton, I., Fitzherbert, E., Foley, C., Hazzah, L., Henschel, P., Hilborn, R., Hopcraft, J. G. C., Ikanda, D., Jacobson, A., Joubert, B., Joubert, D., Kelly, M. S., Lichtenfeld, L., Mace, G. M., Milanzi, J., Mitchell, N., Msuha, M., Muir, R., Nyahongo, J., Pimm, S., Purchase, G., Schenck, C., Sillero-Zubiri, C., Sinclair, A. R. E., Songorwa, A. N., Stanley-Price, M., Tehou, C. A., Trout, C., Wall, J., Wittemyer, G. & Zimmermann, A. (2013): Conserving large populations of lions – the argument for fences has holes. *Ecology Letters* **16**(11): 1413–e3.

Dahiye, Y. M. (2009): Reconnaissance survey for the hirola antelope in northern parts of Garissa district, Kenya. *African Journal of Ecology* **47**(3): 452–453.

De Merode, E., Smith, K. H., Homewood, K., Pettifor, R., Rowcliffe, M. & Cowlishaw, G. 2007. The impact of armed conflict on protected-area efficacy in Central Africa. *Biology letters* **3**(3): 299–301.

De Wit, M. & Stankiewicz, J. (2006): Changes in surface water supply across Africa with predicted climate change. *Science* **311**(5769): 1917–1921.

Demment, M. W. & Van Soest, P. J. (1985): A nutritional explanation for body-size patterns of ruminant and nonruminant herbivores. *American Naturalist*: 641–672.

Dobson, A., Borner, M. & Sinclair, T. (2010): Road will ruin Serengeti. *Nature* **467**(7313): 272–273.

Du Toit, J. & Olff, H. (2014): Generalities in grazing and browsing ecology: using across-guild comparisons to control contingencies. *Oecologia* **174**(4): 1075–1083.

Durant, S. M., Becker, M. S., Creel, S., Bashir, S., Dickman, A. J., Beudels-Jamar, R. C., Lichtenfeld, L., Hilborn, R., Wall, J. & Wittemyer, G. (2015): Developing fencing policies for dryland ecosystems. *Journal of Applied Ecology* **52**(3): 544–551.

Eldridge, D. J., Bowker, M. A., Maestre, F. T., Roger, E., Reynolds, J. F. & Whitford, W. G. (2011): Impacts of shrub encroachment on ecosystem structure and functioning: towards a global synthesis. *Ecology Letters* **14**(7): 709–722.

Ezenwa, V. O. (2004): Parasite infection rates of impala (*Aepyceros melampus*) in fenced game reserves in relation to reserve characteristics. *Biological Conservation* **118**(3): 397–401.

Ford, A. T. & Goheen, J. R. In press. An experimental study on risk effects in a dwarf antelope, *Madoqua guentheri*. *Journal of Mammalogy*.

Ford, A. T., Goheen, J. R., Augustine, D. J., Kinnaird, M. F., O'Brien, T., Palmer, T. M., Pringle, R. M. & Woodroffe, R. In press. Recovery of African wild dogs suppresses prey but does not trigger a trophic cascade. *Ecology*.

Ford, A. T., Goheen, J. R., Otieno, T. O., Bidner, L., Isbell, L. A., Palmer, T. M., Ward, D., Woodroffe, R. & Pringle, R. M. (2014): Large carnivores make savanna tree communities less thorny. *Science* **346**(6207): 346–349.

Frank, D. A., Mcnaughton, S. J. & Tracy, B. F. (1998): The ecology of the earth's grazing ecosystems. *BioScience*: 513–521.

Fryxell, J. M. & Sinclair, A.R. E. (1988): Causes and consequences of migration by large herbivores. *Trends in Ecology & Evolution* **3**(9): 237–241.

Fryxell, J. M., Wilmshurst, J. F., Sinclair, A. R. E., Haydon, D. T., Holt, R. D. & Abrams, P. A. (2005): Landscape scale, heterogeneity, and the viability of Serengeti grazers. *Ecology Letters* **8**(3): 328–335.

Gadd, M. E. (2005): Conservation outside of parks: attitudes of local people in Laikipia, Kenya. *Environmental Conservation* **32**(1): 50–63.

Gardner, T. A., Caro, T., Fitzherbert, E. B., Banda, T. & Lalbhai, P. (2007): Conservation value of multiple-use areas in East Africa. *Conservation Biology* **21**(6): 1516–1525.

Georgiadis, N. J., Ihwagi, F., Olwero, J. G. N. & Romanach, S. S. (2007a): Savanna herbivore dynamics in a livestock-dominated landscape. II: Ecological, conservation, and management implications of predator restoration. *Biological Conservation* **137**(3): 473–483.

Georgiadis, N. J., Olwero, J. G. N., Ojwang, G. & Romanach, S. S. (2007b): Savanna herbivore dynamics in a livestock-dominated landscape. I: Dependence on land use, rainfall, density, and time. *Biological Conservation* **137**(3): 461–472.

Gereta, E., Wolanski, E., Borner, M. & Serneels, S. (2002): Use of an ecohydrology model to predict the impact on the Serengeti ecosystem of deforestation, irrigation and the proposed Amala Weir Water Diversion Project in Kenya. *International Journal of Ecohydrology & Hydrobiology* **2**(1): 135–142.

Goheen, J. R. & Palmer, T. M. (2010): Defensive plant-ants stabilize megaherbivore-driven landscape change in an African savanna. *Current Biology* **20**(19): 1768–1772.

Goheen, J. R., Palmer, T. M., Charles, G. K., Helgen, K. M., Kinyua, S. N., Maclean, J. E., Turner, B. L., Young, H. S. & Pringle, R. M. (2013): Piecewise disassembly of a large-herbivore community across a rainfall gradient: the UHURU experiment. *PloS one* **8**(2): e55192.

Graham, M. D., Douglas-Hamilton, I., Adams, W. M. & Lee, P. C. (2009): The movement of African elephants in a human-dominated land-use mosaic. *Animal Conservation* **12**(5): 445-455.

Green, D. S., Roloff, G. J., Heath, B. R. & Holekamp, K. E. (2015): Temporal dynamics of the reponses by African mammals to prescribed fire. *Journal of Wildlife Management* **79**(2): 235–242.

Hayward, M. W. & Hayward, G. J. (2009): The impact of tourists on lion Panthera leo behaviour, stress and energetics. *Acta theriologica* **54**(3): 219–224.

Hazzah, L. & Dolrenry, S. (2007): Coexisting with predators. Malyika Singh Seminar, New Delhi, 21.

Hilborn, R., Arcese, P., Borner, M., Hando, J., Hopcraft, G., Loibooki, M., Mduma, S. & Sinclair, A. R. E. (2006): Effective enforcement in a conservation area. *Science* **314**(5803): 1266–1266.

Hoare, R. E. & Du Toit, J. T. (1999): Coexistence between people and elephants in African savannas. *Conservation Biology* **13**(3): 633–639.

Hoffmann, W. A., Geiger, E. L., Gotsch, S. G., Rossatto, D. R., Silva, L. C. R., Lau, O. L., Haridasan, M. & Franco, A. C. (2012): Ecological thresholds at the savanna-forest boundary: how plant traits, resources and fire govern the distribution of tropical biomes. *Ecology Letters* **15**(7): 759–768.

Holdo, R. M., Fryxell, J. M., Sinclair, A. R. E., Dobson, A. & Holt, R. D. (2011): Predicted Impact of Barriers to Migration on the Serengeti Wildebeest Population. *PloS one* **6**(1).

Holdo, R. M., Holt, R. D. & Fryxell, J. M. (2009): Opposing rainfall and plant nutritional gradients best explain the wildebeest migration in the Serengeti. *American Naturalist* **173**(4): 431–445.

Holdo, R. M., Sinclair, A. R. E., Dobson, A. P., Metzger, K. L., Bolker, B. M., Ritchie, M. E. & Holt, R. D. (2009): A disease-mediated trophic cascade in the Serengeti and its implications for ecosystem C. *Plos Biology* **7**(9).

Hopcraft, J. G. C., Anderson, T. M., Perez-Vila, S., Mayemba, E. & Olff, H. (2012): Body size and the division of niche space: food and predation differentially shape the distribution of Serengeti grazers. *Journal of Animal Ecology* **81**(1): 201–213.

Hopcraft, J. G. C., Olff, H. & Sinclair, A. R. E. (2010): Herbivores, resources and risks: alternating regulation along primary environmental gradients in savannas. *Trends in Ecology & Evolution* **25**(2): 119–128.

Hopcraft, J. G. C., Sinclair, A. R. E. & Packer, C. (2005): Planning for success: Serengeti lions seek prey accessibility rather than abundance. *Journal of Animal Ecology* **74**(3): 559–566.

Hulme, M., Doherty, R., Ngara, T., New, M. & Lister, D. (2001): African climate change: 1900–2100. *Climate research* **17**(2): 145–168.

Humavindu, M. N. & Stage, J. (2015): Community-based wildlife management failing to link conservation and financial viability. *Animal Conservation* **18**(1): 4–13.

Jachmann, H. (2008): Illegal wildlife use and protected area management in Ghana. *Biological Conservation* **141**(7): 1906–1918.

Jarman, P. J. (1974): The social organisation of antelope in relation to their ecology. *Behaviour* **48**: 215–267.

Jones, T., Bamford, A. J., Ferrol-Schulte, D., Hieronimo, P., Mcwilliam, N. & Rovero, F. (2012): Vanishing wildlife corridors and options for restoration: a case study from Tanzania. *Tropical Conservation Science* **5**(4): 463–474.

Joubert, D. F., Rothauge, A. & Smit, G. N. (2008): A conceptual model of vegetation dynamics in the semiarid Highland savanna of Namibia, with particular reference to bush thickening by *Acacia mellifera*. *Journal of Arid Environments* **72**(12): 2201–2210.

Kinnaird, M. F. & O'Brien, T. G. (2012): Effects of private-land use, livestock management, and human tolerance on diversity, distribution, and abundance of large African mammals. *Conservation Biology* **26**(6): 1026–1039.

Knapp, E. J. (2007): Who poaches? Household economies of illegal hunters in Western Serengeti, Tanzania. *Human Dimensions of Wildlife* **12**(3): 195–196.

Kock, R., Wambua, J., Mwanzia, J., Wamwayi, H., Ndungu, E., Barrett, T., Kock, N. & Rossiter, P. (1999): Rinderpest epidemic in wild ruminants in Kenya 1993-97. *Veterinary Record* **145**: 275–283.

Laris, P. & Wardell, D. A. (2006): Good, bad or 'necessary evil'? Reinterpreting the colonial burning experiments in the savanna landscapes of West Africa. *Geographical Journal* **172**: 271–290.

Lehmann, C. E. R., Anderson, T. M., Sankaran, M., Higgins, S. I., Archibald, S., Hoffmann, W. A., Hanan, N. P., Williams, R. J., Fensham, R. J., Felfili, J., Hutley, L. B., Ratnam, J., San Jose, J., Montes, R., Franklin, D., Russell-Smith, J., Ryan, C. M., Durigan, G., Hiernaux, P., Haidar, R., Bowman, D. M. J.S.& Bond, W. J. (2014): Savanna vegetation-fire-climate relationships differ among continents. *Science* **343**(6170): 548–552.

Little, P. D. (1996): Pastoralism, biodiversity, and the shaping of savanna landscapes in East Africa. *Africa* **66**(1): 37–51.

Macdonald, I. & Frame, G. (1988): The invasion of introduced species into nature reserves in tropical savannas and dry woodlands. *Biological Conservation* **44**(1): 67–93.

Manser, M. B. & Brotherton, P. N. M. (1995): Environmental constraints on the foraging behavior of a dwarf antelope (*Madoqua kirkii*). *Oecologia* **102**(4): 404–412.

Mbaiwa, J. E. & Stronza, A. L. (2011): Changes in resident attitudes towards tourism development and conservation in the Okavango Delta, Botswana. *Journal of Environmental Management* **92**(8): 1950–1959.

Mbise, A. N., Nsengwa, G. R. M. & Mbasha, E. M. S. (1991): Anthrax epizootic in Tarangire National Park Tanzania: an important disease of impala. *Bulletin of Animal Health and Production in Africa* **39**(1): 45–50.

McKee, J. K., Sciulli, P. W., Fooce, C. D. & Waite, T. A. (2004): Forecasting global biodiversity threats associated with human population growth. *Biological Conservation* **115**(1): 161–164.

McNaughton, S. (1985): Ecology of a grazing ecosystem: the Serengeti. *Ecological monographs* **55**(3): 259–294.

Metzger, K. L., Sinclair, A. R. E., Hilborn, R., Hopcraft, J. G. C. & Mduma, S. a. R. (2010): Evaluating the protection of wildlife in parks: the case of African buffalo in Serengeti. *Biodiversity and Conservation* **19**(12): 3431–3444.

Michel, A. L., Bengis, R. G., Keet, D., Hofmeyr, M., De Klerk, L., Cross, P. C., Jolles, A. E., Cooper, D., Whyte, I. & Buss, P. (2006): Wildlife tuberculosis in South African conservation areas: implications and challenges. *Veterinary microbiology* **112**(2): 91–100.

Milner-Gulland, E. & Leader-Williams, N. (1992): A model of incentives for the illegal exploitation of black rhinos and elephants: poaching pays in Luangwa Valley, Zambia. *Journal of Applied Ecology*: 388–401.

Mosser, A., Fryxell, J. M., Eberly, L. & Packer, C. (2009): Serengeti real estate: density vs. fitness-based indicators of lion habitat quality. *Ecology Letters* **12**(10): 1050–1060.

Newmark, W. D. (2008): Isolation of African protected areas. *Frontiers in Ecology and the Environment* **6**(6): 321–328.

Odadi, W. O., Karachi, M. K., Abdulrazak, S. A. & Young, T. P. (2011): African wild ungulates compete with or facilitate cattle depending on season. *Science* **333**(6050): 1753–1755.

Ogutu, J. O., Owen-Smith, N., Piepho, H. P. & Said, M. Y. (2011): Continuing wildlife population declines and range contraction in the Mara region of Kenya during 1977–2009. *Journal of Zoology* **285**(2): 99–109.

Ogutu, J. O., Piepho, H. P., Dublin, H. T., Bhola, N. & Reid, R. S. (2009): Dynamics of Mara-Serengeti ungulates in relation to land use changes. *Journal of Zoology* **278**(1): 1–14.

Ottichilo, W. K., De Leeuw, J., Skidmore, A. K., Prins, H. H. T. & Said, M. Y. (2000): Population trends of large non-migratory wild herbivores and livestock in the Masai Mara ecosystem, Kenya, between 1977 and 1997. *African Journal of Ecology* **38**(3): 202–216.

Owen-Smith, N. & Mills, M. G. L. (2006): Manifold interactive influences on the population dynamics of a multispecies ungulate assemblage. *Ecological monographs* **76**(1): 73–92.

Packer, C., Loveridge, A., Canney, S., Caro, T., Garnett, S. T., Pfeifer, M., Zander, K. K., Swanson, A., Macnulty, D., Balme, G., Bauer, H., Begg, C. M., Begg, K. S., Bhalla, S., Bissett, C., Bodasing, T., Brink, H., Burger, A., Burton, A. C., Clegg, B., Dell, S., Delsink, A., Dickerson, T., Dloniak, S. M., Druce, D., Frank, L., Funston, P., Gichohi, N., Groom, R., Hanekom, C., Heath, B., Hunter, L., Deiongh, H. H., Joubert, C.J., Kasiki, S. M., Kissui, B., Knocker, W., Leathem, B., Lindsey, P. A., Maclennan, S. D., Mcnutt, J. W., Miller, S. M., Naylor, S., Nel, P., Ng'weno, C., Nicholls, K., Ogutu, J. O., Okot-Omoya, E., Patterson, B. D., Plumptre, A., Salerno, J., Skinner, K., Slotow, R., Sogbohossou, E. A., Stratford, K. J., Winterbach, C., Winterbach, H. & Polasky, S. (2013): Conserving large carnivores: dollars and fence. *Ecology Letters* **16** (5): 635–641.

Packer, C., Swanson, A., Canney, S., Loveridge, A., Garnett, S., Pfeifer, M., Burton, A. C., Bauer, H. & Macnulty, D. (2013): The case for fencing remains intact. *Ecology Letters* **1** (11): 1414–e4.

Packer, C., Swanson, A., Ikanda, D. & Kushnir, H. (2011): Fear of darkness, the full moon and the nocturnal ecology of African lions. *PloS one* **6** (7).

Parr, C. L. & Chown, S. L. (2003): Burning issues for conservation: a critique of faunal fire research in Southern Africa. *Austral Ecology* **28**(4): 384(395):

Pringle, R. M., Goheen, J. R., Palmer, T. M., Charles, G. K., Defranco, E., Hohbein, R., Ford, A. T. & Tarnita, C. E. (2014): Low functional redundancy among mammalian browsers in regulating an encroaching shrub (*Solanum campylacanthum*) in African savannah. *Proceedings of the Royal Society B: Biological Sciences* **281**(1785): 20140390.

Prins, H. H. T. & Vanderjeugd, H. P. (1993): Herbivore population crashes and woodland structure in East Africa. *Journal of Ecology* **81**(2): 305–314.

Prins, H. H. T. & Weyerhaeuser, F. J. (1987): Epidemics in populations of wild ruminants – anthrax and impala, rinderpest and buffalo in Lake Manyara National Park Tanzania. *Oikos* **49**(1): 28–38.

Redfern, J. V., Grant, R., Biggs, H. & Getz, W. M. (2003): Surface-water constraints on herbivore foraging in the Kruger National Park, South Africa. *Ecology* **84**(8): 2092–2107.

Redford, K. H. (1992): The empty forest. *BioScience* **42**(6): 412–422.

Reid, R. S., Thornton, P. K. & Kruska, R. L. (2004): Loss and fragmentation of habitat for pastoral people and wildlife in East Africa: concepts and issues. *African Journal of Range and Forage Science* **21**(3): 171–181.

Riginos, C. & Grace, J. B. (2008): Savanna tree density, herbivores, and the herbaceous community: Bbottom-up vs. top-down effects. *Ecology* **89**(8): 2228–2238.

Riginos, C. & Young, T. P. (2007): Positive and negative effects of grass, cattle, and wild herbivores on Acacia saplings in an East African savanna. *Oecologia* **153**(4): 985–995.

Rohde, R. F. & Hoffman, M. T. (2012): The historical ecology of Namibian rangelands: vegetation change since 1876 in response to local and global drivers. *Science of the Total Environment* **416**: 276–288.

Roques, K. G., O'Connor, T. G. & Watkinson, A. R. (2001): Dynamics of shrub encroachment in an African savanna: relative influences of fire, herbivory, rainfall and density dependence. *Journal of Applied Ecology* **38**(2): 268–280.

Sankaran, M., Hanan, N. P., Scholes, R. J., Ratnam, J., Augustine, D. J., Cade, B. S., Gignoux, J., Higgins, S. I., Le Roux, X., Ludwig, F., Ardo, J., Banyikwa, F., Bronn, A., Bucini, G., Caylor, K. K., Coughenour, M. B., Diouf, A., Ekaya, W., Feral, C. J., February, E. C., Frost, P. G. H., Hiernaux, P., Hrabar, H., Metzger, K. L., Prins, H. H. T., Ringrose, S., Sea, W., Tews, J., Worden, J. & Zambatis, N. (2005): Determinants of woody cover in African savannas. *Nature* **438**(7069): 846–849.

Sensenig, R. L., Demment, M. W. & Laca, E. A. (2010): Allometric scaling predicts preferences for burned patches in a guild of East African grazers. *Ecology* **91**(10): 2898–2907.

Setsaas, T. H., Holmern, T., Mwakalebe, G., Stokke, S. & Roskaft, E. (2007): How does human exploitation affect impala populations in protected and partially protected areas? – A case study from the Serengeti Ecosystem, Tanzania. *Biological Conservation* **136**(4): 563–570.

Sinclair, A. R. (2015): Protected areas are necessary for conservation. In Protecting the Wild: Parks and Wilderness, the Foundation for Conservation: 72–79. Wuerthner, G., Crist, E. & Butler T. (Eds.). Washington DC: Island Press.

Sinclair, A. R. E., Mduma, S. & Brashares, J. S. (2003): Patterns of predation in a diverse predator-prey system. *Nature* **425**(6955): 288–290.

Sinclair, A. R. E., Mduma, S. a. R., Hopcraft, J. G. C., Fryxell, J. M., Hilborn, R. & Thirgood, S. (2007): Long-term ecosystem dynamics in the Serengeti: lessons for conservation. *Conservation Biology* **21**(3): 580–590.

Sinclair, A. R. E., Metzger, K. L., Fryxell, J. M., Packer, C., Byrom, A. E., Craft, M. E., Hampson, K., Lembo, T., Durant, S. M., Forrester, G. J., Bukombe, J., Mchetto, J., Dempewolf, J., Hilborn, R., Cleaveland, S., Nkwabi, A., Mosser, A. & Mduma, S. a. R. (2013): Asynchronous food-web pathways could buffer the response of Serengeti predators to El Nino Southern Oscillation. *Ecology* **94**(5): 1123–1130.

Strum, S. C., Stirling, G. & Mutunga, S. K. (2015): The perfect storm: land use change promotes *Opuntia stricta*'s invasion of pastoral rangelands in Kenya. *Journal of Arid Environments* **118**: 37–47.

Thaker, M., Vanak, A. T., Owen, C. R., Ogden, M. B., Niemann, S. M. & Slotow, R. (2011): Minimizing predation risk in a landscape of multiple predators: effects on the spatial distribution of African ungulates. *Ecology* **92**(2): 398–407.

Timmermann, A., Oberhuber, J., Bacher, A., Esch, M., Latif, M. & Roeckner, E. (1999): Increased El Niño frequency in a climate model forced by future greenhouse warming. *Nature* **398**(6729): 694–697.

Underwood, R. (1982): Vigilance behaviour in grazing African antelopes. *Behaviour* **79**(2–4): 8–107.

Vermeulen, S. & Cotula, L. (2010): Over the heads of local people: consultation, consent, and recompense in large-scale land deals for biofuels projects in Africa. *Journal of Peasant Studies* **37**(4): 899–916.

Vitousek, P. M., Mooney, H. A., Lubchenco, J. & Melillo, J. M. (1997): Human domination of Earth's ecosystems. *Science* **277**(5325): 494–499.

Vuorio, V., Muchiru, A., Reid, R. S. & Ogutu, J. O. (2014): How pastoralism changes savanna vegetation: impact of old pastoral settlements on plant diversity and abundance in south-western Kenya. *Biodiversity and Conservation* **23**(13): 3219–3240.

Wall, J., Wittemyer, G., Klinkenberg, B. & Douglas-Hamilton, I. (2014): Novel opportunities for wildlife conservation and research with real-time monitoring. *Ecological Applications* **24**(4): 593–601.

Ward, D. (2005): Do we understand the causes of bush encroachment in African savannas? *African Journal of Range and Forage Science* **22**(2): 101–105.

Watson, F. G. R., Becker, M. S., Milanzi, J. & Nyirenda, M. (2015): Human encroachment into protected area networks in Zambia: implications for large carnivore conservation. *Regional Environmental Change* **15**(2): 415–429.

Watson, J. E. M., Dudley, N., Segan, D. B. & Hockings, M. (2014): The performance and potential of protected areas. *Nature* **515**(7525): 67–73.

Western, D., Groom, R. & Worden, J. (2009): The impact of subdivision and sedentarization of pastoral lands on wildlife in an African savanna ecosystem. *Biological Conservation* **142**(11): 2538–2546.

Wigley, B. J., Bond, W. J. & Hoffman, M. T. (2010): Thicket expansion in a South African savanna under divergent land use: local vs. global drivers? *Global Change Biology* **16**(3): 964–976.

Wittemyer, G., Elsen, P., Bean, W. T., Burton, A. C. O. & Brashares, J. S. (2008): Accelerated human population growth at protected area edges. *Science* **321**(5885): 123–126.

Woodroffe, R. (2011): Ranging behaviour of African wild dog packs in a human-dominated landscape. *Journal of Zoology* **283**(2): 88–97.

Woodroffe, R. & Frank, L. G. (2005): Lethal control of African lions (*Panthera leo*): local and regional population impacts. *Animal Conservation* **8**: 91–98.

Woodroffe, R. & Ginsberg, J. R. (1998): Edge effects and the extinction of populations inside protected areas. *Science* **280**(5372): 2126–2128.

Woodroffe, R. & Ginsberg, J. R. (1999): Conserving the African wild dog *Lycaon pictus*. I: Diagnosing and treating causes of decline. *Oryx* **33**(2): 132–142.

Woodroffe, R., Prager, K. C., Munson, L., Conrad, P. A., Dubovi, E. J. & Mazet, J. a. K. (2012): Contact with domestic dogs increases pathogen exposure in endangered African wild dogs (*Lycaon pictus*). *PloS one* **7** (1).

Woodroffe, R., Thirgood, S. & Rabinowitz, A. (2005): People and wildlife: conflict of co-existence? In People and wildlife: conflict of co-existence? Woodroffe, R., Thirgood, S. & Rabinowitz, A. (Eds.). Cambridge: Cambridge University Press.

3

Population Regulation and Climate Change: The Future of Africa's Antelope

J. Grant C. Hopcraft

Institute of Biodiversity, Animal Health and Comparative Medicine, University of Glasgow, Scotland, United Kingdom

Introduction

The impacts of global climate change on Africa's antelope represent a major threat to their long-term survival, yet remain largely unaddressed. Current conservation efforts focus on mitigating the burgeoning human requirements for natural resources, including water, wood, bushmeat, agricultural land, and minerals, which are fracturing once-contiguous habitats and are the root of the most immediate threats such habitat loss and interruptions of gene flow. However, the implications of climate change on antelope are rarely considered primarily because we lack a clear synthetic framework as to how climate change might impact this incredibly diverse lineage. Fossil records show that the sequential radiations of African antelope species have coincided with periods of widespread climate shifts (deMenocal, 2004) during which the quality and abundance of forage likely drove the evolution of niche differentiation. This led to the extensive diversity we observe today in which antelope occupy virtually every biome on the continent from the desert gemsbok (*Oryx gazella*) to the Afro-alpine mountain nyala (*Tragelaphus buxtoni*). Therefore, the story of antelope diversity is closely tied to climate. This chapter focuses on how environmental gradients such as rainfall determine the abundance and distribution of different African herbivores, and proposes a framework to logically project how climate change may affect antelope populations in the future. This can be used to arm our strategies and help prioritize our objectives for the conservation of African antelope in the 21st century.

Antelope Conservation: From Diagnosis to Action, First Edition. Edited by Jakob Bro-Jørgensen and David P. Mallon.

A major concern for conservation is the increasing isolation of protected areas, leading to natural processes such as population regulation and dispersal becoming progressively more disconnected from regional trends. Species and communities in the future may find it difficult (if not impossible) to move to more suitable locations as would be required under any climate change scenario, because these areas may already have been modified by humans and therefore have become inaccessible. For example, shifts in the vegetation community caused by changes in regional rainfall patterns might necessitate a species to disperse up or down the rainfall gradient. However, if the species is confined to a previously determined protected area and unable to disperse to better habitats, the population would inevitably decline. The relevance for conservation is that *the location of our current protected areas may no longer host the preferred niches of many rare, endangered, or economically important species in the future, particularly if there are significant changes in the global climate.* Therefore, understanding the factors that regulate herbivores and determine their distributions is critical to the long-term management and conservation of ecosystems.

The central question in this chapter is how might long-term climate changes affect the future of African antelope? The approach is a logical progression of questions: how do common environmental gradients such as rainfall alter the quality and quantity of forage? What determines and/or mediates the impact of predation on populations of herbivores? If a niche space is defined by the availability of suitable food and the risk of predation, how is it partitioned between the vast diversity of African herbivores? How might these environmental gradients affect the way animals move between resource patches? How might climate change alter this niche space, and can we anticipate how the distribution of populations of herbivores might respond? This line of questions provides a framework from which we can surmise how shifts in key environmental factors such as rainfall might favour certain species over others, and which species might be most affected by concurrent habitat fragmentation and climate change.

How will climate change affect Africa's habitats?

In sub-Saharan Africa, the primary impact of global climate change is shifting patterns of rainfall, which will have deep and pervasive consequences on ecosystem dynamics (Hulme, 2000). There still remains much debate as to which areas will receive more rainfall and which areas will receive less rainfall; however, all the evidence unanimously agrees that there will be some degree of change in the annual amount of rain and the periodicity with which it falls. Rainfall, rather than temperature or CO_2 concentration, is the primary underlying environmental gradient that determines the structure of all African ecosystems, since virtually all African biomes are water dependent to some degree. An increase in the annual rainfall leads to a greater abundance of grass and trees (Sankaran *et al.*, 2005) while declining rainfall leads to sparse vegetation and desertification (Schlesinger *et al.*, 1990). But aside from this coarse transition of vegetation type that occurs across rainfall gradients, there are more hidden consequences of rainfall on the forage available to herbivores as well as the hunting success of their predators.

Rainfall: the effect on forage quantity and quality

Rainfall alters the quantity and quality of forage available to herbivores through at least two different mechanisms. High annual rainfall leads to the long-term leaching of soils. Many of the key organic molecules that are required by plants for growth are rare in high-rainfall soils (witness the low availability of nitrogen in tropical rainforests for example). As a result, the vegetation growing in areas with poor-quality soil is typically low in the essential nutrients required by herbivores (Scott, 1962; Hook and Burke, 2000). Second, vegetation tends to grow taller and thicker as rainfall increases. Grasses in high rainfall areas become more lignified with carboniferous and silicate support structures, such as Africa's widespread *Hyparrhenia* species of grasses (thatch grass). Additionally, plants tend to invest more resources into physical defence against herbivory, such as thorns of *Acacia* species, or into secondary chemical compounds like tannins in the *Terminalia* species of Africa's miombo woodlands, which are astringent and unpalatable to many herbivores (Shrader *et al.*, 2012).

The coarse and well-defended vegetation in high-rainfall areas effectively dilutes the nutritional content of the forage and makes it difficult for herbivores to extract sufficient energy from it (Botkin *et al.*, 1981; East, 1984; Fritz and Duncan, 1994; Bell, 1982). Therefore, the most nutritious grasslands in Africa occur in semi-arid landscapes on rich unleached soils such as the volcanic-ash plains of the Serengeti where over 1.3 million wildebeest, 250,000 gazelles, and 180,000 zebra reside. The stoloniferous grasses in these areas (such as *Andropogon greenwayi*, which was first described in this ecosystem) grow rapidly into thick grazing lawns that rarely exceed 15 cm in height. These grasses have a very high leaf-to-stem ratio with few support structures. In addition, the short growing season prevents these grasses from developing extensive secondary defence mechanisms and results in high-quality forage (Anderson *et al.*, 2006; Anderson *et al.*, 2007) that is only available to herbivores for a few months every year creating a natural highly productive rotational grazing system (Holdo *et al.*, 2009).

The crucial points when considering the question of how climate change will affect Africa's antelope are that rainfall is (a) positively correlated with the quantity of vegetation available to herbivores, but (b) negatively correlated with its nutritional content. This inverse relationship is central to understanding the current distribution of antelope in Africa; certain species can process large amounts of low-quality forage and thrive in high-rainfall areas while other species depend on small quantities of high-quality forage and tend to be confined to fertile semi-arid areas (Botkin *et al.*, 1981; Fritz and Duncan, 1994; Jarman, 1974; Bell, 1970). The following sections of this chapter explore these patterns of species distributions in further detail.

Rainfall: the effect on predation

Predators face the challenge of capturing and consuming prey and converting this energy into the next generation of offspring in the most efficient manner. Studies illustrate that predators do not move randomly through their home range and serendipitously encounter prey, but instead select specific hunting areas that are characterised by landscape features that increase their success of encountering and capturing their prey (Hilborn *et al.*, 2012; Hopcraft *et al.*, 2005). For instance, stalk and ambush predators such as lions are more successful at catching prey in areas with more than 20% woody cover and less than 500 meters from a drainage system (even if it has no water) (Hopcraft *et al.*, 2005; Valeix *et al.*, 2009b). Prides with these landscape features in their home range show stronger recruitment rates than their neighbours and are difficult to displace by their competitors (Mosser *et al.*, 2009). The key finding from these studies is that the success of a predator does not necessarily depend on the abundance of the prey but rather on the prey's catchability. Therefore, the net effect of rainfall on the structure of the vegetation community not only alters the bottom-up regulation of herbivores by altering the quality and quantity of forage, but it also determines the amount of cover that conceals stalking predators and the efficiency with which a predator can kill its prey (Anderson *et al.*, 2010; Valeix *et al.*, 2009b). This suggests that both the degree of top-down regulation and bottom-up regulation are determined by rainfall and cannot be considered as independent processes (Hopcraft *et al.*, 2010).

Rainfall: the effect on disturbance patterns

Fires are the primary disturbance in Africa (Bond and Keeley, 2005). The periodicity, intensity, and frequency of fires can rapidly switch ecosystems back and forth from an open grassland-state to a closed woodland-state (Dublin *et al.*,

1990; Higgins *et al.*, 2007; Sankaran *et al.*, 2005). However, only regions with intermediate amounts of rainfall actually burn: fires cannot persist at either extreme of the rainfall gradient because either the vegetation is too wet, or the vegetation is too sparse, and there is insufficient fuel-load to sustain a fire (Bond and Keeley, 2005). Therefore, the degree of disturbance is partially determined by rainfall with intermediate rainfall areas capable of exhibiting two alternatively stable states (Staver *et al.*, 2011): open fast-growing grasslands with frequent disturbance more suitable for grazers, or slow-growing closed woodlands with infrequent disturbance more suitable for browsers.

How do ungulates divide their niche space?

The amount of rainfall not only alters the quantity and quality of forage available to herbivores, but also the hunting efficiency of their predators, and the extent to which disturbances such as fires can switch an ecosystem from a grassland-state to a woodland-state or vice versa. Given the pervasive effects of rainfall on the structure of African ecosystems, how do ungulates divide this niche space? And, more to the point, what might be the implications of changing rainfall patterns on the conservation of antelope in the future?

Forage and digestive constraints

From a herbivore's perspective, not all that is green is edible. For antelope to digest cellulose, they require a complex symbiotic fermentation process that occurs in the rumen. This process is relatively time consuming and requires a specialized gastro-intestinal tract. Small antelope such as dik-dik (*Madoqua* spp.) or Thomson's gazelle (*Eudorcas thomsonii*) need more energy per kilogram of body mass than large antelope but face severe space restrictions in their body cavity, which reduces the ingesta retention times. These restrictions preclude them from processing coarse forage that requires a long time to ferment. Essentially small antelope have to extract energy from a small amount of ingesta that is rapidly passed through the digestive system, and therefore they tend to be very selective for the highest quality food. The digestive constraints experienced by small antelope means they select only the most nutritious and highest-energy vegetation such as browsing the apical meristems or grazing the freshest young leaves (known as the Jarman-Bell principle) (Wilmshurst *et al.*, 2000; Peters, 1983; Jarman, 1974; Bell, 1970). The jaw musculature and dental arcade (especially the length of the incisor row and tooth curvature) of small antelope are specifically adapted to maximize their ingestion rates on small and dispersed food items (witness the dik-dik nipping specific shoots in the undergrowth browse as opposed to the buffalo (*Syncerus caffer*) taking large mouthfuls of mixed homogenously dispersed grass) (Gordon and Illius, 1988). Large ungulates on the other hand are relatively unconstrained by the size of their gastrointestinal tract and have long ingesta retention times with a broad dental arcade (Owen-Smith, 1988). The slow fermentation of large quantities of forage enables large ungulates to gain sufficient energy even from low-quality food such as stems, awns, and mature leaf matter. The only restriction large herbivores face is accessing enough forage biomass, and not necessarily the quality of the forage itself (Hopcraft *et al.*, 2010).

Body size and the effects of predation

The body size of a herbivore not only constrains the quality or quantity of forage an ungulate can process, but it also alters their risk of predation. Several studies have shown that large predators tend to focus on large prey and small predators tend to focus on small prey (Radloff and Du Toit, 2004; Owen-Smith and Mills, 2008) such that the prey base is partitioned according to the predators body size. However, if large predators are opportunists and consume small prey as well as large prey whenever they are encountered, then the prey base of small predators is nested within that of large predators, and this exposes small herbivores to many more enemies (Sinclair *et al.*, 2003). Therefore, the degree of regulation of herbivore populations by predation depends both on the relative body size of predators and prey, and the way in which the prey base is partitioned by carnivores of different sizes (Hopcraft *et al.*, 2010). Large herbivores tend to be less regulated by predation because they are difficult to capture and are only consumed by the largest predators, while small herbivores are potentially susceptible to many predators and tend to be more regulated by predation (Sinclair *et al.*, 2003).

Population regulation across body size

Years of ecological research on African antelope briefly summarized in the previous sections raise three key processes that underpin our current understanding of population regulation: (1) the quality and biomass of forage available to herbivores is determined by environmental gradients (such as rainfall) as well as the plants' structural architecture (such as prostrate grasses versus woody vegetation). (2) The quality and abundance of forage affect small and large antelope differently, owing to the different physiological constraints on their digestive systems and morphology. In addition, (3) small antelope are exposed to more predators than large antelope because large predators supplement their diet with small prey. When these findings are considered in combination, the net result is that populations of small herbivores tend to be regulated by the quality of food and by predation. The relative importance of predation in regulating a population decays as body size increases such that populations of intermediate-sized herbivores tend to be regulated by the quality of food available. The largest antelope are too big for most predators and their slow fermentation in large rumens means these populations are regulated by the quantity of food available (Hopcraft *et al.*, 2010) (Plate 11 [i]). Therefore small antelope tend to be top-down regulated whereas large antelope tend to be bottom-up regulated.

Population regulation across rainfall gradients

The division of niche space in antelope is most likely the result of evolutionary selection for alternative strategies to decrease the competition for food while simultaneously avoiding predators. Up to now the narrative has focused on how rainfall, disturbances, and body size all modify the way in which a population can be regulated by food or predation, but how do these factors interact with each other to determine the overall abundance and distribution of different antelope species? Combining the effects of rainfall, disturbances, and body size into a single framework not only improves our understanding about population regulation, but also leads to testable hypotheses as to how climate change might impact the future of antelope in Africa. These insights can help us devise proactive management strategies and targeted research programs for the conservation of antelope, especially in light of the rapid and large-scale changes occurring to their habitats.

Bottom-up regulation

To begin with, consider the simplest case in which bottom-up regulation (i.e., the availability of food) determines the overall abundance of the population. From previous research we know that rainfall affects both the quality and quantity of forage simultaneously but in opposite ways by changing the architecture of the vegetation and its nutritional content, as discussed previously. This implies that the maximum abundance of any antelope in any ecosystem could either be limited by the quantity of forage in very low rainfall conditions or the quality of forage in very high rainfall conditions (Botkin *et al.*, 1981; Fritz and Duncan, 1994). In other words, the way in which bottom-up regulation limits the abundance of antelope varies along a rainfall gradient, with the maximum abundance occurring at some optimal amount of rainfall (Plate 11[ii(a)]). Clearly the slope and intersect of the 'food quality' versus 'food abundance' thresholds in the diagram on Plate 11[ii(a)] will vary by species, which is probably the evolutionary product of inter-specific grazing competition. Therefore each species will converge on its own unique rainfall optimum, but the premise remains consistent between species.

A population could become more abundant than otherwise expected based on the amount of rainfall in an area by seasonally migrating between nutrient-rich patches up and down the rainfall gradient (i.e., migrating between points A and B in the diagram on Plate 11[ii(a)]). This is the strategy of wildebeest (*Connochaetes taurinus*), kob (*Kobus kob*), and some species of gazelle (Fryxell *et al.*, 1988; Mduma *et al.*, 1999; Harris *et al.*, 2009). Interestingly, there are subpopulations of these species, such as herds of resident wildebeest in Serengeti, that do not migrate and as a consequence they remain at much lower densities than the migrant population (Hopcraft *et al.*, 2015).

Top-down regulation

In the real world, however, antelope must contend with predation in addition to the limitations of food supply, and therefore the simplest case scenario that only considers bottom-up regulation does not provide a complete or realistic picture. The impact of predation (or in many cases, the perceived risk of predation rather than predation itself) is inversely correlated with body size such that small antelope are more vulnerable than large antelope (see section 'Body size and the effects of predation'). Furthermore, predators are often facilitated by attributes of the landscape that aid in catching prey, such as vegetative cover that conceals ambush predators (as discussed in 'Rainfall: the effect on predation'), or large open glades that increase the searching efficiency of coursing predators (Hebblewhite *et al.*, 2005; Hilborn *et al.*, 2012; Hopcraft *et al.*, 2005; Valeix *et al.*, 2009a). Many of these landscape attributes are related to vegetation in one way or another, and therefore they are also correlated with rainfall. The increased predation risk due to ambush predators in thick vegetation is illustrated by the negative slope of the predation threshold in the diagram on Plate 11[ii(c)].

A second way in which a population could avoid being regulated by food would be to increase their efficiency at extracting nutrients from low-quality forage. For example, the evolution of the capacious rumen of large grazers such as buffalo allows them to slow the passage of forage in the digestive tract, which results in thorough fermentation and enables them to extract more nutrients from poor-quality food than small ruminants. Therefore physiological adaptations to the digestive system could allow a population of antelope to become more abundant even when the food quality is low (i.e., the light and dark shaded areas in the diagram on Plate 11[ii(b)]). In the extreme case, mega-herbivores such as the elephant (*Loxodonta africana*) can ingest very coarse vegetation in the highest rainfall areas (i.e., low quality branches from rain forest trees). Mega-herbivores continue to be numerous even in areas with exceptionally low food quality (i.e., the hatched area in the diagram on Plate 11[ii(b)]).

In the more realistic case where a population of antelope contends with both top-down and bottom-up forces as illustrated in the diagram on Plate 11[ii(c)], there are at least four different evolutionary solutions by which the effects of predator regulation can be reduced. (1) Herbivores could seasonally migrate up and down a rainfall-induced gradient of vegetative cover to reduce the risk of predation. In other words, when resources are abundant (such as the wet season) herbivores could move into open sparsely vegetated and relatively predator-free refuges (i.e., move from point C to D in the diagram on Plate 11[ii(c)]) independently of the quality of the forage. As the available water decreases and grazing senesces in the dry season herbivores would be forced back into the risky tall-grass woodlands (as predicted by the predator-sensitive foraging hypothesis (Sinclair and Arcese, 1995)). Zebra (*Equus quagga*) exemplify this type of movement; evidence suggests zebra in Serengeti migrate in a similar pattern to wildebeest even though they do not have the same nutritional constraints. Their movement appears to be more calculated in that they avoid risky habitats associated with wet areas in a predator-sensitive manner unlike wildebeest (Hopcraft *et al.*, 2014; Sinclair, 1985). This could be a strategy employed also by some small species of antelope such as Thomson's gazelle and would be an informative angle for further research.

An alternative evolutionary strategy by which a population could reduce the top-down effects of predation is to (2) decrease the chance encounters with predators by becoming exceedingly secretive and hiding in dense vegetation that is associated with more mesic patches within the landscape. This cryptic strategy is exemplified by species such as oribi (*Ourebia ourebi*), dik-dik, and bongo (*Tragelaphus eurycerus*) (Mduma, 1995; Estes *et al.*, 2011). These species are adept at hiding from predators and have evolved to live in wetter parts of the landscape that are dominated by low-quality forage (i.e., they move from point C to E in the diagram on Plate 11[ii(c)]).

The third evolutionary strategy by which antelope could avoid predation is to take the opposite approach and (3) avoid ambush predators by living in very open habitats such as semi-arid savannas and deserts. This is the strategy of species such as besia oryx (*Oryx beisa*), addax (*Addax nasomaculatus*) and dama gazelle (*Nanger dama*) (Gasaway *et al.*, 1996; Owen-Smith and Mills, 2006). These species avoid areas where ambush predators are prevalent, but have evolved adaptations that enable them to live in areas where food and water are spatially and temporally highly variable in open dry landscapes (i.e., they move from point C to F in the diagram on Plate 11[ii(c)]).

A separate strategy by which antelope could reduce the impact of predation is (4) to evolve social structures such as group vigilance (including multiple species associations) that have the dual advantage of simultaneously increasing the detection of predators as well as diluting an individual's risk of being caught (Fryxell *et al.*, 2007). The evolution of vigilant herds could allow a herbivore population that is predator-regulated to become more abundant than otherwise expected based on the amount of rainfall in the area (i.e., expand from the dark shaded area to the moderately shaded area in the diagram on Plate 11[ii(c)]). Vigilant herds of large-bodied animals such as African buffalo avoid virtually all predators, and their populations can expand into the light shaded areas in the diagram on Plate 11[ii(c)] where they become regulated by the availability of sufficient vegetative biomass instead.

Disturbances

Disturbances such as fires introduce heterogeneity and variation into this paradigm and can flip a population between top-down and bottom-up regulation depending on their frequency, duration, intensity, and extent. Fires remove dense vegetation (Bond and Keeley, 2005; Higgins *et al.*, 2000; van Langevelde *et al.*, 2003), which could otherwise conceal predators, as well as changing the quantity and quality of the available forage (Anderson *et al.*, 2007; Eby *et al.*, 2014). Therefore, fires very rapidly switch the mode of regulation from predation to food quantity to food quality and back again, especially in intermediate rainfall areas where they are common. Ecosystems in this intermediate rainfall range, such as savannas, often show multiple stable states (Dublin *et al.*, 1990; Staver *et al.*, 2011) in which the same species occupying two different ecosystems could either be top-down regulated (e.g., wildebeest in Kruger) or bottom-up regulated (e.g., wildebeest in Serengeti) (Hopcraft *et al.*, 2010). Therefore, disturbances change the slope of the predator, food quality, and food quantity thresholds in the diagram on Plate 11[ii(d)] leading to a large amount of variation especially under moderate rainfall conditions.

Body size and the division of niche space

If populations of antelope are regulated either by predation, food quality, or food quantity and these factors are simultaneously influenced by body size and rainfall (Plate 11[i,ii]), then the spatial distribution of different species of antelope should be somewhat predictable. Small herbivores should tend to select the safest areas with the least risk of predation, while large herbivores should select areas with the greatest abundance of forage with little regard for predation (typically high-rainfall areas). Intermediate sized ungulates should select areas based on the quality of the food.

To test the hypothesis that body size leads to the predictable division of niche space, the distribution of five species of grazers in the Serengeti ecosystem was analysed (Hopcraft *et al.*, 2012). The data are from routine aerial surveys collected over a 21-year period from which a recurrence metric was calculated for each 5 × 5 km cell; high values indicate consistent and repeated occurrence of many individuals in the cell across all surveys, while small values indicate infrequent observations of few animals. Landscape features were aggregated into latent variables (Figure 1) representing food quality (measured as grass nitrogen), food abundance (measured as the combination of annual rainfall and the topographic wetting index to estimate the landscape's ability to retain water as a proxy of vegetative biomass), and predation risk (measured as the percent woody cover, distance to drainages, and landscape curvature to estimate the cover and viewsheds available to hunting predators) using structural equation models (SEM). The extent to which the variation in food quality, food abundance, or predation risk explains the variation in recurrence of each species was assessed while simultaneously accounting for correlation between exogenous predictor variables (for more details about SEM analysis see Grace (2008)). The results suggest the distributions of small ungulates such as Thomson's gazelle and Grant's gazelle (*Nanger granti*) are constrained by predation and food quality, whereas the distribution of large grazers such as African buffalo is relatively unconstrained (Figure 1; note that the width and the colour of the arrows are proportional to amount of variation explained by each factor) (Hopcraft *et al.*, 2012). The distributions of intermediate sized ungulates such as hartebeest (*Alcelaphus buselaphus*) and topi (*Damaliscus lunatus*) suggest they optimize access to grass biomass of sufficient quality in relatively predator-safe areas (Figure 1).

These results not only support our understanding of how ungulates divide their niche space, they also allude to the processes shaping the spatial structure of antelope communities. Most importantly, they suggest that if annual rainfall increases in the future this could shift the balance between the availability of food and the risks of predation, which would favour large herbivores over small herbivores, and vice versa.

Do environmental gradients affect animal movement?

The abundance and spatial distribution of African herbivores are largely determined by their evolutionary response to finding food and avoiding predators, but do these factors also affect the movement paths of animals and their ability

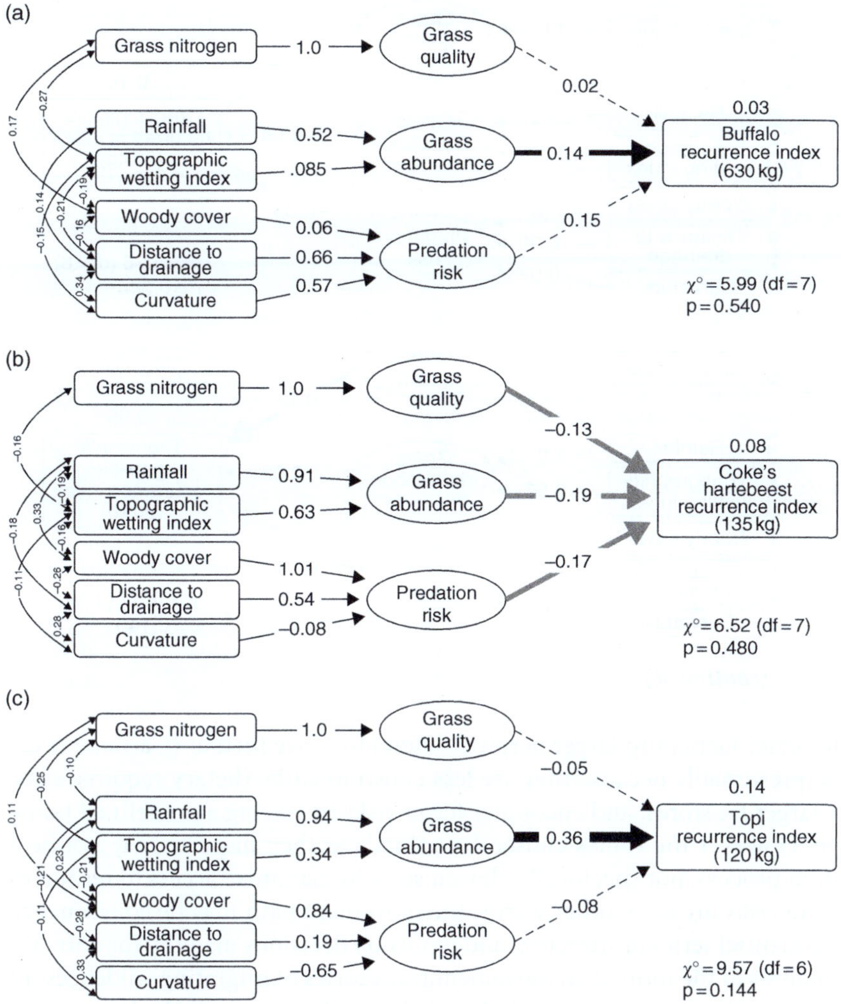

Figure 1 **Structural equation models assessing the spatial recurrence of savanna grazers in the Serengeti ecosystem; (a) African buffalo, (b) Coke's hartebeest, (c) topi, (d) Grant's gazelle, and (e) Thomson's gazelle. Solid lines indicate significant paths (p < 0.05) and are weighted according to their standardized path strength. Positive path strengths are illustrated black lines and increase the probability of recurrence. Negative path strengths in grey decrease the probability of recurrence. Curved double-headed arrows represent correlations between exogenous predictor variables and the overall variation explained by the model is in bold above the recurrence index box. The results illustrate the distribution of small ungulates are constrained by predation and grass quality while large ungulates are weakly constrained by food abundance. (Reproduced from (Hopcraft *et al.*, 2012))**

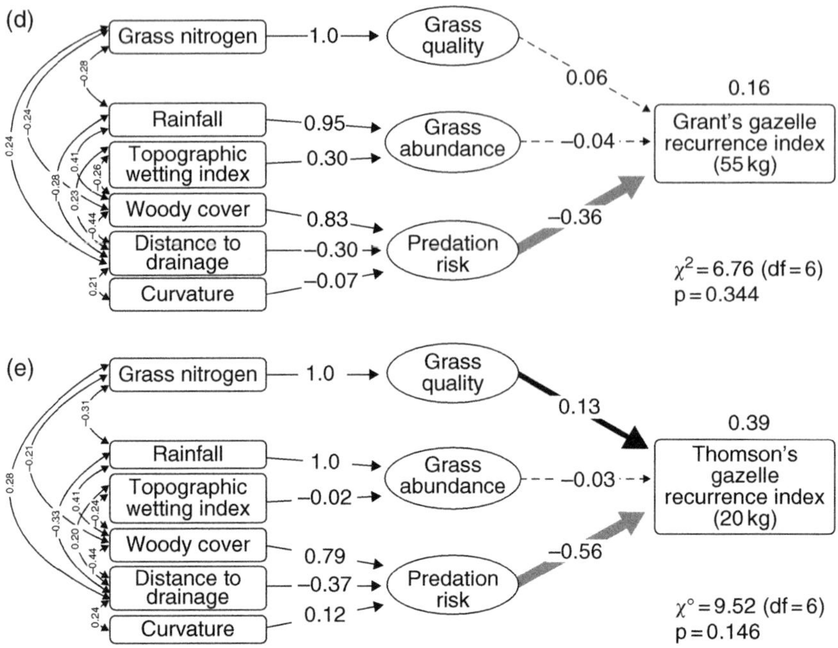

Figure 1 ***(continued)***

to disperse? Generally, larger herbivores tend to move further (Carbone *et al.*, 2005) presumably because they are less constrained by dietary requirements, have larger fat stores, and encounter fewer risks in leaving a predefined home range. However the routes chosen by animals as they disperse may not be a random process, but specifically chosen so as to maximize access to resources or avoid risky areas (in other words, the movement path may be consequence of a sequential series of attraction and repulsion decisions made by the animal). An analysis of the individual movement trajectories of migrating wildebeest in the Serengeti ecosystem suggests the distance they move each day and the direction they choose to turn are almost entirely in response to the quality of food and not the risk of predation (Hopcraft *et al.*, 2014). However, in comparison migratory zebra balance the risk of predation with access to high quality food of sufficient biomass (Hopcraft *et al.*, 2014), which would be more akin to the movement of large bulk-feeding antelope. This finding emphasizes that fact that two species moving through the same landscape at the same time are making decisions about how to move based on completely different cues. This marked difference in movement strategies suggests that the dispersal of herbivores is also determined by gradients of forage quality and abundance as well as the risks of predation (especially humans, whose effects on animal movement are far greater than natural predation (Hopcraft *et al.*, 2014)).

Presumably an animal in a large vigilant group might make different choices about how to move than a solitary individual (Haydon *et al.*, 2008) (wildebeest herds are often in excess of 300,000 animals), which further illustrates how the net effect of top-down and bottom-up forces might be modulated by group size (see section 'Top-down regulation' and Plate 11[ii]).

Conclusion

The structure of ungulate communities in Africa is broadly shaped by the degree of top-down versus bottom-up regulation, which is modified by geographical factors such as rainfall. This has led to the evolution of different biological mechanisms such body size, digestive physiology, and social behaviour that enables each species to exploit specific niches along the predator-food gradient that is itself rainfall dependent. The synergy between water-nutrient cycles and predator-prey interactions selects for different physiological and behavioural responses that collectively determine the general abundance, distributions, and movement of Africa's antelope. In particular, (1) rainfall determines both the quality and quantity of forage that is available to herbivores. (2) Herbivores' ability to process this forage is constrained by the evolutionary trade-offs related to body size such that only large antelope (or hind-gut fermenting herbivores from other clades) can process low-quality food because of the time and space it requires to ferment coarse vegetation in the rumen (Plate 11[i]). Additionally, (3) body size also influences herbivores' sensitivity to predation such that small herbivores are susceptible to many more predators than large herbivores, and consequently they are more ecologically restricted in the habitats they can use (Plate 11[i]). (4) Adaptive behaviours such as migration and group vigilance enable herbivores to escape local regulation (Plate 11[ii(a–c)]) and become more abundant than we would otherwise expect. (5) The extent and frequency of disturbances such as fires, which remove large amounts of vegetation very rapidly, alter the quality and quantity of primary production as well as remove cover for stalking predators, and result in frequent switches between multiple modes of regulation especially in areas with intermediate rainfall (Plate 11[ii(d)]).

The pervasive role of rainfall in African ecology leads us back to the original question of how climate change might affect rainfall patterns across the continent and with it the future of antelope. The magnitude of the effect that rainfall has on both top-down and bottom-up processes is evident across all ecological scales. By simultaneously altering forage and predation, rainfall not only determines the maximum size of populations as a whole, but also affects the spatial distribution of herbivores over the landscape (i.e., the division of niche space), as well as the choices made by individual animals moving through these landscapes. If climate change leads to declining rainfall this would push an ecosystem down the x-axis in the diagram on Plate 11[iii] removing vegetative biomass, opening the visibility, and preventing primary vegetation from investing in lignified support structures and complex defence mechanisms. In essence, based on the current understanding of antelope ecology, a low rainfall scenario would reduce the amount of forage available to herbivores, but increase its nutritional value while simultaneously making predators less effective at capturing prey and would favour herds of small selective antelope such as gazelles over large herbivores such as buffalo. In the extreme case, very low rainfall would lead to desertification and would favour arid-adapted antelope such as addax and oryx living in small highly mobile groups. Conversely, if rainfall increases in the future this would push the ecosystem up the x-axis resulting in hard lignified primary production that is difficult for herbivores to digest and conceals stalking predators. This would favour herds of large herbivores with slow but efficient digestive systems that are too large for most predators to handle. The abundance of small and medium-size antelope would decline because their populations would be become regulated by the quality of forage and greater rates of predation.

Africa's landscapes are becoming increasingly fragmented by humans, and the corridors connecting protected areas are becoming less permeable. What are the implications of habitat fragmentation combined with climate change on antelope populations? A crude forecast based on projecting our current knowledge of antelope ecology suggests that large antelope are better dispersers than small antelope and could more easily relocate to new areas because they are relatively unconstrained by food and predation. Evidence suggests that the movement of selective feeders is closely linked with the distribution of resources while bulk feeders are not, and therefore small-bodied herbivores might have difficulty dispersing large distances into new habitats once they are fragmented. However, large animals are vulnerable to bushmeat hunting and are often the first to be extirpated, an issue that will be considered in more detail in Chapter 5 (see also Macdonald *et al.*, 2011; Topp-Jorgensen *et al.*, 2009). Therefore, it is likely that the natural ability for antelope to disperse in response to climate change could be entirely negated by humans. The challenge for ecologists is to determine how much of the variation we currently observe in antelope abundances is a consequence of climate change and how much is due to direct conflicts with humans. Clearly both of these factors play a role, and once we can pinpoint the causes of decline, then accurate mitigation strategies can be devised, and conservation efforts can be concentrated. International networks of well protected areas linked by corridors that allow natural dispersal are the most logical solution. But does society have the fortitude and political desire to accomplish this?

References

Anderson, T. M., Dong, Y. & Mcnaughton, S. J. (2006): Nutrient acquisition and physiological responses of dominant Serengeti grasses to variation in soil texture and grazing. *Journal of Ecology* **94**: 1164–1175.

Anderson, T. M., Hopcraft, J. G. C., Eby, S., Ritchie, M., Grace, J. B. & Olff, H. (2010): Landscape-scale analyses suggest both nutrient and antipredator advantages to Serengeti herbivore hotspots. *Ecology* **91**: 1519–1529.

Anderson, T. M., Ritchie, M. E., Mayemba, E., Eby, S., Grace, J. B. & Mcnaughton, S. J. (2007): Forage nutritive quality in the Serengeti ecosystem: the roles of fire and herbivory. *American Naturalist* **170**: 343–357.

Bell, R. H. V. (1982): The effect of soil nutrient availability on community structure in African ecosystems. In *Ecology of Tropical Savannas*. Huntley, B. J. & Walker, B. H. (Eds.). Berlin: Springer-Verlag.

Bell, R. H. V. (1970): The use of herb layer by grazing ungulates in the Serengeti. In *Animal populations in relation to their food resources*. Watson, A. (Ed.). Oxford: Blackwell.

Bond, W. J. & Keeley, J. E. (2005): Fire as a global 'herbivore': the ecology and evolution of flammable ecosystems. *Trends in Ecology & Evolution* **20**: 387–394.

Botkin, D. B., Mellilo, J. M. & Wu, L. S.-Y. (1981): How ecosystem processes are linked to large mammal population dynamics. In *Dynamics of large mammal populations*. Fowler, C. W. & Smith, T. D. (Eds.). New York, John Wiley & Sons.

Carbone, C., Cowlishaw, G., Isaac, N. J. B. & Rowcliffe, J. M. (2005): How far do animals go? determinants of day range in mammals. *American Naturalist* **165**: 290–297.

Demenocal, P. B. (2004): African climate change and faunal evolution during the pliocene-pleistocene. *Earth and Planetary Science Letters* **220**: 3–24.

Dublin, H. T., Sinclair, A. R. E. & Mcglade, J. (1990): Elephants and fire as causes of multiple stable states in the Serengeti-Mara Tanzania woodlands. *Journal of Animal Ecology* **59**: 1147–1164.

East, R. (1984): Rainfall, soil nutrient status and biomass of large african savanna mammals. *African Journal of Ecology* **22**: 245–270.

Eby, S. L., Anderson, T. M., Mayemba, E. P. & Ritchie, M. E. (2014): The effect of fire on habitat selection of mammalian herbivores: the role of body size and vegetation characteristics. *Journal of Animal Ecology* **83**: 1196–1205.

Estes, L. D., Mwangi, A. G., Reillo, P. R. & Shugart, H. H. (2011): Predictive distribution modeling with enhanced remote sensing and multiple validation techniques to support mountain bongo antelope recovery. *Animal Conservation* **14**: 521–532.

Fritz, H. & Duncan, P. (1994): On the carrying-capacity for large ungulates of african savanna ecosystems. *Proceedings of the Royal Society of London B Biological Sciences* **256**: 77–82.

Fryxell, J. M., Greever, J. & Sinclair, A. R. E. (1988): Why are migratory ungulates so abundant? *American Naturalist* **131**: 781–798.

Fryxell, J. M., Mosser, A., Sinclair, A. R. E. & Packer, C. (2007): Group formation stabilizes predator-prey dynamics. *Nature* **449**: 1041–1044.

Gasaway, W. C., Gasaway, T. K. & Berry, H. H. (1996): Persistent low densities of plains ungulates in Etosha National Park, Namibia: testing the food-regulating hypothesis. *Canadian Journal of Zoology* **74**: 1556–1572.

Gordon, I. J. & Illius, A. W. (1988): Incisor arcade structure and diet selection in ruminants. *Functional Ecology* **2**: 15–22.

Grace, J. B. (2008): Structural equation modeling for observational studies. *Journal of Wildlife Management* **72**: 14–22.

Harris, G., Thirgood, S., Hopcraft, J. G. C., Cromsigt, J. P. G. M. & Berger, J. (2009): Global decline in aggregated migrations of large terrestrial mammals. *Endangered Species Research* **7**: 55–76.

Haydon, D. T., Morales, J. M., Yott, A., Jenkins, D. A., Rosatte, R. & Fryxell, J. M. (2008): Socially informed random walks: incorporating group dynamics into models of population spread and growth. *Proceedings of the Royal Society B: Biological Sciences* **275**: 1101.

Hebblewhite, M., Merrill, E. H. & Mcdonald, T. L. (2005): Spatial decomposition of predation risk using resource selection functions: an example in a wolf-elk predator-prey system. *Oikos* **111**: 101–111.

Higgins, S., Bond, W.-J. & Trollope, W.-S.-W. (2000): Fire, resprouting and variability: a recipe for grass-tree coexistence in savanna. *Journal of Ecology* **88**: 213–229.

Higgins, S. I., Bond, W. J., February, E. C., Bronn, A., Euston-Brown, D. I. W., Enslin, B., Govender, N., Rademan, L., O'Regan, S., Potgieter, A. L. F., Scheiter, S., Sowry, R., Trollope, L. & Trollope, W. S. W. (2007): Effects of four decades of fire manipulation on woody vegetation structure in savanna. *Ecology* **88**: 1119–1125.

Hilborn, A., Pettorelli, N., Orme, C. D. L. & Durant, S. M. (2012): Stalk and chase: how hunt stages affect hunting success in Serengeti cheetah. *Animal Behaviour* **84**: 701–706.

Holdo, R. M., Holt, R. D. & Fryxell, J. M. (2009): Opposing rainfall and plant nutritional gradients best explain the wildebeest migration in the Serengeti. *American Naturalist* **173**: 431–445.

Hook, P. B. & Burke, I. C. (2000): Biogeochemistry in a shortgrass landscape: control by topography, soil texture, and microclimate. *Ecology* **81**: 2686–2703.

Hopcraft, J. G. C., Anderson, T. M., Perez Vila, S., Mayemba, E. & Olff, H. (2012): Body size and the division of niche space: food and predation differentially shape the distribution of Serengeti grazers. *Journal of Animal Ecology* **81**: 201–213.

Hopcraft, J. G. C., Holdo, R. M., Mwangomo, E., Mduma, S. A. R., Thirgood, S., Borner, M., Fryxell, J. M., Olff, H. & Sinclair, A. R. E. (2015): Why are wildebeest the most abundant herbivore in the Serengeti? In *Serengeti IV*. Sinclair, A. R. E., Metzger, K., Mduma, S. A. R. & Fryxell, J. M. (Eds.). Chicago: University of Chicago Press.

Hopcraft, J. G. C., Morales, J. M., Beyer, H. L., Borner, M., Mwangomo, E., Sinclair, A. R. E., Olff, H. & Haydon, D. T. (2014): Competition, predation and migrations: individual choice patterns of Serengeti migrants captured by hierarchical models. *Ecological Monographs*, 355–372.

Hopcraft, J. G. C., Olff, H. & Sinclair, A. R. E. (2010): Herbivores, resources and risks: alternating regulation along primary environmental gradients in savannas. *Trends in Ecology & Evolution* **25**: 119–128.

Hopcraft, J. G. C., Sinclair, A. R. E. & Packer, C. (2005): Planning for success: Serengeti lions seek prey accessibility rather than abundance. *Journal of Animal Ecology* **74**: 559–566.

Hulme, M., Doherty, R., Ngara, T., New, M. & Lister D. (2000): African climate change: 1900–2100. *Climate Research* **17**: 145–168.

Jarman, P. J. (1974): Social-organization of antelope in relation to their ecology. *Behaviour* **48**: 215–267.

Macdonald, D. W., Johnson, P. J., Albrechtsen, L., Dutton, A., Seymour, S., Dupain, J., Hall, A. & Fa, J. E. (2011): Association of body mass with price of bushmeat in Nigeria and Cameroon. *Conservation Biology* **25**: 1220–1228.

Mduma, S. A. R. (1995): Distribution and abundance of oribi, a small antelope. In *Serengeti II: dynamics, management, and conservation of an ecosystem*. Sinclair, A. R. E. & Arcese, P. (Eds.) Chicago, University of Chicago Press.

Mduma, S. A. R., Sinclair, A. R. E. & Hilborn, R. (1999): Food regulates the Serengeti Wildebeest: a 40-year record. *Journal of Animal Ecology* **68**: 1101–1122.

Mosser, A., Fryxell, J. M., Eberly, L. & Packer, C. (2009): Serengeti real estate: density versus fitness-based indicators of lion habitat quality. *Ecology Letters* **12**: 1050–1060.

Owen-Smith, N. (1988): *Megaherbivores: the Influence of very large body size on ecology*. Cambridge: Cambridge University Press.

Owen-Smith, N. & Mills, M. G. L. (2006): Manifold interactive influences on the population dynamics of a multispecies ungulate assemblage. *Ecological Monographs* **76**: 73–92.

Owen-Smith, N. & Mills, M. G. L. (2008): Predator-prey size relationships in an african large-mammal food web. *Journal of Animal Ecology* **77**: 173–183.

Peters, R. H. (1983): *The ecological implications of body size*. Cambridge: Cambridge University Press.

Radloff, F. G. T. & Du Toit, J. T. (2004): Large predators and their prey in a southern African savanna: a predator's size determines its prey size range. *Journal of Animal Ecology* **73**: 410–423.

Sankaran, M., Hanan, N. P., Scholes, R. J., Ratnam, J., Augustine, D. J., Cade, B. S., Gignoux, J., Higgins, S. I., Le Roux, X., Ludwig, F., Ardo, J., Banyikwa, F., Bronn, A., Bucini, G., Caylor, K. K., Coughenour, M. B., Diouf, A., Ekaya, W., Feral, C. J., February, E. C., Frost, P. G. H., Hiernaux, P., Hrabar, H., Metzger, K. L., Prins, H. H. T., Ringrose, S., Sea, W., Tews, J., Worden, J. & Zambatis, N. (2005): Determinants of woody cover in African savannas. *Nature* **438**: 846–849.

Schlesinger, W. H., Reunolds, J. F., Cunningham, G. L. & Huenneke, L. F. (1990): Biological feedbacks in global desertification. *American Association for the Advancement of Science* **247**: 1043–1048.

Scott, R. M. (1962): Exchangeable bases of mature, well-drained soils in relation to rainfall in East Africa. *Journal of Soil Science* **13**: 1–9.

Shrader, A. M., Bell, C., Bertolli, L. & Ward, D. (2012): Forest or the trees: at what scale do elephants make foraging decisions? *Acta Oecologica* **42**: 3–10.

Sinclair, A. R. E. (1985): Does interspecific competition or predation shape the African ungulate community. *Journal of Animal Ecology* **54**: 899–918.

Sinclair, A. R. E. & Arcese, P. (1995): Population Consequences Of Predation-Sensitive Foraging: The Serengeti Wildebeest. *Ecology* **76**: 882–891.

Sinclair, A. R. E., Mduma, S. & Brashares, J. S. (2003): Patterns Of Predation In A Diverse Predator-Prey System. *Nature* **425**: 288–290.

Staver, A. C., Archibald, S. & Levin, S. A. (2011): The Global Extent And Determinants Of Savanna And Forest As Alternative Biome States. *Science* **334**: 230–232.

Topp-Jorgensen, E., Nielsen, M. R., Marshall, A. R. & Pedersen, U. (2009): Relative Densities Of Mammals In Response To Different Levels Of Bushmeat Hunting In The Udzungwa Mountains, Tanzania. *Tropical Conservation Science* **2**: 70–87.

Valeix, M., Fritz, H., Loveridge, A. J., Davidson, Z., Hunt, J. E., Murindagomo, F. & Macdonald, D. W. (2009a): Does The Risk Of Encountering Lions Influence African Herbivore Behaviour At Waterholes? *Behavioral Ecology and Sociobiology* **63**: 1483–1494.

Valeix, M., Loveridge, A. J., Chamaille-Jammes, S., Davidson, Z., Murindagomo, F., Fritz, H. & Macdonald, D. W. (2009b): Behavioral Adjustments Of African Herbivores To Predation Risk By Lions: Spatiotemporal Variations Influence Habitat Use. *Ecology* **90**: 23–30.

Van Langevelde, F., Van De Vijver, C., Kumar, L., Van De Koppel, J., De Ridder, N., Van Andel, J., Skidmore, A. K., Hearne, J. W., Stroosnijder, L., Bond, W. J., Prins, H. H. T. & Rietkerk, M. (2003): Effects Of Fire And Herbivory On The Stability Of Savanna Ecosystems. *Ecology* **84**: 337–350.

Wilmshurst, J. F., Fryxell, J. M. & Bergman, C. M. (2000): The Allometry Of Patch Selection In Ruminants. *Proceedings of the Royal Society of London B Biological Sciences* **267**: 345–349.

4

Interspecific Resource Competition in Antelopes: Search for Evidence

Herbert H. T. Prins

Resource Ecology Group, Wageningen University, Wageningen, The Netherlands

Introduction

Reviewing the knowledge about competition is an arduous task because modern ecologists are not asking for theoretical musings but demand evidence. 'Evidence-based ecology' is a trend hopefully leading to the same burgeoning of ecological theory and nature conservation as shown by medicine and health care since professionals embraced 'evidence-based medicine'. What is '*evidence*'? Evidence simply is information shown to you or found by you that is able to change your mind concerning some belief you had; evidence is equated with proof, or substantiation, or verification. Since knowledge is defined as '*justified true belief*', it is *evidence* that provides the extraneous justification – without evidence beliefs are merely based on introspection, revelation or cultural transmission.

Most present-day ecologists believe that competition happens and that it was an important force shaping the evolution of species. Murray & Illius (2000), however, outline the rise and possible fall of this particular notion: 'Interspecific competition was perceived initially with confidence as a self-evident axiom of life's evolution soon to be reported in a wide range of organisms, only to recede gradually from view during the slow forward march of empirical research. Opinion today remains divided and ambivalent as to the prevalence and importance of interspecific competition in plant and animal communities with much of the enigma arising from repeated attempts to infer causality from community *patterns*, without first having gained sufficient knowledge of the actual mechanisms by which one species might compete

Antelope Conservation: From Diagnosis to Action, First Edition. Edited by Jakob Bro-Jørgensen and David P. Mallon.

with another.' Especially the last author may have had in mind the work reported later by De Garine-Wichatitsky *et al.* (2004) where they came to 'suggest that both [greater] kudu [*Tragelaphus strepsiceros*] and impala [*Aepyceros melampus*] selected bushes on the basis of the potential number of bites they can provide, and this resulted in different search strategies. Kudu focussed on the larger bushes, which have a larger number of twigs which are out of reach of impala, and kudu also probably directed their path preferentially towards the few larger bushes to maximize encounter rates with this favoured bush category. These differences in bush selection process lead to a low overlap in resource use between the two browsers in this type of savanna' – in other words, the different feeding strategies of these two species led to a diet selection that could have been interpreted as if it had been caused by competition but was not so.

Limits of this review and the sources of information

In this review I only deal with interspecific competition, that is, the competition *between* species. Interspecific competition is simply defined as competition between species for resources (food and water) (see Table 1); antelope species do not compete for shelter or nest holes. I have deliberately left out theoretical modelling studies, because modelling shows what is possible, and not what is happening, even though models may yield important ideas. I also left out instances of indirect competition when, for instance, risk of predation increases for species A due to something that is happening to or is being done by species B. Indirect competition does, however, occur in ungulates (see Prins, 2000).

Electronic data bases (Web of Science, Scopus, CAB and Google Scholar) were searched with key terms 'Competition AND/OR facilitation AND ...'; the term '...' was varied as (1) antelope, (2) gazelle, (3) wildlife and as (a) *Connochaetus*, (b) *Damaliscus*, etc., till all genus names for antelope species had been run through one by one in new searches. The search was restricted to Anglophone literature between 1987 and 2012 and yielded about 110 papers of which some 90 yielded useful data for this review. All retrieved papers were completely read by me.

What is interspecific competition, and what is the evidence one must look for?

Fundamentally there are two ways of looking at interspecific competition: the first merely observes that if a food item is eaten by an individual of one species it cannot be eaten by another individual irrespective of the species it belongs to.

Table 1 **The worldview or paradigm adhered to may depend on the study system of the researcher and make the chances of invoking inter-specific competition less or more likely. For example, most gazelle species are living in arid or semi-arid systems, and most of these areas appear to fall in the domain where non-equilibrium dynamics are thought to be prevalent.**

Worldview	Relationship between animal species within the same guild	Resource limitation	Can inter-species competition be observed?
Equilibrium dynamics	Strong interactions	Present	Competition most likely to be observed, but it requires that the conditions be met.
		Absent	Competition is absent.
	Weak interactions	Present	Competition most likely cannot be observed even if it occurs because the interactions between particular species are diffuse.
		Absent	Competition is absent.
Non-equilibrium dynamics	Strong interactions	Present	Competition most likely will rarely be observed because this is a rare state and will last a short period only.
		Absent	Competition is absent.
	Weak interactions	Present	Competition most likely will never be observed because this is a rare state and the relationship between species makes it near-invisible.
		Absent	Competition is absent.

Intraspecific or interspecific competition is then fundamentally the same: scramble competition is a good example of this type of competition. This occurs when individuals of different species occur in a multi-species assemblage of more-or-less similar species: individuals of an assemblage of three small antelopes (red duiker, *Cephalophus natalensis*, common duiker, *Sylvicapra grimmia*, and suni, *Neotragus moschatus*) foraging on fallen fruits and leaves show this type of competition (Prins *et al.*, 2006). Scramble competition can easily be envisaged: these small antelopes roam about and cannot influence the rate of falling or the place where fruits or leaves will fall; monopolizing a location does not appear to be a good strategy since more fruits can fall somewhere else. Much of the resource use by duiker antelopes in West and Central Africa seems to be of this type (see Newing, 2001). Note that

Belton *et al.* (2008) did not find evidence for competition between suni and other small browsers because they were looking for diet segregation and niche partitioning: scramble competition does not result in these.

This example shows the fundamental issue that is at stake in asking questions about competition, namely, does it matter whether a suni has eaten a fallen fruit if that fruit would not have been consumed by a red duiker? For instance, because ants had found it before the red duiker arrived, or because fungi had taken out all the nutrients already before a red duiker could have digested it? Indeed, some conditions must be met before one can talk about competition; the most important one is that *it must matter* to the individuals that are affected by it, and secondly it must matter *in a particular way*: namely, it must affect their vital rates (mortality, fecundity, longevity). Perhaps it must even affect their fitness. *In addition* one can demand that it must affect the total numbers of the species or population. So, *that* is the evidence one must find.

The other fundamental way of looking at interspecific competition is that it somehow affects or has affected the evolution of the two species that are competing: this is asking whether some form of resource partitioning is either the result of competition (now or in the past) or leads to reduced competition. This latter option I will not cover here, simply because Connell (1980) already pointed out that 'for [this] step, it must be shown that evolution has actually occurred, i.e. that the divergence has a genetic basis.' So the question is then not whether the two species diverged (that is evident) but whether the difference in resource use has a genetic foundation. I am not aware of any publication about antelope ecology that is even scratching the surface of this question.

Knowledge was defined above as '*justified true belief*' and 'evidence' is the information that can provide the justification. So, if one believes in the occurrence of interspecific competition (and one may doubt it following Murray & Illius (2000)) one must search for the evidence that particular actions (like foraging) by individuals of species A negatively affect vital rates of individuals of species B by excluding all other possible causes of depression of these rates (so, excluding predation, disease transfer, droughts, etc.). Simple observational studies nearly always fail to provide the right evidence for this: an excellent example being the study of Melton (1987) who showed how a series of the wrong conclusions and deductions were made from mere observations leading to a believable narrative ('although ticks could have been the main proximate mortality factor, asymmetric resource competition was suggested as the ultimate mortality factor, with impala and nyala [*Tragelaphus angasi*] the principal competitors' (Melton, 1987)). On the basis of his narrative the management decided that there were too many impala and nyala, and started culling. Melton then showed that the whole edifice of implicated inter-specific competition was built on quicksand. In other words, the hypothesis that was based on inter-specific competition between impala and nyala on the one hand, and waterbuck on the other, was soundly rejected.

In this most insightful paper, Melton (1987), then asks himself why the evidence does not bear out what he expects. Interestingly he did not draw the conclusion that inter-specific competition between impala-nyala and waterbuck did not occur, but he points out that the experimental situation was not conducive at all: 'The combined effects of ... two climatic factors [drought, cyclones] make simple testing of the competition hypothesis difficult.' He thus, inadvertently I think, draws attention to the first and last of three fundamental complications, the first being the issue of multi-causality, the second the issue of weak interactions. and the third the issue of non-equilibrium (Jonzén *et al.*, 2002).

Multi-causality is the bane of community ecology. Proper hypothesis-testing is nearly always impossible since it is nearly always impossible to ensure that all other ecological conditions are similar, except in experimentally controlled circumstances (Connell, 1980) and nearly all field studies that try to prove the occurrence of competition fail to deliver the required proof (see below). Next to multi-causality, there is the issue of 'strong interactions' between species versus 'weak interactions'. Weak interactions occur especially in multi-species guilds. It has been observed that if one particular species increases in numbers the negative effect of that increase is exerted diffusely on all other species within the assemblage (e.g., Prins & Douglas-Hamilton, 1990; O'Kane *et al.*, 2011), an issue to a large extent ignored in studies looking for the effects of competition. Thirdly, there is the issue of non-equilibrium. Many rangeland ecologists in tropical savannas adhere to this view so succinctly brought to the fore by Ellis & Swift (1988) namely that animal populations in arid and semi-arid regions rarely if ever reach a situation that conveniently is described as equilibrium. Most plant ecologists and perhaps even animal ecologists in the temperate zone (apart from most North American prairie-ecologists) have the worldview that their systems are in equilibrium, and this has far-reaching consequences (*vide* Illius & O'Connor, 1999; also Rietkerk *et al.*, 1996), but anyone who has followed the saga of the work of Rolf Peterson on the wolves and moose of Isle Royale (see his annual reports), and how he came to the conclusion that the wolves and the moose never reached anything close to equilibrium over a period of half a century, should be acutely aware that resource limitation, if it occurs, is very elusive indeed.

Whether all habitats where antelopes occur are dominated by non-equilibrium dynamics is not certain. Apart from the African savannas, the matter of equilibrium vs. non-equilibrium dynamics of the steppes of Asia is addressed by Kerven (2004) – all these areas harbour gazelles, and our understanding of whether they are resource-limited or not, is not very far developed yet. Nevertheless, it appears that droughts, severe snowstorms, and very severe frosts may prevent any equilibrium between resources and gazelle numbers in these areas (see also Kropotkin, 1904). Likewise, many of the African rainforests appear to provide much less constant food supplies than was thought before (see for instance the work of Caroline Tutin and co-workers in Gabon) (e.g., Tutin & White, 1998). Australian ecologists Andrewartha & Birch (1954) had arrived at this conclusion already long before, perhaps because they were working in a typical boom-and-bust environment. The important point is that it is likely that only rarely are the basic assumptions of resource limitation met for antelopes.

If knowledge is defined as '*justified true belief*', then belief in resource limitation must be investigated first, and very few ecologists nowadays do this: the seminal work on African buffalo (*Syncerus caffer*) by Tony Sinclair (Sinclair, 1977; cf. Prins, 1996) shows how difficult that is, how time-consuming, and how very important for a true understanding of population dynamics and, thus, competition. The only other work that I am aware of that shows a relation between population density, recruitment, and condition in antelopes is about impala (Gaidet & Gaillard, 2008). These fundamental issues are so important that they come close to paradigmatic worldviews, and these have far-reaching implications for the type and quality of the evidence one has to find to convincingly demonstrate interspecific resource competition in, for example, antelopes (Table 1).

Strong evidence, weak evidence, and no evidence

Modern ecology increasingly is faced with the question whether the data or a paper provide evidence for the statement that is made or for testing the hypotheses one believes in (Prins & Gordon, 2014a, b). We know that competition between ungulates is difficult to prove (see Putman, 1996; Prins, 2000). Inspired by the need to base justification on evidence, some ecologists now call for a sifting of the literature to separate beliefs from evidence (Sutherland *et al.*, 2004; Pullin & Knight, 2005). Recently, Spear & Chown (2009) followed this evidence-based approach when they reviewed the available evidence for the biodiversity impacts of non-indigenous ungulates. By-and-large they found very little strong evidence. Spear & Chown (2009) provide a table in which they present the criteria and rate the strength of evidence, on basis of which I constructed Table 2. This table shows how evidence has to be weighed concerning competition or facilitation ('negative competition'). Both for facilitatory effects as for competitive ones, the strength of the evidence is arranged from 'weak' to 'strong', from mere observations on resource partitioning (i.e., the absence of perfect overlap in resource use between two species) and

Table 2 **Strength of evidence for the effects of resource competition and resource facilitation, based on Spear & Chown (2007: Table 1) but modified (I have added the requirement for resource limitation to the facilitation rows). If 1d or 2c reveal non-significant relationships it is likely that diffuse competition or facilitation takes place. If two species have no perfect resource use overlap ('resource partitioning') then this is insufficient information to draw the conclusion that this points at competition (see text).**

Criteria and methodology	Strength of evidence
1. Exploitative competition	
a. Two species use the same resource:	insufficient
• Food: observation, stomach analysis, faecal analysis, literature review, feeding morphology (worst)	
• Habitat: observation, telemetry, habitat suitability modelling (worst)	
b. Effects from species A decreases the resource intake of individuals of species B	weak
c. The resource is limiting for population growth + 1a & 1b	moderate
d. The species which undergoes the competition is significantly negatively affected: the population declines + 1c	strong
e. Direct relationship between population decline and additional species + 1d	very strong
• Replicated field experiment (best)	
• Correlational (less good)	
2. Exploitative facilitation:	
a. Effects from species A promotes the resource intake of individuals of species B	
a.1 Correlational study with randomly stratified selection of replicated sites	moderate
a.2 Islands with and without introduced ungulates	moderate
a.3 Exclosures/enclosures with randomly stratified selection of replicated sites	moderate
b. The resource is limiting for population growth + 2a	moderately strong
c. The species which benefits from facilitation is significantly positively affected: the population increases + 2b	strong
d. Direct relationship between population decline and additional species + 2c	very strong
• Replicated field experiment (best)	
• Correlational (less good)	

habitat segregation, to depressed or elevated levels of food intake, the additionally required proof of resource limitation, to, finally, significant effects on population numbers and the proof of direct effects.

It should be clear from this list and from Table 2 that the type and quality of the data one needs to prove the existence of competition or facilitation increases with the strength of the evidence – the strongest evidence is most difficult to produce.

Cases of insufficient evidence for competition in antelopes

In many papers the authors drew the conclusion that if 'their' two species do not fully and completely have the same diet, then that would mean that there is resource partitioning, and that this would point at the possibility of competition. This is strictly speaking fallacious reasoning for at least some reasons. Indeed, it may mean (a) that the two species have evolved a (slightly) different morphology, or physiology, or set of requirements – such evolution can have been caused by a myriad of factors, and since scientists were not observing the forces that shaped the present-day species, no falsifiable deductions can be made. It may also mean (b) that, say, giving up densities are different because individuals of the two species experience danger differently. It could mean, perhaps, (c) that habitat requirements that are not related to food intake but to energy expenditure or the use of another resource, like water, determine the animals' time budgets differently. I refer to the essay written by Connell (1980) to better appreciate the problems of how to separate competition and evolution.

In antelope studies, there are many papers that show partial resource use and/or differential use of space. The latter is often described as spatial segregation but the verb 'to segregate' has an active connotation while these studies did not compare a situation from before the competitor was around to after the competitor was introduced. In other words, although these studies may yield important insights for nature conservation and protected area planning, they do not prove the existence of competition. Good examples are Wood (1989); Namgail, Fox & Bhatnagar (2004); or Namgail *et al.* (2008). The relative high diet overlap of Przewalski's gazelle (*Procapra przewalskii*) and Tibetan gazelle (*P. picticaudata*) reported by Li *et al.* (2008) only shows that, and does not allow for a conclusion concerning competition because lack of food was not shown to limit the population numbers of these species (see Tables 1 and 2). Interestingly, the diets of these two antelopes did not differ between areas depending on whether they lived together or not (Li *et al.*, 2008), emphasizing that competition between the two species is under the current circumstances of no importance. Nevertheless, their recommendations make complete sense from a nature conservation point of view! Leslie (2010), in my view correctly, observes 'because of the rarity of many of the [wild ungulate species] throughout the Tibetan Plateau, competitive interactions among them are now reduced.' In the same vein, the high degree of diet overlap between introduced greater kudu and native white-tailed deer (*Odocoileus virginianus*) on a ranch in Texas (Gray *et al.*, 2007) merely shows a high degree of diet overlap, but that does not allow for any conclusion concerning competition. In their review of the effect of exotic ungulates, Spear & Chown (2009) could only find one study that met their criteria; this in fact described an effect on white-tailed deer (Harmel 1978) but even that study suffered from lack of replication.

A study of grey rhebok (*Pelea capreolus*) and southern mountain reedbuck (*Redunca fulvorufula fulvorufula*) did not provide evidence for inter-specific competitive exclusion and showed complete habitat overlap; but since both antelopes have a totally different diet, with rhebok being a browser and mountain reedbuck a grazer, these results cannot serve as proof for competition (Taylor *et al.*, 2007). The review provided by Newing (2001) nicely shows differences in diet between the seven duiker species from West Africa and the Congo rainforest, but this cannot be accepted as evidence for competition (for criteria see Table 2) because she catalogues a non-exhaustive listing of other factors that can explain dietary differences, and, indeed, the shade of the competition past may be there, but it is equally possible that these duiker species are generally not food limited. She suggests that food resources were limited in certain seasons and that species were in direct competition – indeed, it could suggest this but many other things can be suggested too, so I do not consider this 'evidence'.

Newing (2001) describes the work of F. Feer (published in French) as 'a detailed comparison between the bay duiker [*C. dorsalis*]and the similar-sized diurnal Peter's duiker [*Cephalophus callipygus*], and concluded that the difference in activity patterns reduced not only direct competition at food patches but also indirect competition, by allowing exploitation of different resources. He argued that, since most fruits fall by day, a smaller quantity and variety of fruits are available by night, and the bay duiker has adapted by developing the ability to use poorer quality fruit and more foliage.' I, however, ask myself what fitness advantage bay duikers get from – apparently – 'avoiding competition' with Peter's duiker? I do not accept this as evidence for competition. In the literature on competition, one (still) can encounter teleological forms of reasoning as if individuals of different species partition their diets because they 'want' or 'try' to avoid competition. In this review, I am not addressing this fundamental flaw, but for an examination of the problem of teleological reasoning, see Depew (2010): the 'goal' of avoiding competition simply cannot be used as a causal explanation.

However, Newing's (2001) reports that 'geographical differences in habitat use in the presence or absence of congenerics would support the hypothesis that habitat use is limited by interspecific competition, and evidence of this kind was found for the yellow-backed duiker [*Cephalophus silvicultur*] and Jentink's duikers [*C. jentinkii*]. Where Jentink's duiker is absent, the yellow-backed duiker is common in closed canopy forest. In contrast, where it is sympatric with Jentink's duiker the yellow-backed duiker has been recorded only in secondary vegetation. Thus it appears that in areas of sympatry, Jentink's duiker displaces the yellow-backed duiker from forest habitats. Since the yellow-backed duiker is present throughout the range of Jentink's duiker, there is no evidence as to whether the reverse is true in secondary habitats.' So perhaps this indeed could be taken as evidence, but if it can be taken as such, it is at its best indirect evidence because neither food intake rates or population limitation were studied.

Knoop & Owen-Smith (2006) considered possible competition between roan antelope (*Hippotragus equinus*) and Burchell's zebra (*Equus quagga*) and blue wildebeest (*Connochaetus taurinus*) as cause of decline of the roan antelope (as of other 'rare antelopes'). They postulate that the 'ability of the roan to feed on the tall vlei grasses provides them with a partial refuge from competition from these two ungulate species. However, the greater dependency of the roan on upland savanna grasses during the dry year raises the potential for greater competitive overlap under such conditions.' However, since in the same paper they conclude that 'it seems that the prime factor limiting the abundance of roan in the region is predation pressure, accentuated by an influx of staple prey species for lion into the roan range' the issue of competition in this study is irrelevant, and this thus does not provide evidence for competition.

Earlier, roan antelope had been subject to another investigation into their decline (Harrington *et al.*, 1999) which posited six possible causes of decline (drought and its effects, resource competition between roan and zebra, a predator-mediated interaction between these two species, diseases, immunisation stress, and effects of artificial waterpoints), and with 18 years of annual counts the researchers tried to disentangle these. Their conclusion was that 'the ultimate cause was the provision of numerous artificial water points in the roan range, which attracted the influx of zebra and wildebeest, particularly during drought conditions', but I have not been able to find the proof for that statement in the paper of Harrington *et al.* (1999) other than a narrative description of grassland conditions becoming 'less favourable for roan as a result of the persistent concentrations of zebra and wildebeest'. I do not take this as evidence for competition to occur in this system (see also the section below 'Multi-causality and untestability'). I feel strengthened in my conclusion by the work of Codron *et al.* (2009) who on basis of the use of stable carbon isotopes could show that roan antelope, sable antelope (*Hippotragus niger*), and tsessebe (*Damaliscus lunatus lunatus*) had little niche overlap with bulk grazers such as Burchell's zebra; they ascribed it to partitioning of resources (interpreted mainly as tall- versus short-grass grazing, respectively) and concluded there was no competition. One illustration of how difficult it is to reach a causal explanation from descriptive studies is the interpretation by Dunham *et al.* (2004) that the decline of the tsessebe in the same area followed 'from rainfall-induced changes in food availability during the dry season, to a period of low rainfall' – in other words, another explanation again. For another population again, in nearby Zimbabwe, the tsessebe decline was ascribed to a combination of drought and competition with cattle (Dunham *et al.*, 2003).

A similar series of papers can be found in which the competing species are livestock, and the mechanisms how livestock can outcompete wildlife (cf. Prins, 1992), thus antelopes are by now well studied by, for instance, Mishra *et al.* (2002); Bagchi, Mishra & Bhatnagar (2004); Mishra *et al.* (2004) (all in the dry side of the Himalayas of India); Campos-Arceiz, Takatsuki & Lhagvasuren (2004); Yoshihara *et al.* (2008); Li, Jiang & Li (2008) (all three studies in the Mongolian area), and Wronski (2010) (in Saudi Arabia). Nearly all these studies face the same methodological problem, namely that they ignore all other effects herders can have, such as hunting, chasing away wild competitors, poisoning, or preventing the wild animals, the use of water sources. This is well illustrated by Wronski's (2010) description of the decline of mountain gazelles (*Gazella gazella*) in Saudi Arabia. It is thus quite questionable whether negative interactions between gazelles and livestock can be taken as indicative of competition simply because the human factor may be completely overriding. A case in point is the devastating effect herders have on wildlife when they use motorbikes for pursuing and killing antelopes as currently happens in the Sahara and in the Gobi Desert area. But it also has been shown how fences and agriculture threaten antelopes' survival in the Asian steppes: for instance the Przewalski's gazelle (Leslie *et al.*, 2010), Tibetan antelope (*Pantholops hodgsonii*), and Tibetan gazelle (Fox *et al.*, 2009).

A study by Bonnington *et al.* (2007) which shows a negative relation between spoor density of puku (*Kobus vardoni*) and cattle density is suggestive but does not meet any of the criteria concerning evidence for competition. The study by Stapelberg *et al.* (2008) on the possible competitive interaction between blue wildebeest and springbok (*Antidorcas marsupialis*), even though it shows that these two Kalahari species do not feed in the same fashion and do not feed on the same forage, cannot be taken as evidence for competition because the conclusion is based on analogy with another ecosystem (*viz.*, Burchell's zebra, white-bearded wildebeest, and Thomson's gazelle (*Eudorcas thomsonii*) in the Serengeti). Indeed, declaring interactions between species as analogous does not make them functionally identical, and even if the wildebeest were regulating the Thomson's gazelle in the Serengeti (Sinclair & Norton-Griffiths, 1982), this is no proof that springbok are resource-limited in the Kalahari. In other words, this study does not provide evidence for competition.

Multi-causality and untestability – more cases without evidence for competition

Much of the research on antelopes and ungulates in general suffers from an exuberance of possible causes to explain numerical changes in study populations. Of course, this reflects reality: few complicated patterns can be explained by a single cause. Listing all possible causes without an endeavour to separate them will, however, not yield any scientific insight in a proper cause-and-effect chain. Without that insight, management becomes a slow process of learning by doing, forgetting again, trying something else, and so on. In engineering, this is called 'tinkering', which should not be confused with adaptive management. So, Mamo *et al.* (2010) may be right when they believe that the decline of mountain nyala (*Tragelaphus buxtoni*) in the Bale Mountains National Park (Ethiopia) is caused by a myriad of reasons but 'many of the demographic parameters of mountain nyala did not change significantly over time'. So, even if mountain nyala suffer from livestock, this study cannot count as evidence because it does not meet the criteria of Table 2.

Likewise, in their study on Mongolian saiga antelope (*Saiga tatarica*), Lushchekina *et al.* (1999) state as possible causes of decline many other possibilities besides competition with livestock. In their study of four-horned antelope (*Tetracerus quadricornis*), Krishna *et al.* (2009) mention that the species is negatively affected by a plethora of factors including other ungulates especially chital (*Axis axis*). There is little doubt in my mind, that all these factors partly or in combination explain the decline of this antelope, but before this can be taken as evidence for competition, quite complicated long-term studies are needed. Also Breebaart *et al.* (2002) show dietary overlap between Boer goats and southern African browsers (namely, greater kudu and black rhinoceros (*Diceros bicornis*)) and thus suggest resource competition, but again the critical tests have not been executed, so this cannot count as evidence either.

A recurrent difficulty in the literature on possible competition is problems with methodology and the desire to see more in the data than statistical analysis or logical inference allows. The most important error is to not apply correction factors for multiple testing; Bonferroni-corrections or some other form of allowing for multiple testing increases the threshold for rejecting a null hypothesis considerably, but in some instances a Šidák-correction would be more appropriate. A case in point is a paper by Borner *et al.* (1987) on the possible decline of the Thomson's gazelle in the Serengeti (Tanzania). They explored five factors (predation, disease, poaching, emigration, and interspecific competition). In the current review, I have not taken this publication as evidence for anything, since I do not believe that any of these conclusions was drawn on the basis of proper inference: on the basis of five censuses (with their very large confidence intervals) one cannot test five possible causes, not even with any Bonferroni-correction.

The Serengeti results are even more difficult to interpret when Dublin *et al.* (1990) state: 'Recent censuses in the 1980s have allowed us to test the hypotheses that: (1) there was competition between wildebeest and Thomson's gazelle. This predicted that gazelle numbers should have declined in the 1980s when wildebeest were food limited. Census figures show no change in gazelle numbers between 1978 and 1986, a result contrary to the interspecific competition hypothesis' while Borner *et al.* (1987) report that 'in the early 1970s, numbers [of the Thomson's gazelle] stood at 0.66 million animals but had decreased to less than 0.25 million animals in 1985'. Dublin *et al.* (1990) write that they 'could not repeat the findings of Borner *et al.* (1987) and we find little evidence for a change in population. Since there may have been differences in bias error between the count of 1972 and those later, it is doubtful that there has been any decline in population since 1972.' I have shied away from drawing a conclusion for this review because if the world's specialists on Serengeti ungulates can have such a divergent opinion (~400,000 Thomson's gazelles!), then how, as a 'Serengeti-outsider', can I say who were right?

The problem of multi-causality or even webs of causality, competing null hypotheses, and conclusion drawing is a very serious problem in ecology. The way forward seems to be as formulated by Jonzén *et al.* (2002). Johnson & Omland (2004) amongst other have been advocating model selection on the basis of the Akaike Information Criterion, but in their paper the sting is in the tail – not to pay meticulous attention to their three caveats highlighted by Jonzén (2002) plays havoc with the presently supposed superiority over the testing of null hypotheses. I found one paper that starts by very precisely defining the biological underpinning of the competing hypotheses (as advised later by Jonzén *et al.* (2002)), namely the paper by Sinclair & Norton-Griffiths (1982). However, their paper suffers from ignoring the effects of many antelope species and other ungulates interacting with each other (the weak interaction, referred to above), because they reduce that plurality to a set of only five competing hypotheses. Even though they did not apply formal statistical decision rules (at that time the Akaike Information Criterion method had not been applied in ecology to allow formal model selection; however, they do not even provide a sign-test), they reject the facilitation hypothesis, and find support for the competition from wildebeest on Thomson's gazelle (in contrast to Dublin *et al.* (1990), see above, which makes reviewing not easy). But to put it bluntly, I think hard evidence was not provided by Sinclair & Norton-Griffiths (1982).

Multi-hypothesis testing appears quite often in the literature on antelope ecology. Also Dunham *et al.* (2003) applied this technique and investigated the effects of rainfall, cattle management, wildlife numbers, captures, hunting, and predation (six hypotheses) on a data set about tsessebe numbers from 1991 to 1999 (five censuses). I do not believe that the way they did it was according to methodologically sound rules, but the paper was accepted by the journal, and has been cited 28 times (26 May 2015: Google Scholar, which is more often than many papers written by me). Similarly, O'Connor & Krüger (2003) tried to explain the decline of reedbuck (*Redunca arundinum*), oribi (*Ourebia ourebi*), grey rhebok, and mountain reedbuck by increased competition from increasing populations of blesbok (*Damaliscus pygargus phillipsi*) and blue wildebeest, but apart from the effect of these other antelope species, they also tested for the effect of precipitation and [cold, wet] weather, habitat quality, habitat area, predation and poaching. The number of data points was only 12 to 15 and they did not apply a Bonferroni-correction. For the purpose of this review, I agree with them, namely, 'it is possible [that] the increase of blesbok and wildebeest after 1999 had a negative effect on the recovery of reedbuck and especially oribi populations, but this cannot be determined from available data.' As outlined, many studies suffer from the problem of too many hypotheses, which then are used to test the same dataset. Mysterud *et al.* (2011) present four not mutually exclusive hypotheses formulated with different but partly identical predictions, which makes testing extremely difficult. They concluded 'that more detailed predictions are needed'.

Areas where much research has been done on antelopes also include the Masai-Mara in Kenya, which in some ways can be seen as a continuation of the Serengeti although it does not share all migratory populations (see, e.g., Ottichilo, de Leeuw & Prins, 2001). Here, too, multi-causality is fleshed out in much detail by Ogutu *et al.* (2011), but they do apply a different statistical technique to overcome the problem of multi-causality. Contrariwise, they use model selection on the basis of the Akaike Information Criterion. They conclude that recruitment rates declined for all species but giraffe, and they test four hypotheses (effects of rainfall, competition, bomas, and huts). This at first sight is a good approach, but then they do not follow the advice of Johnson & Omland (2004) or Jonzén *et al.* (2002) to have precisely formulated a priori hypotheses that are biologically sound (see below): they start searching for patterns in their data and look at lagged rainfall, monthly rainfall, mention interaction with lions, find humps for one species, and others for other species; cohort effects; susceptibility to predation; population density; disease outbreaks. They observe that the effects of predation can interact with many other factors of interest and that decadal oscillations in rainfall can interact with predation. Let me be clear: I think they are right – there is a causal web. But I do not believe that such webs can be understood by blasting a series of census results against all, or many, or some of these causal factors without seriously violating statistical inference rules. I think a path-analysis has to be done, and I doubt that using the Akaike Information Criterion is sufficient: indeed, the AIC method is meant to select between statistical models but not to test the validity of null-models, and Jonzén *et al.* (2002) even remark that the 'interplay between demography and environmental stochasticity may ruin the interpretation at a more fundamental level.... So, what should ecologists do? Some authors suggest that time-series analysis should be used for diagnosis rather than modelling.... The results should then be considered as *clues* instead of facts' [emphasis added]. Anyhow, even with their unique data set with unparalleled precision of numbers of individuals per age class, and unparalleled precision in allocating individuals to age classes (meaning that all error in their models could be ascribed to process error), Ogutu *et al.* (2011) did not conclude whether competition was important or not, even though it was one of the four possible explanatory factors.

Cases of weak evidence for competition in ungulates

Table 2 gives conditions for when evidence is provided for competition: that is when the effects from species A decrease the resource intake of individuals of species B. However, if the study has not shown that resources are limiting population growth, then this only provides weak evidence for competition. To be able to prove that resource intake is reduced by other animals nearly always necessitates an experimental approach, and that is what Cameron & du Toit (2007) did in a well-executed experiment with giraffe (*Giraffa camelopardalis*). Even though giraffe is not an antelope, these animals compete with antelopes, and I thus include them in my review. A more important reason, though, is that this case of competition is one of the few that is underpinned by experimentation. I quote: 'the evolution of the giraffe's elongated neck is widely considered a classic example of evolutionary biology based on foraging competition. This competition hypothesis had been implicitly accepted because it seemed so obvious until Simmons and Scheepers (1996) [see also Simmons & Altwegg, 2010] questioned the evidence for it and suggested that alternative explanations, such as sexual selection, could be equally plausible. The problem was that no study had been designed to explicitly test whether giraffes achieve a foraging advantage by foraging above the reach of smaller browsers' (Cameron & du Toit, 2007; see also Woolnough & du Toit 2001). They then experimentally excluded shorter browsers: steenbok (*Raphicerus campestris*) and impalas that forage relatively low in the trees, and greater kudu that forage at medium height while giraffes normally feed even higher. They showed that their findings are consistent with the hypothesis that competition drives feeding height stratification. However, since multi-causality is quite likely, their experimental evidence for resource competition does not lead to the rejection of the alternative hypothesis that sexual selection is the cause of the long necks. This well-executed study, by the way, discloses other difficulties too: the competitors of the giraffe are a group of species (steenbok, impala, and greater kudu) thus making it quite difficult to find a strong significant negative correlation between the numbers of giraffe and the numbers of one of the competing species. It is very likely that we are dealing here with weak interactions (cf. Prins & Douglas-Hamilton 1990), and the likelihood that in this assemblage strong evidence for competition will be found is thus not large. The study thus provides *weak evidence* for competition.

The difficulty of obtaining firm answers from observational studies is illustrated by a study on the same guild of herbivores in the close by Sabi Sand Wildtuin (South Africa), in which O'Kane *et al.* (2011) reached a different conclusion, namely, that resource depletion rarely had a significant effect on either inter-specific overlap in plant species browsed, overlap in browsing heights, actual browsing heights, or plant-parts utilized, but 'giraffe may have reduced the available forage in the higher branches, competitively displacing elephant lower, but still above nyala and impala'. Woolnough & du Toit (2001) concluded that 'when resources are restricted smaller ungulates competitively displace larger ones from shared feeding sites' but I deduce from their paper that they did not find *evidence* for competition.

In a grazing experiment, Murray & Illius (2000) used topi *Damaliscus lunatus* and blue wildebeest in the migratory Serengeti population (compare with the same idea in Stapelberg *et al.* (2008) where springbok and blue wildebeest were contrasted). Their 'experiments showed that small differences in the structure of grass swards, perhaps barely noticeable to the casual observer, can produce large differences in food intake rate.... But whereas a constraint on bite quantity can be compensated for by an increase in the daily grazing time, no such compensation is possible for a limitation in bite quality, as daily intake is limited by digestive constraints rather than consumption constraints. Hence bite quality competition may be just as exacting as bite quantity competition.' Now many an ecologist would draw the conclusion that Murray & Illius had proven that there is competition between topi and wildebeest. Yet they themselves are, completely correctly so, much less convinced. In line with the criteria of Table 2 I am sure they would draw the same conclusion as me – this is *weak evidence* only, but so beautifully executed!

In another experimental study, Lagendijk *et al.* (2012) found how large browsers affect the diet of smaller ones; the study is thus the opposite to the one of Cameron & du Toit (2007), thus showing the difficulties of generalization.

In a nested design, Lagendijk *et al.* fenced an area to exclude African elephant (*Loxodonta africana*) and another area to exclude both nyala and elephant, but accessible to small-size herbivores (i.e., duiker and suni). They write that 'the study has demonstrated that browsing release by larger size herbivores (e.g., either elephant or nyala) affects resource utilization by their smaller counterparts (e.g., either nyala or small-size herbivores).' Lagendijk *et al.* (2012) did not find evidence of facilitation; rather they show, I think, quite convincing evidence that food intake of one species decreases that of others. I think they drew the right conclusion: this is what the study shows, and not more. Hence they provided *weak evidence* for competition because the study did not show that the affected herbivores were food limited and the researchers did not investigate population effects.

The study of Prins (1992) shows overlap in the use of resources between different species of livestock and different antelopes; that study also shows a negative relationship between livestock biomass and wildlife. At first sight, one may conclude that this is a case of strong evidence (see Table 2 under 1d: 'the species which undergoes the competition is significantly negatively affected: the population declines'); however, for a proper proof condition 1c of Table 2 applies, which in its turn depends on condition 1b ('the resource is limiting for population growth of the antelopes' and 'effects from livestock decrease the resource intake of antelopes', respectively). In other words, the causation of the significant negative correlation is unknown, even though it may be assumed that livestock and their herders are responsible for the decline of the wildlife. This study may be accepted as providing *weak evidence* for competition, but no more than that. Still supporting evidence for competition comes from observations such as presented by Young, Palmer & Gadd (2005) that Burchell's zebra numbers increased when cattle were removed, and that buffalo increased when cattle were removed from the Ngorongoro Crater (Prins, 2000).

A similar study was published by Khan *et al.* (1996); they studied the effect of the removal of the migratory cattle into the Gir Forest in India. The ungulate community in Gir comprises of two large (nilgai, *Boselaphus tragocamelus*; sambar, *Cervus unicolor*), two medium (chital; wild pig, *Sus scrofa*) and two small-sized ungulate species (four- horned antelope, *Tetracerus quadricornis*; chinkara, *Gazella bennettii*). The cessation of migratory livestock into the reserve had an enormous impact especially on the chital (axis deer), but not on the other species because they are browsers. Khan *et al.* (1996) also state that they believe that the presence of excessive domestic livestock was actually limiting the chital population in Gir in the 1960s and 1970s. Whether this study serves as a good example of competition needs a bit more attention. Khan *et al.* (1996) write that 'the increase in chital density provides conclusive but indirect evidence of the cattle population limiting the growth of chital population (a principally selective grazer) as a result of competition for common food resources [but] information on food habits of wild ungulates from cafeteria trials during the dry season, where all species fed on 20 offered browse species, led S. H. Berwick (1974) to believe that no forage competition existed between domestic livestock and wild ungulates'. I have not been able to access Berwick's unpublished PhD thesis (note that Khan *et al.* (1996) did not provide evidence that antelope numbers were positively affected by the removal of the cattle).

A similar result was found by Madhusudan (2004) where gaur *Bos gaurus* (but also chital and Asian elephant *Elephas maximus*) showed a strong increase in numbers along a gradient of decreasing cattle and other livestock. Since he also measured standing crops of food and studied diets, habitat, and space overlap, this study shows perhaps *moderate evidence* for competition although strictly speaking there was no proof that food was limiting population numbers of gaur, and, of course, one can doubt whether gaurs should be considered antelopes. Moreover, because the gradient in livestock numbers was not experimentally controlled, other factors may have played a role too, so that is why I do not rate this study as strong evidence.

Fritz *et al.* (1996) found that 'interspecific competition occurred between cattle and impala, especially during the wet season and the hot dry season'. Yet, this study may be taken as *weak evidence* for competition according the criteria of Table 2.

There is evidence for competition between African elephants and African buffalo (which may be regarded as a very large antelope). De Boer & Prins (1990) report a clear negative influence of the grazing of buffalo on that of elephant. On a time scale of several days, high grazing pressure by elephant in a specific patch on a given day takes place only in the absence of high grazing pressure by buffalo on preceding days, suggesting that elephant react to the grazing by buffalo. The study did not investigate whether elephant were resource limited, but the elephants had reached an enormous density of about 6 per km^2 (Prins & Douglas-Hamilton, 1990) so it is not unlikely that they were resource-limited, and perhaps this provides *weak evidence* for competition (see also Prins, 1996). Evidence also exists of competition between elephant and wildebeest. Dublin (1995) states that the general tendency for elephant to over-utilize woodlands in the dry season may be a direct consequence of competition with migratory wildebeest. Wildebeest can sometimes remove as much as 90% of the herbaceous standing crop during its transient occupancy (McNaughton, 1985; Onyeanusi, 1989; Sinclair, 1975). Immediately following cessation of the rains and prior to the arrival of the wildebeest, elephant show an obvious preference for grasses (Beekman & Prins 1989; Dublin, 1995). However, upon the arrival of the wildebeest, a distinct change in diet occurs. Elephant immediately decrease feeding on grasses and, in the later stages of wildebeest occupancy, stop feeding on grasses altogether, despite the production of new grass following intermittent local thunderstorms (Dublin, 1995). Wildebeest graze the grasses down to lawn height (McNaughton, 1985), a height at which grasses may become difficult for elephant to eat, even though they can use their feet to kick up small plants (Beekman & Prins, 1989). However, I do not know whether it was proven that the Serengeti elephant were resource-limited.

Cases of facilitation in ungulates

The issue of facilitation of antelopes by zebra has been a long-running theme in African savanna ecology (since Vesey-FitzGerald, 1960). The careful experiments by Odadi *et al.* (2011) in which they used donkey as substitute-zebra and cattle as substitute-antelopes showed that when herded together, both species gained more weight, had higher bite rates (especially at low stocking density), and selected diets with a more favourable balance between digestible organic matter and crude protein, than when herded separately. This study showed no evidence of competition between the two species but rather a facilitative interaction between them. But this cannot count as evidence for or against competition or facilitation between *antelopes*.

In their study of the Gir Forest in India referred to above, Khan *et al.* (1996) observed a decline of the nilgai when the cattle were removed. They postulate that the cattle grazers or their livestock had kept the vegetation open and that because of the denied access of the migratory cattle the Gir became more wooded.

A beautiful example of facilitation is provided by Veblen & Young (2010) between zebra and cattle and/or African elephant: the zebra by heavily grazing the grass *Panicum stramineum* shift the competitive balance between the grasses towards *Cynodon plectostachyus*, which is grazed more by cattle and elephant. However, a numerical response has not been shown, neither a change in fitness of the cattle or the elephant: this study thus provides only *moderate evidence*. There are more reports of facilitation in the literature. A number of studies have suggested that facilitation may occur between migratory herbivores in savanna systems although on a smaller timescale. Grazing of coarse grasses by Burchell's zebra and wildebeest stimulates the growth of new, high-quality grasses and dicotyledonous material. This, in turn, may promote grazing by Thomson's gazelle (Bell, 1971; McNaughton, 1976), but these studies do not really show a numerical response or a fitness effect either, and are better interpreted as persuasive than as providing evidence. To what extent the foraging of small herbivores is limited in denser vegetation, and whether browsing of larger herbivores facilitates smaller herbivores in more productive woodland savanna, still needs experimental investigation (see the paper of Odadi *et al.*, 2011, referred to above).

Making things really complicated is the insight that smaller browsers like Grant's gazelle (*Nanger grantii*), Thomson's gazelle, Kirk's dik-dik (*Madoqua kirkii*) and impala (Belsky, 1984; Prins & Van der Jeugd, 1993) affect savanna woodland structure much more than is often appreciated, and more perhaps than elephant. As with the studies of Cameron & du Toit (2007) and O'Kane *et al.* (2011) referred to above, the species that exert this influence are more than one and operate as a guild, which makes it very unlikely that strong statistical correlations can be found to provide very strong evidence for facilitation (see Table 2). However, a study could be set up to prove strong evidence. For example, several authors had suggested that elephant and, to a minor extent, giraffe may facilitate small herbivores by opening up dense thickets (see for references Van de Koppel & Prins, 1998), and Kohi *et al.* (2011) set up experiments to prove that African elephant can change the vegetation to such an extent that greater kudu, impala and duikers benefit. Even though they did this experimentally, they could not, however, prove that the population numbers of the affected species increased: in other words, also this study only provides moderate evidence for facilitation and not by *antelopes*.

Conclusion: there is no evidence that different antelope species compete with each other

I have been able to identify a few studies that provide (weak or moderate) evidence for competition and a few that provide (moderate) evidence for facilitation. The cases of facilitation are of elephant for antelopes (Lagendijk *et al.*, 2012; Kohi *et al.*, 2011) and perhaps of cattle for nilgai (Khan *et al.*, 1996), and zebra facilitating cattle and elephant (Veblen & Young, 2010; cf. Odadi *et al.*, 2011). I have not been able to find evidence that antelopes are facilitating other antelopes, even though one can easily envisage how it could work (Prins & Olff, 1998; Olff *et al.*, 2002). The cases of competition are as follows: small antelopes competing with giraffe (Cameron & du Toit, 2007), cattle that compete with gaur (Madhusudan, 2004), cattle that compete with deer (but not with antelopes) (Khan *et al.*, 1996), cattle that compete with impala (Fritz *et al.*, 1996), cattle that compete with buffalo (Prins, 2000), livestock that compete with wildlife (Prins, 1992), nyala that compete with other antelopes (Lagendijk *et al.*, 2012), and buffalo that compete with African elephant (De Boer & Prins, 1990) and possibly blue wildebeest that compete with African elephant (Dublin, 1995). So the evidence for competition between different species of antelope is very weak given the large number of studies that have been carried out; the same conclusion had been drawn by Gurevitch *et al.* (1992) for herbivorous species in general. So all in all, there is very little evidence for inter-specific competition between antelopes and perhaps the conclusion must be that there is no competition: in 1990 Fred de Boer and I published a paper citing Shakespeare's 'strive mightily but eat and drink as friends', which catches the problem well: of course, each individual is trying to maximize its fitness but to prove that individuals of different species are competing is perhaps just too difficult. Or perhaps it is just unimportant (see Connell, 1980) and niche differentiation is not the result of competition but arises because species evolved separately and already became adapted to different resources or parts of the habitat in which they evolved (Connell, 1980). Thereafter competition may keep them apart, as has been demonstrated in several instances. This was also the view of Grant (1975): 'adaptations already possessed by the species at the time of meeting are the principal determinants of coexistence.' And Peter Grant still believes this (pers. comm. January 2013). Different antelopes are definitely 'apart' and different as Murray and Illius (2000) so elegantly showed, but whether it is *competition* that keeps them apart is very questionable indeed – perhaps each species is sitting on its own Wrightian peak and slightly worse fitness for the slightly less adapted individuals is what keeps each species on its own adaptive peak. For that scenario, one does not have to study putative competition, but fitness differences between individuals.

I think that presently, after some 60 years of work on antelopes, not a single study has presented strong evidence that is solid enough to prove competition, only a bit of weak evidence emerges: presently the most frugal stance is thus to believe that there is no evidence for competition between antelope species.

References

Andrewartha, H. G. & Birch, L. C. (1954): *The distribution and abundance of animals.* Chicago: University of Chicago Press.

Bagchi, S., Mishra, C. & Bhatnagar, Y. V. (2004): Conflicts between traditional pastoralism and conservation of Himalayan ibex (*Capra sibirica*) in the Trans-Himalayan mountains. *Animal Conservation* **7**(2): 121–128.

Beekman, J. H. & Prins, H. H. T. (1989): Feeding strategies of sedentary large herbivores in East Africa, with emphasis on the African buffalo. *African Journal of Ecology* **27**: 129–147.

Bell, R. H. V. (1971): A grazing ecosystem in the Serengeti. *Scientific American* **232**: 54–65.

Belsky, A. J. (1984): Role of small browsing mammals in preventing woodland regeneration in the Serengeti National Park, Tanzania. *African Journal of Ecology* **22**: 271–279.

Belton, L. E., Dalerum, F. & Van Rensburg, B. J. (2008): Factors associated with suni distribution in Tembe Elephant Park, South Africa: implications for management and conservation of a small antelope. *African Journal of Ecology* **46**: 631–636.

Berwick, S. H. (1974): The community of wild ruminants in the Gir forest ecosystem, India. PhD Thesis, Yale University.

Bonnington, B., Weaver, D. & Fanning, E. (2007): Livestock and large wild mammals in the Kilombero Valley, in southern Tanzania. *African Journal of Ecology* **45**: 658–663.

Borner, M., FitzGibbon, C. D., Borner, M., Caro, T. M., Lindsay, W. K., Collins, D. A. & Holt, M. E. (1987): The decline of the Serengeti Thomson's gazelle population. *Oecologia* **73**: 32–40.

Breebaart, L., Bhikraj, R. & O'Connor, T. G. (2002): Dietary overlap between Boer goats and indigenous browsers in a South African savanna. *African Journal of Range & Forage Science* **19**: 13–20.

Cameron, E. Z. & du Toit, J. T. (2007): Winning by a neck: tall giraffes avoid competing with shorter browsers. *American Naturalist* **169**: 130–135.

Campos-Arceiz, A., Takatsuki S. & Lhagvasuren B. (2004): Food overlap between Mongolian gazelles and livestock in Omnogobi, southern Mongolia. *Ecological Research* **19**: 455–460.

Codron, D., Codron, J., Lee-Thorp, J. A., Sponheimer, M., Grant, C. C. & Brink, J. S. (2009): Stable isotope evidence for nutritional stress, competition, and loss of functional habitat as factors limiting recovery of rare antelope in southern Africa. *Journal of Arid Environments* **73**: 449–457.

Connell, J. H. (1980): Diversity and the coevolution of competitors, or the ghost of competition past. *Oikos* **35**: 131–138.

De Boer, W. F. & Prins, H. H. T. (1990): Large herbivores that strive mightily but eat and drink as friends. *Oecologia* **82**: 264–274.

De Garine-Wichatitsky, M., Fritz, H., Gordon, I. J. & Illius A. W. (2004): Bush selection along foraging pathways by sympatric impala and greater kudu. *Oecologia* **141**: 66–75.

Depew, D. J. (2010): Is evolutionary biology infected with invalid teleological reasoning? *Philosophy & Theory in Biology* 2: e105. http://quod.lib.umich.edu/cgi/t/text/idx/p/ptb/6959004.0002.005?rgn=main;view=fulltext;q1=Evo-devo (as 22 February 2016).

Dublin, H. T. (1995): Vegetation dynamics in the Serengeti-Mara Ecosystem: the role of elephants, fire, and other factors. In *Serengeti II: dynamics, management, and conservation of an ecosystem*. Sinclair, A. R. E. & Arcese, P. (Eds.). Chicago: University of Chicago Press.

Dublin, H. T., Sinclair, A. R. E., Boutin, S., Anderson, E. Jago, M. & Arcese P. (1990): Does competition regulate ungulate populations? Further evidence from Serengeti, Tanzania. *Oecologia* **82**: 283–288.

Dunham, K. M., Robertson, E. F. & Grant, C.C. (2004): Rainfall and the decline of a rare antelope, the tsessebe (*Damaliscus lunatus lunatus*), in Kruger National Park, South Africa. *Biological Conservation* **117**: 83–94.

Dunham, K. M., Robertson, E. F. & Swanepoel C. M. (2003): Population decline of tsessebe antelope (*Damaliscus lunatus lunatus*) on a mixed cattle and wildlife ranch in Zimbabwe. *Biological Conservation* **113**: 111–124.

Ellis, J. E. & Swift, D. M. (1988): Stability of African pastoral ecosystems: alternate paradigms and implications for development. *Journal of Range Management* **41**: 450–459.

Fox, J. L., Dhondup, K. & Dorji T. (2009):Tibetan antelope *Pantholops hodgsonii* conservation and new rangeland management policies in the western Chang Tang Nature Reserve, Tibet: Is fencing creating an impasse? *Oryx* **43**: 183–190.

Fritz, H., de Garine-Wichatitsky, M. & Letessier G. (1996): Habitat use by sympatric wild and domestic herbivores in an African savanna woodland: the influence of cattle spatial behaviour. *Journal of Applied Ecology* **33**: 589–598.

Gaidet N. & Gaillard, J. M. (2008): Density-dependent body condition and recruitment in a tropical ungulate. *Canadian Journal of Zoology* **86**: 24–32.

Grant, P. (1975): The classical case of character displacement. *Evolutionary Biology* **8**: 237–337.

Gray, S. S., Simpson, T. R., Baccus, J. T., Manning R. W. & Schwertner T. W. (2007): Seasonal diet and foraging preference of greater kudu *Tragelaphus strepsiceros* in the Llano Uplift of Texas. *Wildlife Biology* **13**: 75–83.

Gurevitch, J., Morrow, L. L., Wallace, A. & Walsh, J. S. (1992): A meta-analysis of competition in field experiments. *American Naturalist* **140**: 539–572.

Harmel, D. E. (1978): The influence of exotic artiodactyls on white-tailed deer production and survival. *Job No. 20. Federal aid project No. W-109-R-1.* Austin: Texas Parks and Wildlife Department.

Harrington, R., Owen-Smith, N., Viljoen, P. C., Biggs, H. C., Mason, D. R. & Funston P. (1999): Establishing the causes of the roan antelope decline in the Kruger National Park, South Africa. *Biological Conservation* **90**: 69–78.

Illius, A. W. & O'Connor T. G. (1999): On the relevance of nonequilibrium concepts to arid and semiarid grazing systems. *Ecological Applications* **9**: 798–813.

Johnson, J. B. & Omland, K. S. (2004): Model selection in ecology and evolution. *Trends in Ecology & Evolution* **19**: 101–108.

Jonzén, N., Lundberg, P., Ranta E. & Kaitala V. (2002): The irreducible uncertainty of the demography–environment interaction in ecology. *Proceedings of the Royal Society of London. Series B: Biological Sciences*, **269**(1488): 221–225.

Kerven, C. (2004): The influence of cold temperatures and snowstorms on rangelands and livestock in northern Asia. In *Rangelands at equilibrium and non-equilibrium recent developments in the debate around rangeland ecology and management:* 41–54. Vetter, S. (Ed.). Bellville: Programme for Land and Agrarian Studies (PLAAS) School of Government, University of the Western Cape (South Africa).

Khan, J. A., Chellam, R., Rodgers, W. A. A. & Johnsingh J. T. (1996): Ungulate densities and biomass in the tropical dry deciduous forests of Gir, Gujarat, India. *Journal of Tropical Ecology* **12**: 149–162.

Knoop, M. C. & Owen-Smith N. (2006): Foraging ecology of roan antelope: key resources during critical periods. *African Journal of Ecology* **44**: 228–236.

Kohi, E. H., de Boer, W. F., Peel, M. J. S., Slotow, R. C., van der Waal, I. M. A., Heitkönig, Skidmore A. K. & Prins H. H. T. (2011): African elephants *Loxodonta africana* amplify browse heterogeneity in African savannas. *Biotropica* **43**: 711–721.

Krishna, Y. C., Clyne, P. J., Krishnaswamy J. & Kumar N. S.(2009): Distributional and ecological review of the four horned antelope, *Tetracerus quadricornis*. *Mammalia* **73**: 1–6.

Kropotkin, P. (1904): *Mutual Aid: A Factor of Evolution*. London: William Heinemann.

Lagendijk, G., Page, B. R. & Slotow, R. (2012): Short-term effects of single species browsing release by different-sized herbivores on sand forest vegetation community, South Africa. *Biotropica* **44**: 63–72.

Leslie, D. M. (2010): *Procapra picticaudata* (Artiodactyla: Bovidae): *Mammalian Species* **42**: 138–148.

Leslie, D. M., Groves, C. P. & Abramov, A. V. (2010): *Procapra przewalskii* (Artiodactyla: Bovidae): *Mammalian Species* **42**: 124–137.

Li, Z., Jiang, Z. & Li, C. (2008): Dietary overlap of Przewalski's gazelle, Tibetan gazelle, and Tibetan sheep on the Qinghai-Tibet plateau. *Journal of Wildlife Management* **72**: 944–948.

Lushchekina, A. A., Dulamtseren, S., Amgalan, L. & Neronov, V. M. (1999): The status and prospects for conservation of the Mongolian saiga *Saiga tatarica mongolica*. *Oryx* **33**: 21–30.

Madhusudan, M. D. (2004) Recovery of wild large herbivores following livestock decline in a tropical Indian wildlife reserve. *Journal of Applied Ecology* **41**: 858–869.

Mamo, Y., Pinard, M. A. & Bekele, A. (2010): Demography and dynamics of mountain nyala *Tragelaphus buxtoni* in the Bale Mountains National Park, Ethiopia. *Current Zoology* **56**: 660–669.

McNaughton, S. J. (1976): Serengeti migratory wildebeest: facilitation of energy flow by grazing. *Science* **191**: 92–94.

McNaughton, S. J. (1985): Ecology of a grazing ecosystem: the Serengeti. *Ecological Monographs* **55**: 259–294.

Mduma, S. A. R. & Sinclair, A. R. E. (1994): The function of habitat selection by oribi in Serengeti, Tanzania. *African journal of Ecology* **32**: 16–29.

Melton, D. A. (1987): Waterbuck (*Kobus ellipsiprymnus*) population dynamics: the testing of an hypothesis. *African Journal of Ecology* **25**: 133–145.

Mishra, C., van Wieren, S. E., Heitkonig, I. M. A. & Prins, H. H. T. (2002): A theoretical analysis of competitive exclusion in Trans-Himalayan large herbivore assemblage. *Animal Conservation* **5**: 251–258.

Mishra, C., van Wieren, S. E., Ketner, P., Heitkonig, I. M. A. & Prins, H. H. T. (2004): Competition between livestock and bharal *Pseudois nayaur* in the Indian Trans-Himalaya. *Journal of Applied Ecology* **41**: 344–354.

Murray, M. G. & Illius, A. W. (2000): Vegetation modification and resource competition in grazing ungulates. *Oikos* **89**: 501–508.

Mysterud, A., Loe, L. E., Zimmermann, B., Bischof, R., Veiberg, V. & Meisingset, E. (2011): Partial migration in expanding red deer populations at northern latitudes – a role for density dependence? *Oikos* **120**: 1817–1825.

Namgail, T., Fox, J. L. & Bhatnagar, Y. V. (2004): Habitat segregation between sympatric Tibetan argali *Ovis ammon hodgsoni* and blue sheep *Pseudois nayaur* in the Indian Trans-Himalaya. *Journal of Zoology* **262**: 57–63.

Namgail, T., Bagchi, S., Mishra C. & Bhatnagar Y. V. (2008): Distributional correlates of the Tibetan gazelle *Procapra picticaudata* in Ladakh, northern India: towards a recovery programme. *Oryx* **42**: 107–112.

Newing, H. (2001): Bushmeat hunting and management: implications of duiker ecology and interspecific competition. *Biodiversity and Conservation* **10**: 99–118.

O'Connor, T. & Krüger S. (2003): Synchronous decline of populations of small antelope in a reserve in the southern Drakensberg, South Africa. *South African Journal of Wildlife Research* **33**: 97–107.

Odadi, W. O., Jain, M., Van Wieren, S. E., Prins, H. H. T. & Rubenstein, D. I. (2011): Facilitation between Bovids and Equids in an African Savanna. *Evolutionary Ecology Research* **13**: 237–252.

Ogutu, J. O., Piepho, H. P., Dublin, H. T., Bhola, N. & Reid, R. S. (2011): Dynamics of births and juvenile recruitment in Mara–Serengeti ungulates in relation to climatic and land use changes. *Population Ecology* **53**: 195–213.

O'Kane, C. A. J., Duffy, K. J., Page, B. R. & Macdonald, D. W. (2011): Overlap and seasonal shifts in use of woody plant species amongst a guild of savanna browsers. *Journal of Tropical Ecology* **27**: 249–258.

Olff, H., Ritchie, M. E. & Prins, H. H. T. (2002): Global environmental controls of diversity in large herbivores. *Nature* **415**: 901–904.

Onyeanusi, A. E. (1989): Large herbivore grass offtake in Masai Mara National Reserve: implications for the Serengeti-Mara migrants. *Journal of Arid Environments* **16**: 203–209.

Ottichilo, W. K., de Leeuw, J. & Prins, H. H. T. (2001): Population trends of resident wildebeest [*Connochaetus taurinus hecki* (Neumann)] and factors influencing them in the Masai Mara ecosystem, Kenya. *Biological Conservation* **97**: 271–282.

Owen-Smith, N., Chirima, G. J., Macandza, V. & Le Roux, E. (2011): Shrinking sable antelope numbers in Kruger National Park: what is suppressing population recovery? *Animal Conservation* doi:10.1111/j.1469-1795.2011.00504.x.

Prins, H. H. T. (1992): The pastoral road to extinction: competition between wildlife and traditional pastoralism in East Africa. *Environmental Conservation* **19**: 117–123.

Prins, H. H. T. (1996): *Behaviour and ecology of the African buffalo: social inequality and decision making*. London: Chapman & Hall.

Prins, H. H. T. (2000): Competition between wildlife livestock in Africa. In *Wildlife conservation by sustainable use:* 51–80. Prins, H. H. T, Grootenhuis, T. J. G. & Dolan, T. T. (Eds.). Boston: Kluwer Academic Publishers.

Prins, H. H. T., de Boer, W. F., van Oeveren, H., Correia, A., Mafuca, J. & Olff, H. (2006): Co-existence and niche segregation of three small bovid species in southern Mozambique. *African Journal of Ecology* **44**: 186–198.

Prins, H. H. T. & Douglas-Hamilton, I. (1990): Stability in a multi-species assemblage of large herbivores in East Africa. *Oecologia* **83**: 392–400.

Prins, H. H. T. & Gordon, I. J. (2014a): Testing hypotheses about biological invasions and Charles Darwin's two-creators ruminations. In *Invasion Biology and Ecological Theory: insights from a continent in transformation*: 1–19. Prins, H. H. T. & I. J. Gordon (Eds.). Cambridge: Cambridge University Press.

Prins, H. H. T. & Gordon, I. J. (2014b): A critique of ecological theory and a salute to natural history. In *Invasion Biology and Ecological Theory: insights from a continent in transformation:* 497–516. Prins, H. H. T. & I.J.Gordon (Eds.). Cambridge: Cambridge University Press.

Prins, H. H. T. & Olff, H. (1998): Species richness of African grazer assemblages: towards a functional explanation. In *Dynamics of tropical communities*: 449–490. Newbery, D. M., Prins, H. H. T. & Brown. N. D. (Eds.) *British Ecological Society Symposium* **37**. Oxford: Blackwell Science.

Prins, H. H. T. & van der Jeugd, H. P. (1993): Herbivore population crashes and woodland structure in East Africa. *Journal of Ecology* **81**: 305–314.

Pullin, A. S. & Knight, T. M. (2005): Assessing conservation management's evidence base: a survey of management-plan compilers in the United Kingdom and Australia. *Conservation Biology* **19**: 1989–1996.

Putman, R. J. (1996): *Competition and resource partitioning in temperate ungulate assemblies*. London: Chapman & Hall.

Rietkerk, M, Ketner, P., Stroosnijder, L. & Prins, H. H. T. (1996): Sahelian rangeland development: a catastrophe? *Journal of Range Management* **49**: 512–519.

Simmons, R. E. & Altwegg, R. (2010): Necks-for-sex or competing browsers? A critique of ideas on the evolution of giraffe. *Journal of Zoology* **282**: 6–12.

Simmons, R. E. & Scheepers, L. (1996): Winning by a neck: sexual selection in the evolution of the giraffe. *American Naturalist* **148**: 771–786.

Sinclair, A. R. E. (1975): The resource limitation of trophic levels in tropical grassland ecosystems. *Journal of Animal Ecology* **44**: 497–520.

Sinclair, A. R. E. (1977): *The African buffalo: a study of resource limitation of populations.* Chicago: University of Chicago Press.

Sinclair, A. R. E. & Norton-Griffiths, M. (1982): Does competition or facilitation regulate migrant ungulate populations in the Serengeti? A test of hypotheses. *Oecologia* **53**: 364–369.

Spear, D. & Chown, S. L. (2009): Non-indigenous ungulates as a threat to biodiversity. *Journal of Zoology* **279**: 1–17.

Stapelberg, H., Van Rooyen, M. V., Bothma, J., Van der Linde, M. J. & Groenewald, H. T. (2008): Springbok behaviour as affected by environmental conditions in the Kalahari. *Koedoe* **50**: 145–153.

Sutherland, W. J., Pullin, A. S., Dolman, P. M. & Knight, T. M. (2004): The need for evidence-based conservation. *Trends in Ecology and Evolution* **19**: 305–308.

Taylor, W., Skinner, J. D. & Krecek, R. C. (2007): Home ranges of sympatric grey rhebok and mountain reedbuck in a South African highveld grassland. *African Zoology* **42**: 145–151.

Tutin, C. E. G. & White, L. J. T. (1998): Primates, phenology and frugivory: present, past and future. In *Dynamics of tropical communities*: 309–337. Newbery, D. M., Prins, H. H. T. & Brown, N. D. (Eds.). British Ecological Society Symposium **37**. Oxford: Blackwell Science.

Van de Koppel, J. & Prins, H. H. T. (1998): The importance of herbivore interactions for the dynamics of African savanna woodlands: an hypothesis. *Journal of Tropical Ecology* **14**: 565–576.

Veblen, K. E. & Young, T. P. (2010): Contrasting effects of cattle and wildlife on the vegetation development of a savanna landscape mosaic. *Journal of Ecology* **98**: 993–1001.

Vesey-FitzGerald, D. F. (1960): Grazing succession among East-African game animals. *Journal of Mammalogy* **41**: 161–172.

Wood, A. K. (1989): Comparative distribution and habitat use by antelope and mule deer. *Journal of Mammalogy* **70**: 335–340.

Woolnough, A. P., & du Toit, J. T. (2001): Vertical zonation of browse quality in tree canopies exposed to a size-structured guild of Africa browsing ungulates. *Oecologia* **129**: 585–590.

Wronski, T. (2010): Factors affecting population dynamics of re-introduced mountain gazelles (*Gazella gazella*) in the Ibex Reserve, central Saudi Arabia. *Journal of Arid Environments* **74**: 1427–1434.

Young, T. P., Palmer, T. M. & Gadd, M. E. (2005): Competition and compensation among cattle, zebras, and elephants in a semi-arid savanna in Laikipia, Kenya. *Biological Conservation* **122**: 351–359.

Yoshihara, Y., Ito, T. Y., Lhagvasuren, B. & Takatsuki, S. (2008): A comparison of food resources used by Mongolian gazelles and sympatric livestock in three areas in Mongolia. *Journal of Arid Environments* **72**: 48–55.

5

Importance of Antelope Bushmeat Consumption in African Wet and Moist Tropical Forests

John E. Fa

Division of Biology and Conservation Ecology, School of Science and the Environment, Manchester Metropolitan University, United Kingdom; Center for International Forestry Research (CIFOR), CIFOR Headquarters, Bogor, Indonesia

Introduction

Human consumption of wild animals ('bushmeat') is escalating in many parts of sub-Saharan Africa, and may not be sustainable for many species. Increased commercial hunting, especially within the moist forest areas, has brought this about (Abernethy *et al.*, 2013). Hunting for trade rather than for subsistence has been promoted by increased penetration of remote forests by logging roads and by the greater access to modern weapons by hunters. Ungulates, particularly antelopes, contribute a very large proportion of the bushmeat biomass hunted in many sub-Saharan countries. Duikers are the main prey for hunters in moist forests whereas other antelopes are hunted in more open habitats (Plate 10).

The meat of wild animals has long been a part of the staple diet of forest-dwelling peoples in tropical areas worldwide. Wild meat is often the most accessible and sustainable source of protein (Fa, Currie & Meeuwig, 2003). Additionally, bushmeat is an open access resource, so cost of production is generally lower than cost of raising livestock. However, in recent years, there has been an important transition in the scale of commercial hunting and trading of wildlife because of accelerating population growth, modernisation of hunting techniques, and greater accessibility of remote forest areas.

Antelope Conservation: From Diagnosis to Action, First Edition. Edited by Jakob Bro-Jørgensen and David P. Mallon.

Wildlife and the meat derived from it, termed 'bushmeat', is sold for public consumption either fresh or smoked. Bushmeat may provide the primary source of animal protein for forest families, and many urban families (particularly those many without access to refrigeration), and can also constitute a significant source of revenue. Moreover, taste and cultural preference may play a role in driving the demand for wild meat, since it is commonly believed that people in tropical forest countries often prefer the taste of wild meat over that of domestic animals (Schenck *et al.*, 2006). That wild meat consumption may be a deeply rooted tradition that is highly resistant to change is supported by the continuing demand for wild meat from formerly rural people now living in middle-class urban or even overseas environments (Chaber *et al.*, 2010). In these cases, wild meat is consumed as a luxury item to maintain a link to a former lifestyle and is not a staple source of animal protein. However, taste studies have suggested that taste is not the primary determinant of consumer demand for wild meat (Schenck *et al.*, 2006).

The high demand for bushmeat and the lucrative trade associated with it are the main reasons for the high extraction rates estimated for many West and Central African countries (Fa & Peres, 2001). There is considerable evidence that commercial hunting has been growing in importance for some time, and thus increasing numbers of hunters currently supplement their incomes with the sale of meat (Abernethy *et al.*, 2013). Such commercial hunting increases the number of animals extracted from the forest and reduces the sustainability of offtake for numerous prey populations (Bennett & Robinson, 2000), ultimately threatening the persistence of some taxa, especially the large-bodied species (Fa & Brown, 2009). Furthermore, increases in hunting have been linked to the rise in commercial hunters, who can access heavier-calibre weapons than those available to villagers (Stiles, 2011).

Mammals are the main source of bushmeat protein throughout Africa (Fa & Brown, 2009). In particular, ungulates, primates, and rodents make up most of the meat consumed in tropical forests throughout the world (Robinson & Bennett, 2004). Amongst these, ungulates are by far the most important, both in species numbers and biomass. In this chapter, I first describe the factors determining abundance and biomass of ungulates in African wet and moist forests. I then review levels of extraction of antelope bushmeat in the different habitats and by species. Using data from the estimated potential production of the different African antelope species I raise several discussion points arising from this analysis that are relevant to the broader debate about improving conservation outcomes for species of importance to many people in the continent as a protein source.

Antelope richness and abundance in tropical ecosystems

Understanding the factors influencing antelope biomass and productivity of populations is essential to assess the sustainability of offtake. Plant biomass decreases steadily as rainfall declines in the tropics and under conditions of low seasonality and at low altitudes, 'wet forests' can be supported only in areas with rainfall above 4000 mm, 'moist forests' between 2000 and 4000 mm, and 'dry forests' between 1000 and 2000 mm. Between 100 mm and 1000 mm of rainfall, savannas, scrub and even dry woodlands thrive, but little plant biomass is possible under arid conditions of <100 mm annual rainfall. Mammalian biomass initially increases with more rainfall, but falls as forest canopy covers over habitat suitable for herbivorous ungulates. As a general pattern, ungulate biomass increases with rainfall between 500 and 1000 mm; maximum biomass is reached at 1000–1500 mm, and then declines from 1500–3200 mm (Mandujano & Naranjo, 2010). In wet and moist forest, much of the plant biomass is in the form of inedible tree trunks, and most of the leaves are heavily defended by plant secondary compounds, such as lignins and toxins, so are inedible to most mammals (Waterman & McKey, 1989; Robinson & Bennett, 2004). A high proportion of the primary productivity is concentrated in the canopy and only available to relatively small arboreal mammals such as primates, sloths, and rodents; food availability for large ungulates in tropical forests is low (e.g., Hart, 2000).

Previous studies of African savanna ecosystems, in particular, have shown that species diversity and composition, standing biomass and population densities of large herbivores are all strongly influenced by rainfall and soil nutrient contents (Coe *et al.*, 1976; Ogutu & Owen-Smith, 2003; Olff *et al.*, 2002), as well as other factors such as fire or spatial heterogeneity. More recently, Mandujano & Naranjo (2010) estimated that African savannas support significantly higher ungulate biomass (9,935 kg km^{-2}) than tropical dry forests (4,933 kg km^{-2}) or tropical wet forests (995 kg km^{-2}). From figures presented in Robinson & Bennett (2004), the proportion of biomass represented by ungulates is highest (100%) in savanna grasslands like the Serengeti (total biomass 11,606 kg km^{-2}) and lowest in Amazonian wet forests (23%; 1,087 kg km^{-2}). African savannas and palaeotropical dry forests support the highest absolute ungulate biomass, although in the absence of livestock, neotropical and palaeotropical dry and wet forests sustain a relatively similar biomass (Mandujano & Naranjo, 2010); when livestock is present in the neotropical dry forest, total ungulate biomass including livestock is similar to that of the African savannas without livestock, and to palaeotropical dry forests (Figure 1).

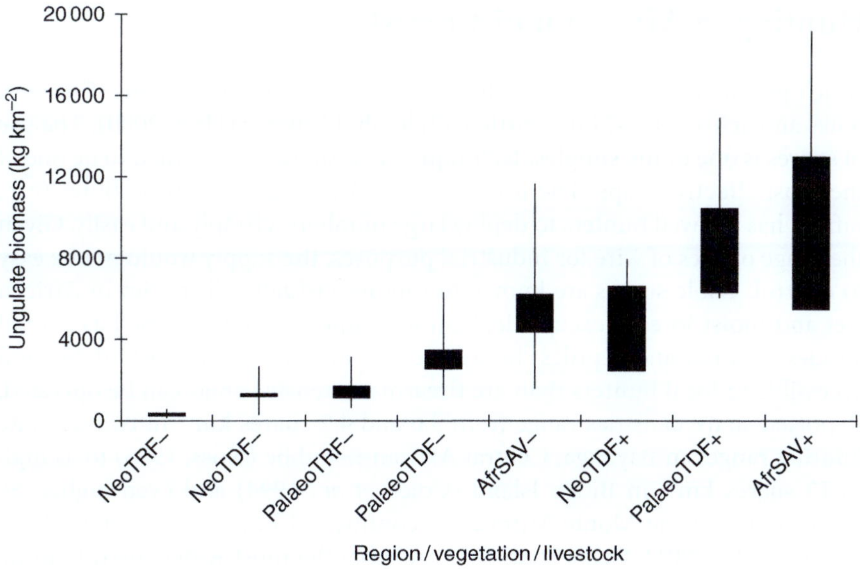

Figure 1 **Comparison of ungulate biomass between tropical dry forests (TDF), tropical rainforests (TRF), and savannas (SAV) in neotopical (Neo), palaeotropical (Palaeo) and African (Afr) regions separately. Localities where livestock were (+) or were not (–) included are shown. Modified from Mandujano & Naranjo (2010).**

Antelopes make up the major group of ungulates present in sub-Saharan Africa. Bovid species diversity, as measured by range overlap, is highest across central and east Africa, where it is entirely dominated by antelopes, with more moderate levels of diversity within western and southern Africa, and south Asia (Traill *et al.*, 2011). Low levels of diversity occur in other parts of Asia and across Europe and America, where antelopes are absent. The latter may partly reflect ecological limitations, but is largely due to the history of bovid evolution during the Miocene (Bibi *et al.*, 2009). Within forests in central and west Africa, bovid species richness includes 17 small to medium-sized frugivorous forest duikers (two duiker genera, *Philantomba* and *Cephalophus*, subfamily Cephalophinae), three dwarf antelopes (*Neotragus* spp., subfamily Antilopinae), three large-sized spiral-horned antelopes (bushbuck *Tragelaphus scriptus*, sitatunga *T. spekeii* and bongo *T. eurycerus*; subfamily Bovinae), as well as the forest buffalo (*Syncerus caffer nanus*; subfamily Bovinae). Duikers, more than other forest antelopes, are the most important source of food and income throughout central and West Africa (Fa & Brown, 2009).

Hunting in African moist forests

Antelopes in African forests are hunted with a variety of weapons (firearms, bows and arrows, nets) but mostly with leg-hold snares (Hart, 2000). The use of snares is one of the simplest techniques, and snares are claimed to be one of the most effective traps. The use of wire cable, replacing natural fibres (e.g., raffia), has allowed hunters to deploy large numbers, cheaply and easily. Given the range of uses of wire for industrial purposes, the supply would not be easy to control. Cable snares are known to capture virtually all species in African wet and moist forests, except elephant and hippopotamus, as well as several species of birds and reptiles. Because cable snares are more affordable and accessible to local hunters than are firearms, extensive areas can be operated. Reported snare densities range from 3.6 and 4.2 snares km^{-2} in two separate hunting ranges in Bayanga, Central African Republic (Noss, 1995) to as high as 13 snares km^{-2} in Bioko Island (Colell *et al.*, 1994) and even higher, 56 snares km^{-2}, in the Monte Mitra area, continental Equatorial Guinea (Fa & Garcia Yuste, 2001). Cable snaring is probably the most widely used hunting method in African moist forests (Noss, 1998). Because snaring is largely indiscriminate, prey captured by hunters using snares tend in general to reflect the relative abundance of the prey in the targeted forests (Fa *et al.*, 2005).

Although hunters who use snaring outnumber those who hunt with guns, the introduction of firearms has replaced more traditional weapons such as the bow and arrow, and consequently has contributed to increased impact of some groups of forest animals that are less vulnerable to snaring, such that arboreal primates (Mittermeier, 1987). Moreover, even if an animal has become scarce, on the verge of local extinction, hunters will tend to pursue it every time they encounter it, regardless of how seldom that is, particularly if the animal is large. However, the use of firearms is often secondary to snaring because of the greater start-up and running costs and greater skill involved in gun-hunting (Kümpel, 2006).

Antelope hunting levels in African moist forests

Fa *et al.* (2005) summarised information from a comprehensive search of studies undertaken in central and west Africa, including unpublished reports and other grey literature located by direct contact with researchers. Studies included were sites (hunting camps or villages) exclusively located within tropical moist forest, with available records of number of individuals and identification of all mammal species hunted for at least one year. Data from a total of 36 sites were available for analysis; these had an average of about 19 hunters operating from settlements of less than 1,000 inhabitants, from seven west and central African countries (Cameroon, Equatorial Guinea, Gabon, Congo, Democratic Republic of Congo, Central African Republic, and Ghana). Annual harvest rate for each site and per species (carcasses yr^{-1}) were calculated using annual harvest figures for all sites in which the species appeared. Average number of hunters active per 100 days in each study site was used as a proxy of hunting pressure. The species' general ecological characteristics (body size, abundance), as well as their level of arboreal habit, and speed of movement were employed in analyses to assess susceptibility to being hunted. A total of 71 mammal species [18 ungulates (4 families) as well as 22 primates (5 families), 13 rodents (4 families), 12 carnivores (4 families), 3 pangolins, and elephant, hyrax and aardvark] were recorded as hunted in wet and moist African forests. The three most common taxa reported in most hunting studies were the blue or Maxwell's duiker (*Philantomba* spp.), the bay duiker (*Cephalophus dorsalis*), and the brush-tailed porcupine (*Atherurus africanus*). Annual harvest rates of bushmeat, calculated for different sites in central and West Africa (Fa *et al.*, 2005), can exceed 12,000 carcasses yr^{-1} (average of about 2,000 carcasses yr^{-1}), equivalent to as much as 84,000 kg yr^{-1} (average of about 18,000 kg yr^{-1} per site; however, catchment sizes were in most cases unreported). Per hunter, mean extraction rate was around 100 carcasses yr^{-1}, close to 1,000 kg yr^{-1} in weight. In those study sites where mammalian biomass hunted per unit area has been calculated, such as the Ituri Forest, in the Democratic Republic of Congo, extraction figures can vary from as low as 50.0 kg/km^2 to as high as 209.7 kg/km^2 even where the human population density is comparable, at less than 2.00 persons/km^2 (Hart, 2000).

Data from most hunting sites indicate that it is the small (2.0–4.9 kg) and medium-sized (5–14.9 kg) mammals that provide the largest proportion of animals killed (32% and 30% respectively) with larger-bodied (15.0–99.9 kg) species only contributing 22%. However, large mammals alone can make up 55% of the total bushmeat biomass extracted in a year. Larger ungulate species are hunted less than smaller species; the same applies to carnivores whereas larger rodents and medium-sized primates were hunted more. Abundance rather than size of the animal was the main predictor of harvest level (although, particularly for primates highly social and vocal species were also more vulnerable to hunting; Fa *et al.*, 2005). Overall, antelopes accounted for 38.4% of the harvest rate if measured as biomass/year and 36.6% if measured as carcasses/year.

Published data on animal groups contributing to bushmeat trade in central and west Africa show that antelopes are the most important group (Fa, 2007). For example, in bushmeat markets in the Cross-Sanaga Rivers region in Nigeria and Cameroon, Fa *et al.* (2006) found that antelopes contributed over 70% of all individual animals sold in 89 markets, with blue duikers representing more than 50%. Data obtained from bushmeat markets and village hunts

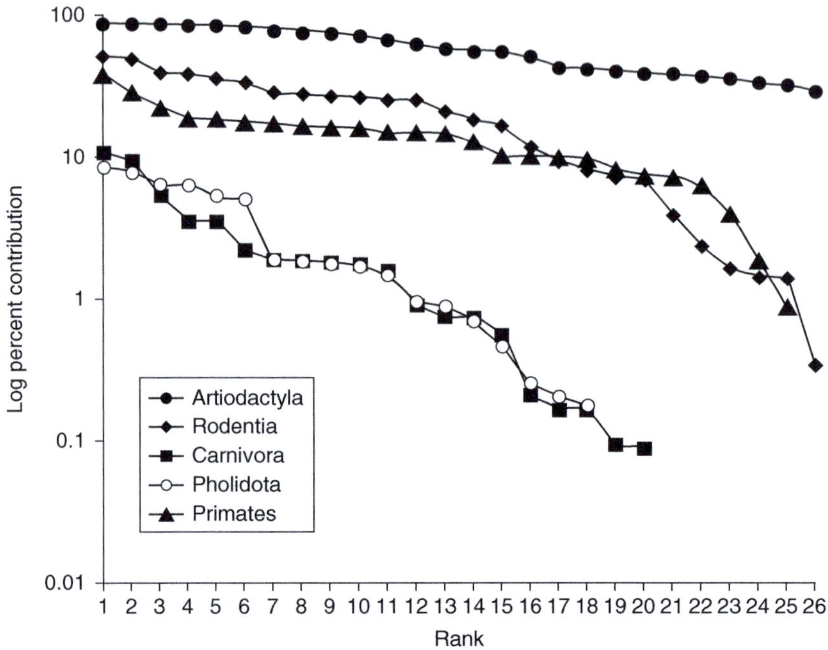

Figure 2 **Rank order diagram indicating the percentage contribution of the main mammal groups to hunting in 26 tropical moist forest sites in sub-Saharan Africa. Data from Abugiche (2008); Albrechtsen *et al.* (2007); Coad (2008); Dethier & Ghuirghi (2000); Fa & Garcia Yuste (2001); Fa *et al.* (1995); Kümpel (2006); Maisels (1996); Colell *et al.* (1994); Muchaal (1999); Noss (2000); Okouyi (2006); Poulsen *et al.* (2009); Rieu (2003); Rist (2007); van Vliet (2007); Willcox & Nambu (2007).**

in the Congo Basin further shows the importance of ungulates as a source wild meat. Ungulates, of which antelopes and the water chevrotain (Tragulidae) contributed the vast majority of carcasses and biomass per year (>95% in all sites), averaged 59.56 ± 4.04% in all sites, followed by rodents (20.31 ± 3.06%) and primates (13.42 ± 1.68%) (Figure 2).

Impact of bushmeat hunting on forest antelopes

At a given site, hunting will not be sustainable when harvest of wildlife resources consistently exceeds the maximum sustainable offtake. Generalised supply and demand curves for wildlife resources indicate that at high rainfall sites, that is, moist forests, observed harvests tend to be close to or to exceed supply (Robinson & Bennett, 2004). This accords with Bennett & Robinson's (2000) conclusion, based on studies across tropical forests, that hunting rates for many species generally were clearly not sustainable; a situation likely to have worsened in more recent times given the known increases in demand for wild meat in line with the rise in human populations (Abernethy *et al.*, 2013). At intermediate rainfall levels between 500 and 1500 mm, characterised typically as dry forests and open savannas, the high standing biomass and preponderance of ungulates can allow a much higher maximum sustainable offtake (Fa *et al.*, 2014). However, human population density is often high and hunting easy, and evidence of extreme overexploitation is emerging also from these habitats (Lindsey *et al.*, 2013). In more arid areas (below 500 mm/year), the often low human population densities may reduce harvest demands and favour sustainable offtake; however, lower ungulate biomass in arid areas also reduces the maximum sustainable offtake. Not all species are affected equally by exploitation. Because the intrinsic rate of natural increase (r_{max}) of small and medium-sized ungulates is relatively high (Robinson & Redford, 1986), and certain smaller taxa are often abundant, higher offtakes (Robinson & Bennett, 2000) are possible for these groups, such as the blue duiker (*P. monticola*) and Maxwell's duiker (*P. maxwelli*).

Actual measurements of hunting sustainability of African forest antelopes are largely unavailable, though the general consensus is that current levels of hunting are probably unsustainable, at least near large settlements (Mockrin *et al.*, 2011). As a rule, if hunting pressure is not too heavy, and large neighbouring tracts of undisturbed forest are able to buffer and replenish hunted areas, game populations can readily bounce back after exploitation, provided that source areas are allowed to exist (Novaro *et al.*, 2000). Research on the impact of hunting on *Cephalophus* species indicates that intensive hunting can cause dramatic drops in numbers, as reported for the Ogilby's duiker (*C. ogilbyi*) in the Oban Hills Region, Nigeria (Jimoh *et al.*, 2012), or even the disappearance of species such as the bay duiker (*C. dorsalis*) in the Ipassa Reserve (north-east Gabon) (van Vliet *et al.*, 2007). Even in the case of the blue duiker, there is some evidence that hunting can reduce numbers considerably (Hart, 2000; Jimoh *et al.*, 2012), but complete extirpation is often less likely because of source-sink dynamics (Van Vliet & Nasi, 2008), as long as connectivity in the landscape permits enough dispersal and immigration, and hunting offtake is not too high (Hart, 2000).

Fa *et al.* (2005) showed that, as expected, hunters are most likely to be active in areas with most game. Heavily hunted areas had, on average, smaller prey, with evidence of depletion of larger-sized species, as clearly shown for hunted prey in a parallel system in the Amazonian forests (Jerozolimski & Peres, 2003) and also indicated in the bushmeat species traded in African markets (Fa *et al.*, 2015). In African moist forest areas, there is often a higher proportion of ungulates extracted in lightly hunted areas, but more rodents are taken in the more heavily hunted sites (see Fa, 2000; Cowlishaw *et al.*, 2005). Rodents only become important prey items in disturbed areas (Eves & Ruggiero, 2000). Therefore, increases in rodents hunted suggest a reduction in the availability of more favoured bushmeat species (Fa *et al.*, 2015). A lower reproduction rate coupled with lower density of large-bodied ungulate species renders these more vulnerable to over-exploitation. In the case of African forest antelopes, species heavier than 60 kg, such as the yellow-backed duiker (*Cephalophus silvicultor*), spiral horned antelopes (members of the genus *Tragelaphus*), and forest buffalo (*Syncerus caffer nanus*) are more at risk from overhunting.

Information required to measure hunting sustainability includes the extent and variability of hunting, population densities and production of exploited species, as much as the response of hunters to exploitation. Few studies are available that have been able to collect such level of detail. Exceptionally, Hart's (2000) study of the impact of human hunting on forest antelopes in the Ituri Forest, DRC, is unprecedented in assessing population characteristics as well as the impact of human and natural predation on hunted populations of six duiker species and the water chevrotain. The importance of this study is the fact that data were gathered from hunted and unhunted populations as well from radio-collaring individual animals. The study showed that annual hunter offtake of red duikers (such as *C. dorsalis.*) and of blue duikers was between 12% and 45% respectively of standing biomass compared with approximately 9% offtake removed by leopards (*Panthera pardus*) and golden cats (*Profelis aurata*) in the unhunted sites.

Conclusion

The foundations of the bushmeat trade in African tropical forests are the forest antelopes, which provide the main source of animal protein. Although the impact of current hunting levels may be unsustainable for some species, especially in areas of high hunting pressure, there is no doubt that the forest ungulate community as a whole can provide long-term benefits to countless numbers of people if adequately managed. However, sustainable hunting of forest-dependent species such as duikers requires the protection of large forest areas from settlement and fragmentation. This prerequisite is unlikely, given the expansion of human populations and more so with the increase of the commercial wild meat trade in remote areas that would otherwise remain lightly hunted or even unhunted. In particular, our understanding of the different impacts of hunting and habitat change on individual species is still very limited. Some duiker species are able to survive or even flourish in heavily disturbed habitats, and therefore there is scope to manage them for bushmeat in such areas. However, heavy hunting is likely to affect prey populations negatively unless hunters can operate in large catchments, with low intensities of hunting and long periods before hunted areas are reused. Protecting such areas from forest clearance is a challenge but one that needs to be accepted. Provided that protected areas are not permeable to commercial bushmeat hunting (see Fa *et al.*, 2006; Macdonald *et al.*, 2011, 2012) these sites can ensure source populations of hunted species outside parks as well as a safe haven for those species that are highly endangered (Tranquilli *et al.*, 2014).

In general, antelope species are hunted relative to their abundance in a locality, which is often a reflection of the species' biological attributes (see above) as well as the hunting history of that area. The vulnerability of a species to extirpation, however, is greatest for larger-bodied bovids; these are likely to disappear first whilst *Philantomba* duikers under the same hunting regimen can persist for longer periods. However, even among similar species such as medium-sized duikers (10–18 kg), some like the white-bellied duiker (*Cephalophus leucogaster*) are known to become more vulnerable to overhunting than other duiker species. In the Ituri Forest, DRC, Hart (2000) argues that it is this species' wide-ranging movements over large home ranges that expose it to higher risk by snare capture than more sedentary species. Such species may also be more sensitive to habitat disturbance (logging and farming activities). Hence, sustainable hunting of antelope species will require systems that are able to discriminate between the different species in the hunted locality. This remains the ultimate challenge in African tropical moist forests.

Applying hunting regimes where rural communities are capable of regulating their pressure on a species is still to be widely implemented. Such systems require effective monitoring methods that have technical validity but are practically feasible. Alternatives to professional techniques, based on the involvement of local people, are emerging. Such locally based monitoring methods may be more sustainable over time, allow greater spatial coverage and quicker management decisions, lead to increased compliance, and help encourage attitude shifts toward more environmentally sustainable practices (Rist *et al.*, 2010). Nevertheless, regulation of the use of snares in particular will allow a greater approximation to sustainable hunting practices. This is because snare hunting does not allow the level of selectivity required to be able to discriminate between rare and abundant species. Even though snares can be set to trap smaller or larger taxa by altering the strength of the cable used, a broad range of species (ungulates, but also large rodents, and even some primates) is still caught. Determining the optimal snare density of a certain type (e.g., those targeting smaller antelopes and large rodents) that can deployed in an area may be a potential solution, but one that presently requires empirical testing. Moreover, determining no-take zones that can provide refuge sites for species targeted in hunted sites needs adequate study. Based on such data, it will then be possible to advocate rules of access and hunting rights to defined populations of hunters, closed seasons if applicable, as well as the clear demarcation of hunting zones and refuge areas.

References

Abernethy, K. A., Coad, L., Taylor, G., Lee, M. E. & Maisels, F. (2013): Extent and ecological consequences of hunting in Central African rainforests in the twenty-first century. *Phil. Trans. R. Soc. B* **368**: 20120303. 10.1098/rstb.2012.0303 (as 10 February 2016).

Abugiche S. (2008): *Impact of hunting and bushmeat trade on biodiversity loss in Cameroon: a case study of the Banyang-Mbo Wildlife Sanctuary*. PhD Thesis, Brandenburg University of Technology, Cottbus.

Albrechtsen, L., Macdonald, D. W., Johnson, P. J., Castelo, R. & Fa, J. E. (2007): Faunal loss from bushmeat hunting: empirical evidence and policy implications in Bioko Island. *Environmental Science & Policy*, **10**: 654–667.

Bennett, E. L. & Robinson, J. G. (2000): Hunting of wildlife in tropical forests: implications for biodiversity and forest peoples. *Global Environment Division*. Washington, DC: World Bank.

Bibi, F., Bukhsianidze, M., Gentry, A. W., Geraads, D., Kostopoulos D. S. & Vrba, E. S. (2009): The fossil record and evolution of Bovidae: state of the field. *Palaeontologia Electronica*, **12**.

Chaber, A.-L., Allebone-Webb, S., Lignereux, Y., Cunningham, A. A. & Rowcliffe, J. M. (2010): The scale of illegal meat importation from Africa to Europe via Paris. *Conservation Letters* **3**: 317–321.

Coad, L. (2007): Bushmeat hunting in Gabon: Socio-economics and hunter behaviour. PhD Thesis, University of London and Institute of Zoology, London.

Coe, M. J., Cumming D. H. M. & Phillipson J. (1976): Biomass and production of large African herbivores in relation to rainfall and primary production. *Oecologia* **22**: 341–354.

Colell, M., Maté, C. & Fa, J. E. (1994): Hunting among Moka Bubis in Bioko: dynamics of faunal exploitation at the village level. *Biodiversity and Conservation* **3**: 939–950.

Cowlishaw, G., Mendelson, S. & Rowcliffe, J. M. (2005) Evidence for post-depletion sustainability in a mature bushmeat market. *Journal of Applied Ecology*, **42**: 460–468.

Eves, H. E. & Ruggiero, R. G. (2000): Socio-economics and the sustainability of hunting in the forests of Northern Congo (Brazzaville). In *Hunting for sustainability in tropical forests*: 427–454. Robinson, J. G. & Bennett E. L. (Eds.). New York: Columbia University Press.

Fa, J. E. (2000): Hunted animals in Bioko Island, West Africa: sustainability and future. In *Hunting for sustainability in tropical forests*: 168–198. Robinson, J. G. & Bennett, E. L. (Eds.). New York: Columbia University Press.

Fa, J. E. (2007): Bushmeat markets—white elephants or red herrings? In *Bushmeat and livelihoods: wildlife management and poverty reduction*: 47–60. Davies, G. & Brown, D. (Eds.). Oxford: Blackwell.

Fa, J. E. & Brown, D. (2009): Impacts of hunting on mammals in African tropical moist forests: a review and synthesis. *Mammal Review* **39**: 231–264.

Fa, J. E. & Garcia Yuste, J. E. (2001): Commercial bushmeat hunting in the Monte Mitra forests, Equatorial Guinea: extent and impact. *Animal Biodiversity and Conservation* **24**: 31–52.

Fa, J. E. & Peres, C. A. (2001): Game vertebrate extraction in African and Neotropical forests: an intercontinental comparison. In *Conservation of exploited species*: 203–241. Reynolds, J. D., Mace, G. M., Redford, K. H. & Robinson, J. G. (Eds.). Cambridge: Cambridge University Press.

Fa, J. E., Ryan, S. L. & Bell, D. J. (2005): Hunting vulnerability, ecological characteristics and harvest rates of bushmeat species in Afrotropical forests. *Biological Conservation* **121**: 167–176.

Fa J. E., Juste J., Del Val J. P. & Castroviejo J. (1995): Impact of market hunting on mammal species in Equatorial Guinea. *Conservation Biology* **9**: 1107–1115.

Fa, J. E., Seymour, S., Dupain, J., Amin, R., Albrechtsen, L. & Macdonald D. (2006): Getting to grips with the magnitude of exploitation: bushmeat in the Cross-Sanaga rivers region, Nigeria and Cameroon. *Biological Conservation* **129**: 497–510.

Fa, J. E., Olivero, J., Farfán, M. Á., Márquez, A. L., Vargas, J. M., et al. (2014): Integrating sustainable hunting in biodiversity protection in central Africa: hot spots, weak spots, and strong spots. *PLoS ONE* **9**(11): e112367. doi:10.1371/journal.pone.0112367.

Fa, J. E., Olivero, J., Farfan, M. A., Marquez, A. L., Duarte J., Nackoney, J., Hall, A., Dupain, J., Seymour, S., Johnson, P. J., Macdonald, D. W., Real, R. & Vargas, J. M. (2015): Correlates of bushmeat in markets and depletion of wildlife. *Conservation Biology Online.*

Fa, J. E., Currie, D. & Meeuwig, J. (2003): Bushmeat and food security in the Congo Basin. *Environmental Conservation* **30**: 71–78.

Hart, J. A. (2000): Impact and sustainability of indigenous hunting in the Ituri Forest, Congo-Zaire: a comparison of hunted and unhunted duiker populations. In *Hunting for sustainability in tropical forests*: 106–153. Robinson, J. G. & Bennett, E. L. New York: Columbia University Press.

Jerozolimski, A. & Peres, C. A. (2003): Bringing home the biggest bacon: a cross-site analysis of the structure of hunter-kill profiles in neotropical forests. *Biological Conservation* **111**: 415–425.

Jimoh, S. O., Ikyaagba, E. T., Alarape, A. A., Adeyemi, A. A. & Waltert, M. (2012): Local depletion of two larger duikers in the Oban Hills Region, Nigeria. *African Journal of Ecology* **51**: 228–234.

Kümpel, N. (2006): *Incentives for sustainable hunting of bushmeat in Rio Muni, Equatorial Guinea*. PhD Thesis. Imperial College, London.

Lindsey P. A., Balme, G., Becker, M., Begg, C., Bento, C., Bocchino C., Dickman A., Diggle, R. W., Eves H., Henschel P., Lewis, D., Marnewick, K., Mattheus J., McNutt, W., McRobb, R., Midlane, N., Milanzi, J., Morley R., Murphree, M., Opyene, V., Phadima J., Purchase G., Dennis Rentsch, Christopher Roche, Joanne Shaw, Hugo van der Westhuizen, Nathalie Van Vliet, Patience Zisadza-Gandiwa (2013): The bushmeat trade in African savannas: impacts, drivers, and possible solutions. *Biological Conservation* **160**: 80–96.

Macdonald, D. W., Johnson, P. J., Albrechtsen, L., Dutton, A., Seymour, S., Dupain, J., Hall, A. & Fa, J. E. (2011): Association of body mass with price of bushmeat in Nigeria and Cameroon. *Conservation Biology* **25**: 1220–1228.

Macdonald, D. W., Johnson, P. J., Albrechtsen, L., Seymour, S., Dupain, J., Hall, A. & Fa, J. E. (2012): Bushmeat trade in the Cross–Sanaga rivers region: evidence for the importance of protected areas. *Biological Conservation*, **147**: 107–114.

Maisels F. (1996): *Synthesis of information concerning the Park National d'Ozala, Congo*. Projektbericht: ECOFAC-Composante Congo.

Mandujano, S. & Naranjo, E. J. (2010): Ungulate biomass across a rainfall gradient: a comparison of data from neotropical and palaeotropical forests and local analyses in Mexico. *Journal of Tropical Ecology* **26**:13–23.

Mittermeier, R. A. (1987): Effects of hunting on rainforest primates. In *Primate conservation in the tropical rainforest*: 109–146. Marsh, C. W. & Mittermeier, R. A. (Eds.). New York: Alan R. Liss.

Mockrin, M. H., Rockwell, R. F., Redford, K. H. & Keuler N. S. (2011): Effects of landscape features on the distribution and sustainability of ungulate hunting in Northern Congo. *Conservation Biology* **25**: 514–525.

Muchaal P. K. & Ngandjui G. (1999): Impact of village hunting on wildlife populations in the Western Dja Reserve, Cameroon. *Conservation Biology* **13**: 385–396.

Noss, A. J. (1995): Duikers, cables and nets: a cultural ecology of hunting in a Central African Forest. Ph.D Thesis. University of Florida.

Noss, A. J. (1998): The impacts of cable snare hunting on wildlife populations in the forests of the Central African Republic. *Conservation Biology* **12**: 390–398.

Novaro, A. J., Redford, K. H. & Bodmer R. E. (2000): Effect of hunting in source-sink systems in the neotropics. *Conservation Biology* **14**: 713–721

Ogutu, J. O. & Owen-Smith, N. (2003): ENSO, rainfall and temperature influences on extreme population declines among African savanna ungulates. *Ecology Letters* **6**: 412–419.

Okouyi, J. (2006): *Savoirs locaux et outils modernes cynégétiques: développement de la filière commerciale de viande de brousse à makokou (gabon)*. PhD dissertation. University of Orléans. Orléans.

Olff, H., Ritchie, M. E. & Prins, H. H. T. (2002): Global determinants of diversity in large herbivores. *Nature* **415**: 901–904.

Poulsen J. R., Clark, C. J., Mavah G. & Elkan P. W. (2009): Bushmeat supply and consumption in a tropical logging concession in Northern Congo. *Conservation Biology* **23**: 1597–1608.

Rieu L. (2003): *Du chasseur au consommateur: organisation de la filière viande de brousse dans un site industriel forestier d'Afrique Centrale*. Société SEFCA, Mambélé, République Centrafricaine. Projektbericht CIRAD/Université Montpellier.

Rist, J., Milner-Gulland, E. J., Cowlishaw, G. & Rowcliffe, M. (2010): Hunter reporting of catch per unit effort as a monitoring tool in a bushmeat-harvesting system. *Conservation Biology* **24**: 489–499.

Robinson, J. G. & Bennett, E. L. (2000): Carrying capacity limits to sustainable hunting in tropical forests. In *Hunting for sustainability in tropical forests*: 13–30. Robinson John G. & Bennett Elizabeth L. (Eds.). New York: Columbia University Press.

Robinson, J. G. & Bennett, E. L. (2004): Having your wildlife and eating it too: an analysis of hunting sustainability across tropical ecosystems. *Animal Conservation*, **7**: 397–408.

Robinson, J. G. & Redford, K. H. (1986): Body size, diet, and population density of neotropical forest mammals. *American Naturalist* **128**: 665–680.

Schenck, M., Effa, E. N., Starkey, M., Wilkie, D., Abernethy, K., Telfer, P., Godoy, R. & Treves A. (2006): Why people eat bushmeat: results from two-choice, taste tests in Gabon, Central Africa. *Human Ecology* **34**: 433–445.

Stiles D. (2011): Elephant meat and ivory trade in Central Africa. *Pachyderm* **50**: 26–36.

Traill, L. W., Wanger, T. C. & Kidd, D. (2011): Large, particular bovids may require localised conservation effort to prevent extinction. *Raffles Bulletin of Zoology*. 2012 Supplement **25**: 111–116.

Tranquilli, S., Abedi-Lartey, M., Abernethy, K., Amsini, F., Asamoah, A., et al. (2014): Protected areas in tropical Africa: Assessing threats and conservation activities. *PLoS ONE* **9**(12): e114154. doi:10.1371/journal.pone.0114154

Van Vliet (2007): Variabilité spatiale et temporelle au sein du systeme. Chasseur-animal-territoire de chasse villageois. Pour une approche geographique de l´estimation de la durabilité de la chasse en Afrique Centrale. Cas de *Cephalophus* spp. du Nord-est Gabon. PhD Thesis. Université Toulouse le Mirail. Toulouse

Van Vliet, N. & Nasi, R. (2008): Hunting for livelihood in northeast Gabon: patterns, evolution, and sustainability. *Ecology and Society* **13**: 33. (Online). http://www.ecologyandsociety.org/vol13/iss2/art33/(as at 10 February 2016).

Van Vliet, N., Nasi, R., Emmons, L., Feer, F., Mbazza, P. & Bourgare, M. (2007): Evidence for the local depletion of bay duiker *Cephalophus dorsalis*, within the Ipassa Man and Biosphere Reserve, north-east Gabon. *Afr. J. Ecol.* **45**: 440–443.

Waterman, P. G. & McKey, D. (1989): Herbivory and secondary compounds in rain-forest plants. In *Tropical rain forest ecosystems: biogeographical and ecological studies*: 513–536. Lieth, H. & Werger, M. J. A. (Eds.). Amsterdam and New York: Elsevier.

Willcox A. S. & Nambu D. M. (2007): Wildlife hunting practices and bushmeat dynamics of the Banyangi and Mbo people of Southwestern Cameroon. *Biological Conservation* **134**: 251–261.

6

Opportunities and Pitfalls in Realising the Potential Contribution of Trophy Hunting to Antelope Conservation

Nils Bunnefeld[1] and E. J. Milner-Gulland[2]

[1]*School of Natural Sciences, Biological and Environmental Sciences, University of Stirling, Scotland, United Kingdom*
[2]*Department of Zoology, University of Oxford, United Kingdom*

Trophy hunting as a conservation tool for antelopes

Trophy hunting is widespread around the world and includes a diverse list of species. For example, Palazy *et al.* (2012b) analysed 427 trophy-hunted species and subspecies listed on the website of Safari Club International, of which 202 were ungulates. Here we focus on antelopes, a group of species within the family Bovidae. Geographically, we focus on Africa since this is where most research on trophy hunting of ungulates has been carried out. We define trophy hunting as the selective hunting of wild animals. Parts of the animal are often kept as a trophy (skin, antlers, head), and the carcass may be consumed as food. We exclude purely recreational hunting which is more common among local people in Europe, North America, and Australia (Sharp & Wollscheid, 2009).

Trophy hunting has been advocated in the last ten years as a conservation tool because it puts a monetary value on natural resources and has the potential for a sustainable offtake that provides income into the long term (Lindsey *et al.*, 2007). Furthermore, over 1.3 million km^2 in sub-Saharan Africa are now used for trophy hunting, which is more than state land provided for wildlife (Lindsey *et al.*, 2006), and it has financed re-introductions, for example black wildebeest (*Connochaetus gnou*) and southern white rhino (*Ceratotherium*

Antelope Conservation: From Diagnosis to Action, First Edition. Edited by Jakob Bro-Jørgensen and David P. Mallon.

simum simum) in South Africa (Leader-Williams *et al.*, 2005). However, trophy hunting is also subject to controversy in terms of its effects on the ecology of the target species, on the wider ecosystem, and the degree to which, in practice, it contributes to the wider economy and to improving the lives of local people who bear the cost of living with wildlife (Loveridge *et al.*, 2007). Others have considered the ethical and moral issues around trophy hunting (Dickson, 2009). Rather than retread this ground, we firstly review more recent research on trophy hunting with a focus on the ecological and evolutionary effects of hunting on target species, and on the socio-economic drivers of trophy hunting and its impact on poverty alleviation. We then move on to consider ways to improve the sustainability of trophy hunting. If sustainability can be improved through an understanding of the holistic nature of the hunting system, then the case for trophy hunting as a conservation tool can be considered based on its intrinsic merits, rather than with respect to issues surrounding its implementation in practice.

What effect does trophy hunting actually have?

Effects on the target species

There is controversy about the overall effect of trophy hunting on the conservation status of target species. The most complete analysis of the population trends of trophy hunted species was carried out by Packer *et al.* (2011) who found that lion (*Panthera leo*) and leopard (*Panthera pardus*) populations had decreased over the last decade in many parts of Tanzania, and trophy hunting was a significant factor explaining this trend in a regression analysis. So far no such paper has been written for antelopes, summarising studies on population trends of antelopes subject to trophy hunting, but some studies exist, which indicate the complexity and uncertainty surrounding assessments of the sustainability of antelope trophy hunting.

Based on life history theory and modelling, one can expect that hunting very few individuals for trophies, especially males of polygynous group-living species with no paternal care, has negligible effects on population size and thus can be sustainable. Caro *et al.* (2009) assess the ecological effects of trophy hunting in the Selous Game Reserve in south-eastern Tanzania, where a large range of species can be hunted, including antelopes and carnivores. They show that theoretically up to 6% of the population (male-only quotas) can be hunted for species that show no paternal care, no infanticide, and live in middle-size to large harems (5–10 individuals); these characteristics fit many antelopes such as impala (*Aepyceros melampus*) and kudu (*Tragelaphus* spp). However, they found that it was difficult to assess the impact of trophy hunting on local target populations because scientific understanding of population status was based on large-scale aerial surveys; ground counts were not available for any of the trophy hunted species. Furthermore, when monitoring data were available, methods differed widely between species and areas, ranging from aerial surveys for larger species to foot transects for more cryptic or smaller species. Together, these factors make it difficult to compare study areas and decompose the relative contribution of trophy hunting on population status compared to other factors, such as differences in habitat and climate, or the demographic processes, which particularly affect small and isolated populations.

While there has been some attention paid to the direct effect of offtake on the population size of trophy-hunted species, less attention has been given to the evolutionary consequences of selective hunting (Mysterud & Bischof, 2010). Crosmary *et al.* (2013) analysed 34 years (1974–2008) of horn-length data of male impalas, greater kudus (*Tragelaphus strepsiceros*), and sable antelopes (*Hippotragus niger*) hunted for trophies in Zimbabwe. Horn length declined by 4% in impalas and by 6% in sable antelopes, whereas in greater kudus, horn length increased by 14%. While monitoring of horn length can be valuable to make inferences about sustainability, the actual mechanisms behind the observed data are hard to infer. Some of these patterns could be attributed to selective hunting, but proof is difficult because of interacting effects with hunting pressure and the proportion of old males in the harvest (Mysterud, 2011). Long-term studies including monitoring both the population demography and the offtake are needed to better understand hunting selectivity and its ecological and evolutionary effects. While long-term data sets exist for common species hunted in Europe and North America, such as red grouse (*Lagopus lagopus scoticus*) and bighorn sheep (*Ovis canadensis*) (Bunnefeld *et al.*, 2009; Bunnefeld *et al.*, 2011b; Pelletier *et al.*, 2012), time series data relating hunting selectivity to population demography and then quantitative traits of trophy hunted species are rare in developing countries.

The lack of high-quality data poses a challenge to understanding the overall impact of trophy hunting on target populations and its role in conservation. Furthermore, there is still very little known about the general trends in hunted populations, the effects of trophy hunting on non-target species, on overall biodiversity and the effects on ecosystems. Only recently have researchers analysed long-term trends in red deer (*Cervus elaphus*) hunting in Europe, finding that differences in offtake numbers and selectivity were linked to cultural differences across European countries and that changes in management regulations were able to reverse declines in trophy size (Milner *et al.*, 2006; Rivrud *et al.*, 2013).

Socio-economic drivers and mechanisms

General economic theory predicts that the realised price of a good is determined by supply and demand, with the demand decreasing when prices increase. However, research on 202 ungulate taxa has suggested that demand is inelastic to price when species are rare, causing prices to increase with rarity. This could encourage overexploitation of these species. Palazy *et al.* (2012b) find that rarer species achieve more than twice the price of similar species that are more common. For example, to obtain a hunting permit for the relatively rare black faced impala (*Aepyceros melampus petersi*), listed by IUCN as Vulnerable, could cost around 2800 US$ whereas the more common, Least Concern, southern impala (*Aepyceros melampus melampus*) only costs a fraction of the price (around 400 US$). Thus trophy hunters, like other types of collectors, are especially interested in rare species. Scarcity-induced price increases could therefore stimulate rather than depress demand (Hall *et al.*, 2008). Prescott *et al.* (2012) showed for 22 genera of bovids that prices increased by 39% (by almost US$900 on average) from 2004 to 2010 for species whose IUCN status deteriorated during that period, for example from Least Concern to Near Threatened, whereas for species with stable or improving IUCN status, the price did not change significantly (2255 prices from 78 operators in 12 countries). However, they were not able to distinguish whether the reason for this difference was to do with changes in perceived rarity (e.g., as a consequence of IUCN uplisting a species) or decreases in the actual population or quota size (Atickem *et al.*, 2011). Further studies also using the Safari Club International data set found that assigning any IUCN status, even Endangered, in a given year had no effect on the trophy numbers in the next five years (Palazy *et al.*, 2012a). Upgrading the status, however, led to a decrease in trophy numbers except for the upgrade to Endangered, where it led to an increase in trophy numbers (Palazy *et al.*, 2012a). All these studies are difficult to interpret, and their effects on actual conservation status hard to predict, in the absence of information on the governance and institutional structures within which populations are managed (Harris *et al.*, 2013).

Patterns of data paucity also occur when examining the evidence concerning whether or not trophy hunting contributes to poverty alleviation. There are few studies that evaluate the opportunities and costs that trophy hunting brings to the poor. Some case studies exist that estimate the income for local people from community-based natural resource management that includes a trophy-hunting element in Southern Africa, but they are mostly one-off assessments (e.g., Jones (2009)). Most of these interventions are not linked to a scientific programme, thus do not get scientifically evaluated for their socio-economic or ecological sustainability, poverty reduction, and biodiversity conservation outcomes. Results and conclusions are usually published in the grey literature and remain difficult to access and unvalidated by peer review (Dickman *et al.*, 2011). Suich (2010) carried out a synthesis of 24 community-based natural resource management (CBNRM) programmes in Namibia, which included trophy hunting as one activity, to assess their livelihood impacts. She found that most research reports focus on the positive effects and only to a lesser extent the negative effects of CBNRM programmes. The CBNRM programmes reduced access to land by zoning it for wildlife and tourism and reduced food availability through more frequent events of crop raiding. Furthermore, employment through trophy hunting was found to be highly seasonal and thus not as important for the community as other year-round activities such as producing and selling crafts. Child (2012) uses the example of rhinos in South Africa to argue that high prices and clear propriety rules for wildlife products could contribute to sustainable use through trophy hunting. Interventions need to have clear objectives and targets based on an understanding of the mechanisms by which the conservation action will improve the status of biodiversity and local wellbeing. They should include scientific evaluation and monitoring as an integral component in order to generate evidence for the impacts of trophy hunting on local people.

There is large uncertainty about the actual price that foreign hunters would be willing to pay for trophy hunting antelopes under different circumstances. There might be an argument that foreign hunters would be willing to pay more if they knew that the money would be invested into conservation or into the local communities. A preliminary study by Lindsey *et al.* (2006) suggests that hunters do have a preference for these attributes, but does not provide an estimate of the potential premium that a conservation label could command. Studies on other conservation issues have estimated these price premiums: for example, the amount consumers would be willing to pay for tiger-friendly margarine (Bateman *et al.*, 2010). These studies use willingness-to-pay methods, an economic survey–based approach (Hanley *et al.*, 1998). A study such as this could inform potential price increases to cover the additional costs of operating trophy-hunting companies which offer conservation and social benefits as well as hunting experiences. However, studies investigating the financial viability of this approach have focused mainly on a few, relatively expensive species such as elephant, lion, leopard, buffalo, and sable antelope (Lindsey *et al.*, 2012), and thus such information remains to be compiled for the wide range of antelopes that are hunted for trophies.

Improving the sustainability of trophy hunting

Management Strategy Evaluation

One approach that can be used to evaluate the sustainability of trophy hunting, both in terms of the conservation of wildlife and the socio-economic circumstances of stakeholders, is Management Strategy Evaluation (MSE).

It was initially developed by fisheries scientists for the International Whaling Commission (Smith *et al.*, 1999) and has recently seen some uptake in terrestrial conservation (Bunnefeld *et al.*, 2011a; Milner-Gulland, 2011; Nuno *et al.*, 2013a). The strength of the approach is that it takes into account three important components of natural resource management systems: the ecological and social dynamics of the system itself, the observation process that includes monitoring data with all its errors and biases, and the assessment and decision-making processes of the managers (Figure 1, Box 1). MSE is especially useful when the system has quantifiable uncertainties, for example not knowing the true population or stock size, which is the case for virtually all fisheries and most hunting and conservation projects, including trophy hunting (Bunnefeld *et al.*, 2013; Edwards *et al.*, 2014).

Despite the lack of robust data and analyses of the mechanisms and outcomes of trophy hunting as a conservation tool, there is a wealth of experience and anecdotal evidence, which suggests that there are two major issues which trophy hunting faces if it is to become a sustainable and positive component of antelope conservation. These are (1) illegal hunting alongside authorised trophy hunting, often for bushmeat; and (2) the lack of clear adaptive management

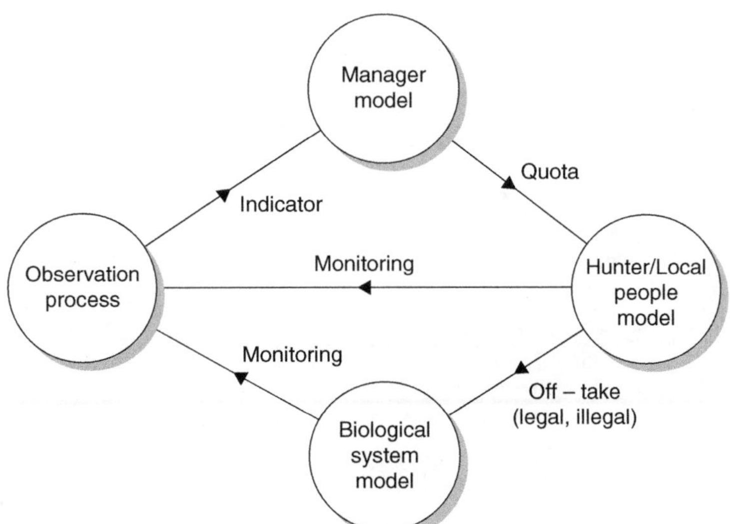

Figure 1 **Management Strategy Evaluation framework including a manager decision-making model (also called assessment model) passing the quota to the hunter and/or local people's decision-making model, who may deviate or not from the quota. These people hunt animals with impacts on their biology described in a biological model. Finally, the observation model translates monitoring data into indicators that can be used by the manager to inform decision making.**

Box 1: Case study: Working towards the sustainability of trophy hunting for an endangered antelope in Ethiopia using Management Strategy Evaluation.

Bunnefeld *et al.* (2013) demonstrate the power of an MSE approach for trophy-hunted species using mountain nyala (*Tragelaphus buxtoni*) as an example. The MSE framework is able to consider the multiple objectives in a system where the conservation of an endangered species is only possible by taking the incentives and the welfare of local people into account. Furthermore, the effect of different forms of uncertainty on the performance of harvest control rules can be assessed. Two different potential management activities to reduce uncertainty were tested for mountain nyala trophy hunting; improving population monitoring and reducing poaching levels. The trophy-hunting model tested the performance of three quota-setting methods: traditional fixed quotas, quotas proportional to the population size, and adaptively setting the trophy quotas in a way similar to data-based empirical harvest control rules in fisheries. The adaptive approach is based on previous quotas, an estimate of the population size in

Box Figure 1 **Male mountain nyala (*Tragelaphus buxtoni*) in the Bale Mountain National Park, Ethiopia (© Nils Bunnefeld).**

previous years and a flexibility parameter that translates the steepness of the population trend into a change in quota size.

The use of an adaptive approach can, to some extent, mitigate the effects of monitoring uncertainty on performance because it takes past offtake and monitoring data into account. However, when poaching is high, both population size and quota drop considerably. This suggests that it would be in the interests of safari companies to support anti-poaching activities (which is in fact what we observe for mountain nyala; Atickem *et al.*, 2011). If the government is interested in maintaining system stability, it would be worthwhile for them to invest enough in monitoring nyala populations in order to ensure that the monitoring coefficient of variation remains below 15%. And indeed, although safari companies currently do not monitor nyala numbers, the government does carry out a minimal level of monitoring. This analysis has therefore suggested one potential mechanism behind the observed behavior of two key stakeholders in the system, illustrating the usefulness of the MSE framework. This enables multiple viewpoints to be modelled and thereby provides a strong tool for making conservation recommendations.

that links monitoring of antelope population trends to changes in management actions. Often, these two are intertwined; because local people do not feel a stake in the trophy-hunting or conservation project, there are no incentives to refrain from unauthorised hunting, and the lack of adaptive management means that these issues are not addressed. For example, in the area west of the Serengeti, bushmeat hunting is a relatively common activity with around 18% of households involved, both for cash and for food, which is likely to be affecting wildlife numbers (Hilborn *et al.*, 2006; Nuno *et al.*, 2013b). A choice experiment by Moro *et al.* (2013) suggests that alternative incomes are not sufficiently attractive to stop illegal hunting. A lack of adaptive management means that there is no clear mechanism by which new insights can influence conservation and management strategies (Nuno *et al.*, 2014).

Improving monitoring of population trends

To be able to make decisions about the sustainable level at which trophy quotas can be set, the state of the resource needs to be assessed. Intuitive or common knowledge is often not enough or can be misleading. Recent scientific advances in monitoring techniques and analytical refinements mean that it is possible to make more precise and accurate estimates of population sizes while also calculating the uncertainty of our knowledge. However monitoring of antelope populations is often still done without the benefit of these new techniques. Singh and Milner-Gulland (2011) review advantages and disadvantages of a range of commonly used monitoring methods for ungulates in Central Asia, such as aerial surveys and ground surveys. They conclude that the challenges faced in Central Asia, which resonate with many monitoring programmes worldwide, include low capacity (lack of expertise in the latest monitoring methods, awareness of biases and errors) as well as logistical and financial constraints.

One of the most widely used modern methods for monitoring ungulates is distance sampling, which has been developed to correct for bias in field survey data. The method is based on the idea that all individuals on a transect line are seen, but observability decreases as distance from the transect increases (Burnham *et al.*, 1985; Thomas *et al.*, 2010). Fitting a curve through the distribution of sightings as distance from the transect increases enables the observability bias to be corrected for. Given that these estimates depend on a fitted curve, the uncertainty around the fit and thus of the abundance estimate can be calculated. However, model-based methods for estimating abundance, such as distance sampling, also have disadvantages, which may explain their limited uptake in developing countries where funding and capacity are especially limited.

Involving local people in monitoring their own resources can increase local involvement in management, hence improving the acceptability of trophy hunting and providing employment opportunities (Danielsen *et al.*, 2009). However, methods such as distance sampling, which require substantial technical input, may be unrealistic for locally based managers (Singh & Milner-Gulland, 2011). Kinahan and Bunnefeld (2012) carried out distance sampling and total counts for mountain nyala (*Tragelaphus buxtoni*), an endangered antelope hunted for its horns in the Bale Mountains of Ethiopia. Total counts required only 12% of the resources of distance sampling while the precision and the power to detect a change was similar for both methods. In addition to the differences in direct costs, total counts had the advantage that they were easily understood and interpreted by local people while complex, model-based monitoring methods need considerable support from scientists with advanced quantitative skills and expensive computer equipment. Gaidet-Drapier *et al.* (2006) compared different count methods for 27 species in the Zambezi Valley (Zimbabwe), including impala and greater kudu, and concluded that foot and bicycle counts delivered similar detection rates while keeping the costs low compared to aerial surveys and car counts. These studies suggest that under limited budgets participatory monitoring is a feasible approach to managing trophy-hunted species.

Understanding local people and improving local engagement

In order to tackle the underlying issues behind the lack of success of trophy hunting as a conservation tool, there is often a need to improve the benefits that local people perceive. Co-management has become a central concept in the management of natural resources (e.g., Evans *et al.*, 2011). This enables several stakeholders (often including the government and local community organisations) to build joint institutions for resource management (Plummer & Armitage, 2007). Often, the government supplies technical expertise and scientific evaluations of resource status, as well as institutional support at the national level, while communities supply traditional knowledge and evaluations, and the local institutions are responsible for ensuring that harvests are sustainable and rules are complied with. For example in Northern Canada, caribou (*Rangifer tarandus*) are managed by co-management organizations composed of government and local representatives deciding on the yearly hunting quota for local communities. Even though co-management has been successful in giving local communities the right to hunt caribou while also enabling the recovery of caribou populations, the co-management organizations face two main challenges; how to make decisions when the population estimates are increasingly uncertain due to funding shortages and how to increase compliance with management decisions (Gunn, 2009). Research indicates that further involvement of the local people, and specifically hunters, in the monitoring process is needed to increase monitoring efforts and that regular communication with local people seems to lead to increased cooperation with management decisions (Hanley & Russell, 2000; Urquhart, 1996).

Revenue sharing is a more top-down approach than co-management, but in both approaches local people get financial benefits from the use of their local natural resources. This approach has challenges, as sometimes the relationship between the revenues sent to local communities and the status of wildlife populations can be indirect and unclear. For example, in a case study on trophy hunting in Ethiopia, Yitbarek *et al.* (2013) identified the following main challenges in the current system of revenue sharing and wildlife tourism: there was a lack of a connection between revenue and wildlife tourism among the people that have received revenues; clear structures for reporting and monitoring were missing, so accountability was compromised; the aims of the project were not entirely clear to all stakeholders, and some had only heard about revenue sharing during meetings discussing the problems which had emerged during the project.

Understanding the decisions people make and the underlying incentives is crucial in order to better structure institutions for managing trophy hunting and engaging local people. In villages around the Serengeti National Park and the Grumeti Game Reserve, people illegally hunt impala and other antelopes for cash and subsistence, which adds further uncertainty when setting sustainable quotas for trophy hunting. Moro *et al.* (2013) used choice experiments to show that on average an additional 1.4 cows in a household or 36,000 TZS (22 USD) increase in income per month would deliver equal utility to respondents as a one-week hunting trip. More importantly, this method is able to quantify the heterogeneity between different groups of users, for example, for people with no livestock at all, being given 2.7 cows would decrease the time spent hunting by just one week. This gives indications of how much livelihoods would have to change to divert people from illegal activities. The drawback of this method is that it only concerns behavioural change at the margins, not overall. Delivering 1.4 cows times the number of households does not mean that all people stop hunting, because the figure of 1.4 cows is an amalgamation of a range of people that have differing willingness to accept. Moreover, behaviour in hypothetical situations is not necessarily connected to choices made when confronted with real-world situations. However, this approach could be used to get an indication of the levels and types of compensation that individuals or types of people (e.g., people with particular livelihoods) might need to incentivise sustainable behaviour. This could then act as a guide for structuring interventions within an adaptive management framework.

Many conservation studies measure the attitudes and behavioural intentions of local people, but these are a poor indicator of actual behaviour, which is affected by a range of different factors including social norms and people's ability to change their actions (Heberlein, 2012). In order to improve the effectiveness of conservation interventions such as antelope reintroductions or community-based hunting areas, it is crucial to understand the drivers of actual behaviour and to monitor indicators of change in behaviour, such as the number of households involved in bushmeat hunting, rather than focussing on improving people's attitudes. Nuno *et al.* (2013b) show how indirect questioning techniques can be used to get better estimates of the prevalence of illegal behaviour while protecting the anonymity of the respondent. It is difficult to use this type of method to answer questions about the volume of illegal offtake, but it can give information on which demographic groups are more likely to hunt and how season or distance to protected area affects the prevalence of hunting. The study suggests that bushmeat hunting is carried out all year round instead of restricted to the dry season as earlier studies had indicated. It also finds that households with seasonal or full-time employment are more likely to hunt than households without any employment, which is interesting in the light of suggestions that higher rates of employment may reduce illegal hunting. The study did not delve into the reasons behind this, but generally there is a lack of evidence and understanding concerning the relationship between hunting and other livelihoods, which is urgently needed if livelihoods-based conservation interventions are to be effective (Wicander *et al.*, 2014).

Conclusion

A lack of high-quality data still remains a major obstacle to the assessment of the ecological and socio-economic sustainability of trophy hunting for antelopes. This holds both for socio-economic data and data on the ecology of trophy-hunted species. The data gaps include information on the behaviour and motivations of trophy hunters as well as local people. These data are crucial to be able to detect trends with confidence and to make evidence-based decisions on the development and modification of interventions, such as hunting quotas, law enforcement, or benefit sharing. While many examples exist that report the failure of benefit sharing as a strategy, there are still some cases that give positive signals; both failures and successes need to be analysed in order to learn from case studies so as to develop guidelines for improving the sustainability of antelope trophy hunting. There are currently few feasible alternatives to trophy hunting and ecotourism for the conservation of large antelopes outside protected areas. Arguments for trophy hunting as a conservation tool could be strengthened by the more transparent dissemination of existing data on offtake, ecological monitoring, and the magnitude and distribution of benefit sharing. These data could then be analysed in an objective way, enabling the development of research programmes that both improve and supplement current data sets, and informing the design of interventions aiming for the ecological and socio-economic sustainability of trophy hunting as a conservation tool.

References

Atickem, A., Loe, L. E., Langangen, O., Rueness, E. K., Bekele, A. & Stenseth, N. C. (2011): Estimating population size and habitat suitability for mountain nyala in areas with different protection status. *Animal Conservation* **14**: 409–418.

Bateman, I. J., Fisher, B., Fitzherbert, E., Glew, D. & Naidoo, R. (2010): Tigers, markets and palm oil: market potential for conservation. *Oryx* **44**: 230–234.

Bunnefeld, N., Baines, D., Newborn, D. & Milner-Gulland, E. J. (2009): Factors affecting unintentional harvesting selectivity in a monomorphic species. *Journal of Animal Ecology* **78**: 485–492.

Bunnefeld, N., Edwards, C. T. T. E., Atickem, A., Hailu, F. & Milner-Gulland, E. J. (2013): Incentivizing monitoring and compliance in trophy hunting. *Conservation Biology* **27**: 1344–1354.

Bunnefeld, N., Hoshino, E. & Milner-Gulland, E. J. (2011a): Management strategy evaluation: a powerful tool for conservation? *Trends in Ecology & Evolution* **26**: 441–447.

Bunnefeld, N., Reuman, D. C., Baines, D. & Milner-Gulland, E. J. (2011b): Impact of unintentional selective harvesting on the population dynamics of red grouse. *Journal of Animal Ecology* **80**: 1258–1268.

Burnham, K. P., Anderson, D. R. & Laake, J. L. (1985): Efficiency and bias in strip and line transect sampling. *Journal of Wildlife Management* **49**: 1012–1018.

Caro, T. M., Young, C. R., Cauldwell, A. E. & Brown, D. D. E. (2009): Animal breeding systems and big game hunting: Models and application. *Biological Conservation* **142**: 909–929.

Child, B. (2012): The sustainable use approach could save South Africa's rhinos. *South African Journal of Science* **108**.

Crosmary, W. G., Loveridge, A. J., Ndaimani, H., Lebel, S., Booth, V., Côté, S. D. & Fritz, H. (2013): Trophy hunting in Africa: long-term trends in antelope horn size. *Animal Conservation* **16**: 648–660.

Danielsen, F., Burgess, N. D., Balmford, A., Donald, P. F., Funder, M., Jones, J. P. G., Alviola, P., Balete, D. S., Blomley, T., Brashares, J., Child, B., Enghoff, M., Fjeldsa, J., Holt, S., Hubertz, H., Jensen, A. E., Jensen, P. M., Massao, J. M., Mendoza, M., Ngaga, Y., Poulsen, M. K., Rueda, R., Sam, M., Skielboe, Stuart-Hill, Topp-Jorgensen, T. G. E. & Yonten, D. (2009): Local participation in natural resource monitoring: a characterization of approaches. *Conservation Biology* **23**: 31–42.

Dickman, A. J., Macdonald, E. A. & Macdonald, D. W. (2011): A review of financial instruments to pay for predator conservation and encourage human-carnivore coexistence. *Proceedings of the National Academy of Sciences of the United States of America* **108**: 19836–19836.

Dickson, B. (2009): The ethics of recreational hunting. Recreational Hunting, *Conservation and Rural Livelihoods*: 59–72.

Edwards, C. T. T. E., Bunnefeld, N., Balme, G. & Milner-Gulland, E. J. (2014): Data-poor management of African lion hunting: how to set quotas when the population size is unknown. *Proceedings of the National Academy of Sciences of the United States of America* **111**: 539–543.

Evans, L., Cherrett, N. & Pemsl, D. (2011): Assessing the impact of fisheries co-management interventions in developing countries: a meta-analysis. *Journal of Environmental Management* **92**: 1938–1949.

Gaidet-Drapier, N., Fritz, H., Bourgarel, M., Renaud, P. C., Poilecot, P., Chardonnet, P., Coid, C., Poulet, D. & Le Bel, S. (2006): Cost and efficiency of large mammal census techniques: comparison of methods for a participatory approach in a communal area, Zimbabwe. *Biodiversity and Conservation* **15**: 735–754.

Gunn, A. (2009): Caribou and musk ox harvesting in the northwest territories. In Conservation of Biological Resources: 314–330. Milner-Gulland, E. J. & Mace, R. (Eds.). Osney Mead, Oxford: Blackwell Science Ltd.

Hall, R. J., Milner-Gulland, E. J. & Courchamp, F. (2008): Endangering the endangered: the effects of perceived rarity on species exploitation. *Conservation Letters* **1**: 75–81.

Hanley, N., Wright, R. E. & Adamowicz, V. (1998): Using choice experiments to value the environment – design issues, current experience and future prospects. *Environmental & Resource Economics* **11**: 413–428.

Hanley, T. A. & Russell, D. E. (2000): Ecological role of hunting in population dynamics and its implication for co-management of the Porcupine caribou herd. *Rangifer* **20**: 71–78.

Harris, R. B., Cooney, R. & Leader-Williams, N. (2013): Application of the anthropogenic Allee effect model to trophy hunting as a conservation tool. *Conservation Biology* **27**: 945–951.

Heberlein, T. A. (2012): Navigating environmental attitudes. Oxford: Oxford University Press.

Hilborn, R., Arcese, P., Borner, M., Hando, J., Hopcraft, G., Loibooki, M., Mduma, S., & Sinclair, A. R. E. (2006): Effective enforcement in a conservation area. *Science* **314**: 1266–1266.

Jones, B. T. (2009): Community benefits from safari hunting and related activities in Southern Africa. In Recreational hunting, conservation and rural livelihoods: science and practice: 157–177. Dickson, B., Hutton, J. & Adams, W. M. (Eds.). Oxford: Wiley-Blackwell.

Kinahan, A. A. & Bunnefeld, N. (2012): Effectiveness and cost efficiency of monitoring mountain nyala in Bale Mountains National Park, Ethiopia. *Endangered Species Research* **18**: 105–114.

Leader-Williams, N., Milledge, S., Adcock, K., Brooks, M., Conway, A., Knight, Mainka, M., Martin, S. E. B. & Teferi, T. (2005): Trophy hunting of black rhino *Diceros bicornis*: proposals to ensure its future sustainability. *Journal of International Wildlife Law and Policy* **8**: 1–11.

Lindsey, P. A., Alexander, R., Frank, L. G., Mathieson, A. & Romanach, S. S. (2006): Potential of trophy hunting to create incentives for wildlife conservation in Africa where alternative wildlife-based land uses may not be viable. *Animal Conservation* **9**: 283–291.

Lindsey, P. A., Balme, G. A., Booth, V. R. & Midlane, N. (2012): The significance of African lions for the financial viability of trophy hunting and the maintenance of wild land. *Plos One* **7**.

Lindsey, P. A., Roulet, P. A. & Romanach, S. S. (2007): Economic and conservation significance of the trophy hunting industry in sub-Saharan Africa. *Biological Conservation* **134**: 455–469.

Loveridge, A. J., Reynolds, J. D. & Milner-Gulland, E. J. (2007): Does sport hunting benefit conservation? In Key topics in conservation biology: 224–240. Macdonald, D. W. & Service, K. (Eds.). Blackwell.

Milner, J. M., Bonenfant, C., Mysterud, A., Gaillard, J. M., Csanyi, S. & Stenseth, N. C. (2006): Temporal and spatial development of red deer harvesting in Europe: biological and cultural factors. *Journal of Applied Ecology* **43**: 721–734.

Milner-Gulland, E. J. (2011): Integrating fisheries approaches and household utility models for improved resource management. *Proceedings of the National Academy of Sciences of the United States of America* **108**: 1741–1746.

Moro, M., Fischer, A., Czajkowski, M., Brennan, D., Lowassa, A., Naiman, L. C. & Hanley, N. (2013): An investigation using the choice experiment method into options for reducing illegal bushmeat hunting in western Serengeti. *Conservation Letters* **6**: 37–45.

Mysterud, A. (2011): Selective harvesting of large mammals: how often does it result in directional selection? *Journal of Applied Ecology* **48**: 827–834.

Mysterud, A. & Bischof, R. (2010): Can compensatory culling offset undesirable evolutionary consequences of trophy hunting? *Journal of Animal Ecology* **79**: 148–160.

Nuno, A., Bunnefeld, N. & Milner-Gulland, E. J. (2013a): Matching observations and reality: using simulation models to improve monitoring under uncertainty in the Serengeti. *Journal of Applied Ecology* **50**: 488–498.

Nuno, A., Bunnefeld, N., Naiman, L. C. & Milner-Gulland, E. J. (2013b): A novel approach to assessing the prevalence and drivers of illegal bushmeat hunting in the Serengeti. *Conservation Biology* **27**: 1355–1365.

Nuno, A., Bunnefeld, N. & Milner-Gulland, E. (2014): Managing social-ecological systems under uncertainty: implementation in the real world. *Ecology and Society* **19**: 52.

Packer, C., Brink, H., Kissui, B. M., Maliti, H., Kushnir, H. & Caro, T. (2011): Effects of trophy hunting on lion and leopard populations in Tanzania. *Conservation Biology* **25**: 142–153.

Palazy, L., Bonenfant, C., Gaillard, J. M. & Courchamp, F. (2012a): On the use of the IUCN status for the management of trophy hunting. *Wildlife Research* **39**: 711–720.

Palazy, L., Bonenfant, C., Gaillard, J. M. & Courchamp, F. (2012b): Rarity, trophy hunting and ungulates. *Animal Conservation* **15**: 4–11.

Pelletier, F., Festa-Bianchet, M. & Jorgenson, J. T. (2012): Data from selective harvests underestimate temporal trends in quantitative traits. *Biology Letters* **8**: 878–881.

Plummer, R., & Armitage, D. R. (2007): Charting the new territory of adaptive co-management: a Delphi study. *Ecology and Society* **12**.

Prescott, G. W., Johnson, P. J., Loveridge, A. J. & Macdonald, D. W. (2012): Does change in IUCN status affect demand for African bovid trophies? *Animal Conservation* **15**: 248–252.

Rivrud, I. M., Sonkoly, K., Lehoczki, R., Csanyi, S., Storvik, G. O. & Mysterud, A. (2013): Hunter selection and long-term trend (1881–2008) of red deer trophy sizes in Hungary. *Journal of Applied Ecology* **50**: 168–180.

Sharp, R. & Wollscheid, K.-U. (2009): An overview of recreational hunting in North America, Europe and Australia. In Recreational Hunting, Conservation and Rural Livelihoods: 25–38. Dickson, B., Hutton, J., Adams, W. A. (Eds.) Oxford: Wiley-Blackwell.

Singh, N. J. & Milner-Gulland, E. J. (2011): Monitoring ungulates in Central Asia: current constraints and future potential. *Oryx* **45**: 38–49.

Smith, A. D. M., Sainsbury, K. J. & Stevens, R. A. (1999): Implementing effective fisheries-management systems – management strategy evaluation and the Australian partnership approach. *ICES Journal of Marine Science* **56**: 967–979.

Suich, H. (2010): The livelihood impacts of the Namibian community-based natural resource management programme: a meta-synthesis. *Environmental Conservation* **37**: 45–53.

Thomas, L., Buckland, S. T., Rexstad, E. A., Laake, J. L., Strindberg, S., Hedley, S. L., Bishop, J. R. B., Marques, T. A. & Burnham, K. P. (2010): Distance software: design and analysis of distance sampling surveys for estimating population size. *Journal of Applied Ecology* **47**: 5–14.

Urquhart, D. (1996): Caribou co-management needs from research: simple questions – tricky answers. *Rangifer* **16**: 263–272.

Wicander, S., Coad, L., Angu, K. A. & Walters, G. (2014): Improving effectiveness of alternative livelihood projects to reduce bushmeat hunting. Technical Brief Environemtal Change Institute, http://www.eci.ox.ac.uk/publications/downloads/coad14-brief-long.pdf.

Yitbarek, T. W., Tadie, D., Timer, G. & Fischer, A. (2013): Evaluating governance processes in the sharing of revenues from wildlife tourism and hunting in Ethiopia. *Environmental Conservation* **40**: 253–265.

7

Antelope Diseases – the Good, the Bad and the Ugly

Richard Kock[1], Philippe Chardonnet[2], and Claire Risley[3]

[1]*Department of Pathology & Infectious Diseases, Royal Veterinary College, Hatfield, United Kingdom*

[2]*International Game Foundation, Paris, France*

[3]*Institute of Biological, Environmental and Rural Sciences (IBERS), Aberystwyth University, Wales, United Kingdom*

Introduction

Despite the long association between humans and antelope, surprisingly little has been written about their diseases. Perhaps our interest has focused more on killing and eating them than preserving them. The veterinary attention has often been negative, as livestock are thought to be threatened by antelope, from contracting infections when in proximity or sharing pasture. Much of this belief is based on opinion rather than science, and it is high time more work was dedicated to understanding antelope diseases, before their numbers dwindle to levels where disease might begin to threaten their very existence. Objectives of the paper were to review the literature and gather and share experience from colleagues, but also to encourage veterinary colleagues to see co-existence as the way forward, rather than exclusion on veterinary grounds. This is vital if we are ever to conserve our remarkable inherited biodiversity on the planet.

The paper is divided into three sections: the first two describe infection in terms of the 'good' and the 'bad' impact of disease on antelope population health and ecosystem dynamics as well as on the health of livestock and people. In the final section, which we provocatively refer to as 'the ugly', we discuss how veterinarians and veterinarian regulations have impacted conservation

Antelope Conservation: From Diagnosis to Action, First Edition. Edited by Jakob Bro-Jørgensen and David P. Mallon.

of antelope. We focus on evidence of disease and its impacts in free-ranging antelope populations. Zoological medicine records many diseases affecting antelope in captivity, but this aspect is not considered here, and readers should refer to zoo medicine texts if interested.

The 'good': disease ecology – what do we know about the role of infectious disease in natural antelope population dynamics and in genetic health?

Diseases can sometimes be tied into the fundamental ecology of an ecosystem. Herbivore disease can be tightly coupled to grassland fires; historical disease impacts on herbivores in the Serengeti changed burning there and fundamentally affected the animal and plant ecology of the ecosystem for the major part of a century (Holdo *et al.*, 2009). However, ecologists observed early on that healthy animal populations also may suffer disease with few ecological consequences. Few ecological consequences overall are experienced where disease impacts only one population of a meta-populational species; for example pinniped population dynamics can be influenced by the frequency and severity of epidemics without this necessarily causing long-term decline (Harwood & Hall, 1990). Pathogens are, by definition, bad for the animal in which disease is caused. Infecting organisms can, however, in addition to being pathogenic, be commensal or mutual, even if they are pathogenic in other animals, or descended from a line of pathogens. For example, medical ecologists have noted that the shared evolutionary fate of humans and their symbiotic bacteria has selected for mutualistic interactions that are essential for human health, and it is only seldom that ecological or genetic changes uncouple this shared fate resulting in disease (Dethlefsen *et al.*, 2007). More fundamentally, the removal of infection can have unintended ecological consequences, the most notable being problems associated with the current massive preponderance of domestic animals over their wild counterparts (e.g., as measured by methane production; Crutzen *et al.*, 1986). Sometimes artificial interventions designed to control disease or increase production can have adverse consequences, in comparison to which diseases themselves can be seen as 'good'. Animal diseases have been controlled to enable production, foster trade, and support markets, and sometimes the control of just one disease (e.g., foot and mouth disease: FMD) has come at the expense of the functioning of whole ecologies (e.g., due to restricting migration; Ferguson *et al.*, 2013). Genetically modified animals appear to be more susceptible to natural infections than their wild cousins, and their farming is therefore dependent on the elimination of their parasites and pathogens. Arguably, treatment of pathogens can have worse consequences than the disease problem it was designed to solve; for example, it can cause antimicrobial resistance in microbial populations in both wild and domestic animals (Laxminarayan *et al.*, 2013). Equally control and elimination of pathogens or vectors of important diseases through pharmaceuticals or chemical means can result in the death of more symbiotic and ecologically valuable taxa as has happened in birds (Fry, 1995).

Geneticists have moreover shown how disease helps maintain the heterozygosity of populations (Acevedo-Whitehouse *et al.*, 2005, 2006). There is increasing evidence that difference in susceptibility to disease has a genetic basis, and this was shown in bovines (Allen *et al.*, 2010; Driscoll *et al.*, 2011). The assumption from these results is that disease (through parasites and pathogens) contributes to maintaining genetic health, through 'selecting' and 'removing' the more homozygous, disease-susceptible individuals and their genes (Acevedo-Whitehouse *et al.*, 2003; Luquet *et al.*, 2012; Paterson *et al.*, 1998; Amos *et al.*, 2009); individuals which may be more in-bred at other loci than their heterozygous conspecifics. Disease may also play a role in preselection of animals for predation through reducing their fitness, and this in turn may determine evolutionary polymorphism and strain diversity of the infective agents and/or pathogens adapted to these hosts (Morozov, 2012). This may be particularly pertinent in antelope, the reason being that they constitute the main food source of large predators and live in large aggregations or herds, thereby providing opportunity for eco-immunological evolutionary processes (Martin *et al.*, 2011).

Disease ecology of natural populations is complex, and there are probably many interactions at the level of the host, pathogen, and the environment, for example co-infection (Ezenwa & Jolles, 2011; Cleaveland *et al.*, 2008) and predation effects (Hethcotea *et al.*, 2004). With some pathogens, such as trypanosomes and avian influenza, regular challenge by the immune system of a wide range of mammalian hosts induces rapid evolution of new antigenic profiles in parasite populations, and a strong selection pressure for heterozygosity in the parasite population. The host community responds with adaptive immunity and subclinical infection and with at worst disease when stressed (Huchzmeyer *et al.*, 2001; van Gils *et al.*, 2007). These findings fundamentally challenge the germ theory, which is based on the idea that disease is caused by infection with a single microorganism (pathogen) and replace it with an ecological theory of disease which is pertinent to antelope disease research.

Often, impacts of disease on antelope can be neutral. Evidence exists of tolerance in wild animal species to endemic infections which are pathogenic to domestic animals. Piroplasm infections, such as trypanosomosis, theileriosis, babesiosis, and anaplasmosis, are well tolerated, and disease only occurs when animals, immunologically naïve, are suddenly introduced into the endemic zone through, for instance, translocation (Hove *et al.*, 1998). Bacterial infections like brucellosis are not infrequently detected in antelope but rarely with any clinical disease (Lundervold, 1994; Rementsova *et al.*, 1987), and endemic virus infections are usually cryptic, such as vector-borne bluetongue (Formenty *et al.*, 1994), and directly transmitted infections like malignant catarrh fever virus are maintained in wildebeest (*Connochaetus taurinus*) without clinical disease (Russell *et al.*, 2009). Internal and external parasites of some African and Asian antelopes are extensively described (Boomker *et al.*, 1983, 1986a, 1986b, 1989, 1991a, 1991b, 1996; Horak *et al.*, 1978, 1979, 1980, 1981, 1982a–c, 1983, 1986, 1988, 1989, 1991, 1992a–b, 1995, 1998; Morgan *et al.*, 2005), but the contribution of these organisms to antelope disease is less well described, except in a few case studies. Besides these few examples, comprehensive studies on infection in free-ranging antelope are relatively rare, given that these animals are relatively difficult to work with in nature, being shy, cryptic, heavily predated, and scavenged. A more holistic and interdisciplinary approach is needed in research to understand the role of epidemic diseases in the population ecology of antelope.

Table 1 gives a summary of the infections recorded in antelope where pathogens only rarely lead to disease.

The 'bad': disease impacts and threats to fragmented and depressed antelope populations – are these significant?

Although infections may sometimes be an integral part of healthy ecosystems, it is well recognised that disease can also have significant negative effects on the ecology of natural ecosystems (Holdo *et al.*, 2009; Table 2). This section will focus on the more negative aspects of disease in relation to antelope conservation and management in general, as well as the impact on the population dynamics of endangered species, on fragmented populations and on landscape change in particular.

Public health authorities and veterinarians are generally more focused on wildlife disease from the perspective of individual human or livestock health (Siembeda *et al.*, 2011), whilst disease ecologists, wildlife veterinarians, and those interested in practicing 'One Health' or Ecohealth consider the pathogen in terms of populations, general ecology, and ecosystems health (Zinsstag *et al.*, 2010). Where there are large populations of animals reproducing effectively, disease is not generally considered a problem – rather the opposite, being integral to the natural ecology. However, vets and others have argued that, given the highly disrupted ecology in most landscapes today, emergence of and exposure to pathogens and/or environmental toxins is not insignificant

Table 1 **Antelope infections (usually asymptomatic carriers or hosts)**

Agent	Species	Disease reported	Reservoir	Reference
Vector Borne Parasites				
Anaplasma	All	No	Yes	Ngeranwa *et al.*, 2008
Trypanosomes	All	No	Yes	Verhulst *et al.*, 1993; Funk *et al.*, 2013
Internal and External Parasites				
Nematodes	All	Yes (*Bronchonema magna* in springbok *Antidorcas marsupialis* and bontebok *Damaliscus pygargus pygargus*; *Haemonchus contortus* in blesbok *D.p.phillipsi*)	Yes	Multiple publ.: Boomker *et al.*, 1983–1996; Horak *et al.*, 1978–1998; Morgan *et al.*, 2005; Magwedere, 2012
Cestodes (*cysticercosis, echinococcosis*)	All	No	Yes	Magwedere, 2012
Trematodes Liver fluke	All	Yes	Yes	Bindernagel, 1972
Warble and nasal bots	Lechwe (*Kobus leche*), Hartebeest (*Alcelaphus buselaphus*)	Yes		Stafford, 1991; Howard, 1977

Filarial worms	Bushbuck (*Tragelaphus scriptus*)	Yes		Personal observation Richard Kock, 1997
Scabies	Various	Yes		Gakuya *et al.*, 2011; Bazargani *et al.*, 2007; Wright *et al.*, 1988
Ticks	All	Yes	Yes	Horak *et al.*, 1991, 1998
Virus				
Blue tongue	Various	No (Africa, Asia)		Formenty *et al.*, 1994
Capripox	Various	Antibody		Hedger and Hamblin, 1983; Young *et al.*, 1970
Akabane	Various	Antibody		Al-Busaidt *et al.*, 1987
Crimean Congo haemorrhagic fever	Various	Antibody		Bengis *et al.*, 1997
Malignant catarrhal fever	Wildebeest	No	Yes	Russell *et al.*, 2009
Bacteria/others				
Brucella	Wildebeest	No		Waghela and Karstad, 1986; Magwedere, 2012
Chlamydia	Various	No		Mansell *et al.*, 1995
Cytocoetes ondiri	Bushbuck, ticks	No	Yes	Davies, 1981

and constitutes a modern threat to species survival (Pounds *et al.*, 2006; Green *et al.*, 2006). Still, to date only a handful of extinctions across all taxa has been ascribed to disease alone. Cases reported as examples are mostly anecdotal and possibly underestimated through lack of a methodological approach to historical data (Macphee & Greenwood 2013; Wyatt *et al.*, 2008). Analysis of 'expert' opinion considers disease as a significant threat to antelope populations (Pedersen *et al.*, 2007); however, this being opinion-based, robust evidence is needed to establish whether this is indeed the case. There are relatively few well documented examples of long-term population impact from infectious disease in antelope.

Many antelopes aggregate in smaller or larger numbers, and can show disease susceptibility: for instance, rinderpest in wildebeest and lesser kudu (*Tragelaphus imberbis*) (Kock, 2006); FMD and pasteurellosis in saiga (*Saiga tatarica*) (Bekenov *et al.*, 1998) and heartwater in springbok (Fourie & Horak, 1987). The ecological immunology of such diseases probably relates to many factors, including the exotic origin of the infection, spillover from domestic species (rinderpest to African antelope and FMD to Asian antelope), the introduction of a species to unsuitable habitat outside the natural range (springbok and heartwater; de Graaff and Penzhorn, 1976) or stress effects.

The critically endangered saiga antelope appears to be an extreme example and possibly an important indicator of environmental stressors or changes. Massive outbreaks of rapidly fatal disease have been reported over the last century (Bekenov *et al.*, 1998). These continue in the present century, in 2010 and 2011 (Kock *et al.*, 2012), and most recently in May 2015, in a catastrophic die-off of more than 130,000 adults and uncountable calves, including entire herds of up to 60,000 individuals, in the core population (Kock *et al.*, 2015). Death appears to be from haemorrhagic septicaemia with *Pasteurella multocida* serotype B, an opportunist commensal bacteria playing the key role (Kock *et al.*, 2015). These events appear to be triggered by environmental factors acting either on the pathogen or the host and, possibly linked to climate, calving aggregations, pasture dynamics, and other seasonal factors. This die-off syndrome is only now beginning to receive research attention on the underlying disease ecology, which is needed to fully understand the causes and consequences.

In wildebeest, introduction of rinderpest virus from the Horn of Africa in the 19th century and the subsequent establishment as an endemic disease was considered to be detrimental to the species' ecology as a result of annual calf mortality and resulting population decline. This continued until eradication of the virus from cattle in Masailand and its disappearance from wildlife in the 1960s, after which time the wildebeest numbers increased fivefold (Holdo *et al.*, 2009). There was a small re-infection of wildlife in the Serengeti ecosystem in the early 1980s but this did not persist. There was also major ecological change observed in the environment over this period ascribed to the infection and its eradication, and shown by shifting vegetation and species composition in response to fluctuating herbivore numbers (Holdo *et al.*, 2009). Similarly, the dramatic decline of wild grazing herbivores like saiga in Central Asia – from disease impacts and poaching over the last two decades – is likely to be associated with changes in the steppe plant ecology, and this is not a surprise since the grasses co-evolved with ungulates affecting an assemblage of associated features (Mack & Thompson, 1982).

Studies on rinderpest have shown the tragelaphine antelopes to be particularly susceptible and mortalities in, for instance, lesser kudu have been suggested to be up to 90% (Kock *et al.*, 2006). An interesting aspect of the tragelaphine susceptibility is phylogenetic relatedness to the maintenance host of the pathogen, in this case domestic cattle (*Bos* sp.). Phylogeny is a measure of susceptibility for some species to some infections, and there is a genetic basis for this related to similarity of cell receptors in closely related species which the parasite recognises and exploits. Interestingly, infection by 'peste des petits ruminants' (PPR) virus (referred to as PPRv) has been confirmed in a wide variety of antelope through serology (Munir, 2013; Kock, 2006), but it has not been reported as a clinical disease in natural free-ranging populations of antelope, despite a current expansion of its range in Asia and Africa. PPRv is adapted to domestic sheep and goats, and it is thus not surprising that disease has only been reported to affect free-ranging mountain caprines of Asia (Abubakar *et al.*, 2011; Bao *et al.*, 2011; Hoffmann *et al.*, 2012). This status may change, however, and Mahapatra *et al.* (2015) report high infection rates in antelope of the Ngorongoro Conservation Area of the Greater Serengeti Ecosystem, but again there is no confirmed clinical disease syndrome in these species as yet. PPR in captive and semi-captive antelopes has been associated with severe mortality, with up to 50% morbidity and 100% case fatality (Elzein *et al.*, 2004; Furley *et al.*, 1987). Here the conditions of captivity might have compromised their natural immunity. All in all, this supports the hypothesis that phylogeny affects disease susceptibility.

FMD impacts on antelope populations can be significant (Anderson, 1975), and high mortalities have been recorded: for instance, ~70% in mountain gazelle (*Gazella gazella*) in a small wildlife reserve in Israel (Frankenberg, 1992; Shimshony, 1988) and a massive die-off of >140,000 Mongolian gazelle (*Procapra gutturosa*) in 1974 (Solokov *et al.*, 1982) with another major outbreak reported in 1998 (Schaller *et al.*, 1998). In Africa, where the buffalo (*Syncerus caffer*) acts as a reservoir, the disease can be considered endemic, and wildlife disease is rarely reported with relatively mild symptoms (Thomson *et al.*, 2003; Weaver *et al.*, 2013). In Central Asia, FMD and pasteurellosis were reported causes of epidemic disease in saiga, with sporadic die-off of a few hundred to several hundreds of thousands of animals (Bekenov *et al.*, 1998), but gaps in infection cycles allowed for full population recovery. Causation of some saiga mortality, ascribed to pasteurellosis, has been questioned (Kock *et al.*, 2012), and historical die-offs cannot always be ascribed for certain to infectious disease. Often, a full epidemiological or pathological picture is missing from the data sets and reports. Where studies are undertaken, they can be enlightening, for example, springbok distribution in South Africa correlates with the presence or absence of heartwater (Fourie & Horak, 1987). Anthrax and rabies are known to be associated with major epidemics and mortalities of antelope, but the ecology and impact of the disease have not been well established except in very few locations: for example, a large epidemic of rabies in greater kudu in Namibia between 1977 and 1983 occurred due to the sudden increase in the population and change in browsing behaviour of the animals (Hubschule, 1988; for other examples, see Hugh-Jones & De Vos, 2002; Bengis *et al.*, 2002; Beyer & Turnbull, 2009; Beyer *et al.*, 2012).

Pathology and serology of some diseases are relatively well described from immobilised, hunted, and culled populations of antelope, with the role of pathogens confirmed, especially where it is a shared with domestic livestock (Hamblin *et al.*, 1990; Lundervold *et al.*, 2004; Morgan *et al.*, 2006; Fourie & Horak, 1987; Peter *et al.*, 1988; Warghela & Karstad, 1986; Anderson & Rowe, 1998; Vosloo *et al.*, 2009; Verhulst *et al.*, 1993; Zieger *et al.*, 1998; Kock *et al.*, 2006). Among other diseases that remain poorly described, are those associated with an insect storm. For example, a pulmonary syndrome has been described from bongo (*Tragelaphus eurycerus eurycerus*) in Ndoki, Congo, in 1997 (Huchzermeyer *et al.*, 2001) and similar epidemic syndromes causing mortalities and sometimes affecting the central nervous system have been associated with mosquito and *Stomoxys* insect storms observed in Ngorongoro (Tanzania), Kibale Forest (Uganda) and Aberdares Forest (Kenya) (personal observation, R. Kock). In Uganda and Kenya, syndromes were observed affecting bushbuck and in Kenya, included histological evidence of nematodes in the brain from material collected; these were possibly filarial worms transmitted by biting flies. Parasitic diseases of the ruminant nervous system have been reported in the Americas (Nagy, 2004) but remain unconfirmed in African wildlife despite this syndromic evidence. Suggestions of disease-associated declines in *Damaliscus* spp. (tiang *D. lunatus tiang* and korrigum *D. lunatus korrigum*) in West and Central Africa and in Dinder, Sudan, are also inadequately described (Sayer, 1982; Whitney & el Moghraby, 1982), and the aetiology remains obscure, although the evidence of rinderpest and PPR antibody in bohor reedbuck (*Redunca redunca*) from Dinder, Sudan, (Kock, 2006; Kock, 2012) suggests that these ruminant plagues may have been a factor.

The main question for conservation is this: how important are diseases in terms of their impact on populations? In most records of significant epidemic disease in antelope populations, they recovered to similar of even higher numbers within a relatively short time period (e.g., in saiga, Bekenov *et al.*, 1998). Even some of the most severe plagues reported in wildlife populations, like rinderpest in buffalo and distemper in lions (*Panthera leo*), were followed by population recovery within a few years (Kock *et al.*, 2006; Munson *et al.*, 2008). For example, in Tsavo National Park in Kenya, buffalo numbers dropped from 11,798 in 1994 to 5,941 in 1999 after a rinderpest epidemic which started in 1993 (Kock *et al.*, 1999; Ngene *et al.*, 2011); the last evidence for infection was in 1998, and by 2008 the population had recovered to 9,371. This epidemic was coincident with drought, which would have exacerbated the effects of disease. Hence it appears that drought is a far greater threat to ungulate populations than disease per se (Duncan *et al.*, 2012).

Epidemic disease is not the only threat that infections present to antelope populations; persistent infection in a population can have apparently greater impacts than the more obvious high-mortality outbreaks described above. In the Serengeti system, at the turn of the 19th century after the great pandemic, rinderpest became endemic, maintained by the large populations of wildlife and cattle, and it kept the wildebeest population depressed at about 250,000. Annual infection of calves reduced the recruitment rate of wildebeest; hence the disease was described as 'yearling disease'. After the disease was eradicated in the 1960s, the population increased to over 1 million with significant impacts on the ecology of the system (Holdo *et al.*, 2007).

From the above it seems that in resilient populations in large landscapes, a combination of infectious disease and drought is countered by host reproductive rate, and populations remain stable or recover. This may not be the case for heavily poached and fragmented populations, and such combined events would be the perfect storm, which is more likely to result in extirpation of populations. This scenario is increasingly likely with climate change and ongoing expanding trends in human and domestic livestock populations, and agricultural land use.

Another viral disease in antelope of concern is malignant catarrhal fever (MCF) of cattle caused by infection with a herpes virus (Alcelaphine herpes virus AHV1) carried by wildebeest. The main area of risk is around the Greater Serengeti ecosystem, and it causes considerable economic loss to the local Masai communities. The livestock herders manage the contact and reduce incidence, but MCF creates an opportunity cost for them, as they cannot access large areas of pastoral rangeland during the wildebeest calving season (Russell *et al.*, 2009). The impact of MCF on antelope is a political issue, and current attitudes of Masai pastoralists indicate a negative trend where costs of wildlife are far outweighing the benefits (Homewood *et al.*, 2012).

As mentioned in Chapter 1, 30% of antelope species are categorized as at risk of extinction by IUCN. Little or no data exist on disease and its impacts on the free-ranging populations of many of these species, partly because the populations are so sparse. For a few species and threatened subspecies, there is some data, but much of the evidence is anecdotal, for instance, the role of disease in the population declines of the vulnerable korrigum in West and Central Africa discussed above. It is also possible that the last population of hirola antelope (*Beatragus hunteri*), a critically endangered genus in the Somali region of Kenya, crashed due to a severe rinderpest epidemic in their range in the 1980s; between 1981 and 1985 their numbers declined from over 13,000 animals to between 1,000 and 2,000 individuals (Butynski, 2000). The disease was not confirmed in this species, as neither trained wildlife veterinarians nor wildlife authority staff were active in the area due to security concerns. Retrospective participatory epidemiological studies showed that large numbers of cattle and wildlife died within the hirola range (Kock, 2008). Significant quantities of data on diseases affecting antelope species originate from captive collections, where significant mortalities from PPR, FMD, and bovine tuberculosis (BTB) have been recorded (Furley *et al.*, 1987; Greth, 1992; Elzein *et al.*, 2004; Kinne *et al.*, 2010). Whether the management conditions in captivity were a factor in the outbreaks is of interest, but care should be taken in extrapolating susceptibilities to free-ranging populations from these data.

Non-infectious disease threats are also of concern. These can be categorised into diseases associated with pasture, browse, and water abundance, and they are related to the quality and composition, resulting in starvation, malnutrition, or toxicity. Pollution is also a concern, and there is evidence for the potential for this in peri-urban protected areas. For example in Kenya, heavy metal toxicity in waterbuck (*Kobus ellipsiprymnus*) and impala (*Aepyceros melampus*) was ascribed to battery factory contaminated waste dumps on the periphery of the Nakuru National Park. Heavy metals were shown to be at toxic levels in pasture, most probably from contamination of the soils from leaching and through natural drainage channels from the dumps. It is assumed that once the soils were contaminated there was subsequent assimilation into plants and waterbuck and impala ingested high quantities of cadmium and lead, so much so that much of the calcium in the animals' skeletons was replaced with heavy metals, causing pathological softening and spontaneous bone fractures (Kisia *et al.*, 2002; Jumba *et al.*, 2007). There is also a risk of disease as a result of supplementary feeding of antelope and there are examples of exposure to contaminants including prions (e.g., Bovine Spongiform Encephalopathy BSE) in captive animals (Kirkwood *et al.*, 1997).

From available data, it is possible to speculate on a susceptibility index for the antelope. For example, herd species and close relatives to domestic animals are likely more at risk from prevalent infectious diseases than solitary forest or bush-dwelling species. Most at risk from infectious diseases might also be the more ancient lineages such as saiga and hirola antelope, assuming these

Table 2 **Major infectious diseases of antelope**

Disease and causative agent	Maintenance hosts	Antelope disease record
Foot-and-mouth disease *Apthovirus*	Cattle and African buffalo	Impala (Keet *et al.*, 1996; Vosloo *et al.*, 2009); saiga antelope (Kindyakov *et al.*, 1970), Mongolian gazelle (Bolortsetseg *et al.*, 2012)
Rinderpest *Morbillivirus*	Cattle, wild ruminants biologically competent, but population insufficient	Tragelaphine antelope (Kock, 2006)
Peste des petits ruminants *Morbillivirus*	Goats and sheep	Only recorded in captive or semi-captive antelope (e.g., Furley *et al.*, 1987; Elzein *et al.*, 2004; Kinne *et al.*, 2010)
Rift Valley fever *Phlebovirus*	Aedine mosquitoes	Various antelope (Olive *et al.*, 2012; Evans *et al.*, 2008)
Rabies *Lyssa virus*	Numerous wild canids, felids, and viverids, domestic dogs	Greater kudu (*Tragelaphus strepsiceros*) (Hubschle, 1988) spread by browse
Ebola virus	Fruit bats (*Hypsignathus monstrosus, Epomops franqueti, Myonyteris torquata)*	Duikers (*Cephalophus sp.*) (Rouquet, 2005)
Heartwater Rickettsia *Cowdria ruminantium*	Certain wild ruminants (e.g., buffalo), chelonians and gallinaceous birds, ticks	Springbok (Fourie & Horak, 1987; Peter *et al.*, 1998)
Babesiosis	Ruminants and ticks *Rhipicephalus evertsi*	Sable antelope (*Hippotragus niger*) (Hove *et al.*, 1998)
Theileriosis	Ruminants and ticks *Rhipicephalus appendiculatus*	Various associated with translocation (Nijhof *et al.*, 2005; Steyl *et al.*, 2012)
Anthrax *Bacillus anthracis*	Soil saprophyte	Large die-offs recorded in impala, kudu, tiang (Grant *et al.*, 2002; Lindeque & Turnbull, 1994; Gainer, 1987)

(*continued overleaf*)

Table 2 (*continued*)

Disease and causative agent	Maintenance hosts	Antelope disease record
Bovine tuberculosis ***Mycobacterium bovis***	Cattle, buffalo	Lechwe (maintenance host), kob (*Kobus kob*), blackbuck (*Antelope cervicapra*), greater kudu and bushbuck (Munyeme *et al.*, 2011; Gupta & Singh, 1987; Weber *et al.*, 1992; Renwick *et al.*, 2007; Zieger *et al.*, 1998)
Pasteurellosis	Ruminants latent carriers	Saiga antelope (Bekenov *et al.*, 1998)
Corynebacterium pseudotuberculosis	Environment	Wildebeest spp., (*Connochaetes gnou/ taurinus*), blesbok red hartebeest (*Alcelaphus buselaphus caama*), springbok (Müller *et al.*, 2011)

species have relatively primitive immune systems. Also, more remote and isolated populations, which are likely to be naïve to many pathogens are probably more susceptible to introduction of pathogens from livestock or human activities. However, these hypotheses need to be tested.

The 'Ugly': veterinary regulations and disease control measures – are these good or bad for antelope conservation?

Most disease examples given above are only one part of a range of risk factors influencing species viability. Sometimes disease is just the last straw in a long line of insults to populations. In this last section the indirect impact of disease on antelope conservation is considered.

In some instances, disease control measures themselves contribute to species decline and habitat disruption, for instance, through use of fencing and animal movement controls, or use of pesticides and population control measures (Kock, 2005; Mbaiwa & Mbaiwa, 2006; Ferguson & Hanks, 2010). Multi-host infection and the risk of spillover of pathogens between wild Bovidae and livestock became a major driving force in decisions on biosecurity and land-use planning from about the 1960s and led to a policy of separation of wildlife resources from commercial livestock, predominantly in the southern African countries. Where separation was not possible, wildlife host eradication measures were sometimes applied, usually without thought to the ecological consequences. For example, during tsetse control many hundreds of thousands of wildlife were destroyed (Smies, 1980). Currently, these draconian approaches are resisted, given their historical failure to resolve disease problems and the overall negative impact on biodiversity and natural resources, which are more highly valued now. More recently, large areas of rangeland in a few southern African countries have reverted to wildlife or wildlife-livestock mixed systems, exploited through consumptive and non-consumptive means for economic gain: in these countries, economic benefit from wildlife can outweigh the contribution of agriculture, and this trend in land use conversion back to wildlife and for opening up previously compartmentalised zones continues (Child *et al.*, 2012; Glew *et al.*, 2010).

Some ecologically insensitive disease control measures still remain as part of veterinary policy, promulgated by national authorities and international organisations. The most potentially damaging is for FMD, which if implemented as planned may eradicate some wild species from large areas. The 'Progressive Control Pathway' for FMD (PCP-FMD; FAO 2012) lays out a five-stage plan resulting in eventual freedom from the disease. However, among conservationists, there is a growing concern that a perceived imperative to progress up this control ladder may lead to interventions, such as veterinary cordon fencing, being implemented without full evaluation of environmental consequences or an assessment of the true risk that wildlife disease reservoirs may pose to livestock development (Ferguson *et al.*, 2013). With FMD, fencing and movement control are proven to reduce infection risk in livestock and in some situations, such as the Americas, Europe, and parts of Asia, it has proven eradicable. However, the full environmental costs of these measures are not accounted for. Where there are wildlife maintenance hosts, as in Africa, the situation is very different when compared to Europe and similar temperate ecosystems, and removal of the virus is unlikely without complete landscape change and destruction of the buffalo and possibly the antelope resource, which is totally unacceptable for conservation, socio-cultural, and economic reasons. In any case, the FMD 'problem', like BTB in many countries, is to some extent an artificial one, related to protection of high-value markets where the infection is absent, achieved through export trade restrictions, a form of hidden tariff – in endemic rangeland settings, the impact of the disease on livestock productivity is relatively minor. There is also a simple solution to this through 'Commodity Based Trade' whereby cattle carcasses can be processed into commodities, to be free of FMD virus (lymph nodes removed and deboned), and this has been embraced by the International Office for Epizootic Diseases (OIE). Surprisingly this approach has yet to be taken up adequately at the national or international levels (Thomson *et al.*, 2003) suggesting resistance to free markets. The actual risk of antelope acting as a source of FMD infection in livestock is also poorly quantified and rarely reported, and where reported the evidence is largely circumstantial (see Hargreaves *et al.*, 2004). African buffalo are a rare source of outbreaks, and recurrent disease in impala is reported and probably represents a possible vector (Vosloo *et al.*, 2009).

Other reasons why the livestock and veterinary sectors tend to discriminate against antelope presence, and which neither fence nor separation can do much about, are the vector-borne diseases including trypansomiasis and tick-borne disease like heartwater, corridor disease, or East Coast Fever. The contribution of antelope to the disease cycles is not well understood, but is considered a lesser concern than in the case of the African buffalo.

If a 'One Health' approach is taken to disease, especially in the context of rangelands, then there is little justification for excluding wildlife under any single domestic species disease control policy (Kock *et al.*, 2010). This approach is supported by increasing evidence that intensification of livestock systems, especially on rangeland, is simply bad economics and wrong policy. Hesse and Cavanna (2010) show productivity differentials in a study of rangeland between the more intensive sedentary (SS) and mobile open (MOS) systems. In East Africa, SS are reported to produce 0.023 and MOS 0.057 kg meat production per kg of animal; calving rates for SS are recorded at 45% and for MOS at 65%, total deaths of SS recorded at 35% and for MOS 15%, and calf mortalities recorded in SS at 40% and in MOS at 11%. Another example is in the northern rangelands of Kenya where integrated multiple-land-use systems exist and have provided tangible benefits. The Northern Rangelands Trust (NRT) initiative and its constituent conservancies are significantly enhancing livelihoods for their participating communities when compared to equivalent, non-participating communities (Glew *et al.*, 2010). These benefits occur at the household and community level, and are derived from revenues produced by conservation-linked enterprises on the back of a livestock system. Another case study in support of integrated systems is from Lake Mburo in Uganda, where no significant difference was detected in livestock morbidity between mixed-species systems compared to ranch livestock – possibly less (Okaido *et al.*, 2009). Some experimental work on the relationship between livestock and wildlife has been carried out showing some positive benefits and some negative impacts of integration depending on the system (Riginos *et al.*, 2012). Largely, evidence is in favour of mixed land-use approaches for both wildlife and livestock communities, for rangeland economics and conservation of species (Butt *et al.*, 2012; Niamir-Fuller, 2012), although this is also disputed (Norton-Griffiths, 2007). When carbon and water benefits and payments from ecosystem services are included, and now being audited more frequently in economic analysis, the margins of benefit will likely be even more obvious (McDermott *et al.*, 2012). The situation is complex, but implementation of a sectored policy on diseases like FMD involving fencing in these areas would impoverish the communities once again, even if a few commercial operators which export might benefit (Kock, 2010).

On the other side of the coin and much less talked about is the disease risk that livestock and people pose to antelope. A number of spillover events have occurred causing considerable impact on antelope populations, including BTB spilling into wildlife most probably from cattle (e.g., in the Kafue, Kruger, and Hluluwe and Umfolozi National Parks in southern Africa; Munyeme *et al.*, 2010; De Garine-Wichatitsky *et al.*, 2010; Gous & Williams, 2009), the rinderpest virus introduction to Africa, and the associated pandemic and subsequent epidemics mentioned earlier. Both these diseases evolved to become major diseases in domestic cattle in Eurasia before their introduction to Africa, rinderpest from an earlier morbillivirus related to BTB, and measles virus and BTB as part of the *Mycobacterium* complex of which *Mycobacterium tuberculosis* of humans is an earlier form. BTB is maintained in domestic animals and can spill over to wildlife where some species have become a reservoir (Brosch *et al.*, 2002). Rinderpest was maintained in cattle without any persistent wildlife reservoir, and this enabled global eradication through vaccination; unfortunately, this has not been possible with BTB partly because of persistent wildlife reservoirs. Other diseases associated with livestock spillover into wildlife are anthrax, pasteurellosis, rabies, and PPR (Pfukenyi *et al.*, 2009; Abubakar *et al.*, 2011; Clegg *et al.*, 2007; Bekenov *et al.*, 1998). Ironically, in some situations, fencing can protect wildlife from this sort of incursion and reduce risk as well as halt or slow down habitat destruction from slash and burn or agricultural expansion. Fences can have conservation benefits but rarely if ever are they ecologically viable, and the concept of a minimum viable population argues against fragmentation of any kind, with resilience being lost in vertebrate populations below 7000 individuals (Reed *et al.*, 2003).

Conclusion

With rising food insecurity, and the predicted increase in livestock production in the antelope ranges in Africa and Asia, the true disease risk in integrated antelope/agricultural/livestock systems and the associated economic costs and benefits need to be determined and quantified. Some attempts at this have been made, but much more research needs to be done, especially on:

- Ecology and epidemiology of economically important infectious disease in antelope hosts
- Diseases arising from the presence of livestock and humans in landscapes with mixed land-use, including the risk of spillover and its impact especially on rare antelope

We know very little about antelope diseases *in situ*, that is, about their role in genetic health, pathogenesis, ecology and epidemiology of free-living animals. Most of our knowledge relates to captive animals and epidemiology for a limited number of diseases in free-living populations, notably rinderpest, tuberculosis, and anthrax. Information on the likely impacts of disease in antelope-livestock systems is vital to equip decision makers to make the best decisions on future disease-control measures. The key policy decision is likely to be whether separation or integration of wild and domestic species on rangelands will optimise the balance between biodiversity conservation, and food and economic security.

References

Abubakar, M., Rajput, Z. I. & Ali, Q. (2011): Evidence of peste des petits ruminants virus (PPRV) infection in Sindh ibex (*Capra aegagrus blythi*) in Pakistan as confirmed by detection of antigen and antibody. *Health* (San Francisco) **43**(4): 745–747. doi:10.1007/s11250-010-9776-y

Acevedo-Whitehouse, K., Gulland, F., Greig, D. & Amos, W. (2003): Inbreeding-dependent pathogen susceptibility in California sea lions. *Nature* **422**: 35.

Acevedo-Whitehouse, K., Spraker, T. R., Lyons, E., Melin, S. R., Gulland, F., Delong, R. L. & Amos, W. (2006): Contrasting effects of heterozygosity on survival and hookworm resistance in California sea lion pups. *Molecular Ecology* **15**(7): 1973–1982.

Acevedo-Whitehouse, K., Vicente, J., Gortazar, C., Höfle, U., Fernández-de-Mera, I.G. & Amos, W. (2005): Genetic resistance to bovine tuberculosis in the Iberian wild boar. *Molecular Ecology* **14**(10): 3209–3217.

Al-Busaidt, S. M., Hamblin, C. & Taylor, W. P. (1987): Neutralising antibodies to Akabane virus in free-living animals in Africa. *Tropical Health and Production* **19**: 197–202.

Allen, A. R., Minozzi, G., Glass, E.J., Skuce, R. A., McDowell, S. W., Woolliams, J. A. & Bishop, S. C. (2010): Bovine tuberculosis: the genetic basis of host susceptibility. *Proceedings Biological Science* **277**(1695): 2737–2745.

Amos, W. & Acevedo-Whitehouse, K. (2009): A new test for genotype-fitness associations reveals a single microsatellite allele that strongly predicts the nature of tuberculosis infections in wild boar. *Molecular Ecology Resources* **9**(4): 1102–1111. doi: 10.1111/j.1755-0998.2009.02560.x

Anderson, E.C., Anderson, J., Doughty, W. J. & Drevmo, S. (1975): The pathogenity of bovine strains of foot and mouth disease virus for impala and wildebeest. *Journal Wildlife Diseases* **11**: 248–255.

Anderson, E. C. & Rowe, L. W. (1998): The prevalence of antibody to the viruses of bovine virus diarrhoea, bovine herpes virus 1, Rift Valley fever, ephemeral fever and bluetongue and to Leptospira sp. in free-ranging wildlife in Zimbabwe. *Epidemiology and Infection* **121**(2): 441–449.

Bao, J., Wang, Z., Li, L., Wu, X., Sang, P., Wu, G., Sang P, Wu G, Ding G, Suo L, Liu C, Wang J, Zhao W, Li J, Qi L. (2011): Detection and genetic characterization of peste des petits ruminants virus in free-living bharals (*Pseudois nayaur*) in Tibet, China. *Research in Veterinary Science* **90**: 238–240.

Bazargani, T. T., Hallan, J. A., Nabian, S. & Rahbari, S. (2007): Sarcoptic mange of gazelle (*Gazella subguttarosa*) and its medical importance in Iran. *Parasitological Research* **101**(6): 1517–1520.

Bekenov, B., Grachev A. & Milner Gulland, E. J. (1998): The ecology and management of the Saiga antelope in Kazakhstan. *Mammal Review* **28**(1): 1–52.

Bengis, R. G. & Veary, C. M. (1997): Public health risks associated with the utilization of wildlife products in certain regions of Africa. *Revue Scientifique et Technique Office Internationale Epizooties* **16**: 586–593.

Bengis, R. G., Kock, R. A. & Fischer, J. (2002): Infectious animal diseases : the wildlife/livestock interface. *Revue Scientifique et Technique Office Internationale Epizooties* **21**(1): 53–65.

Beyer, W. & Turnbull, P. C. B. (2009): Anthrax in animals. *Molecular Aspects of Medicine*, **30**(6): 481–489. doi:10.1016/j.mam.2009.08.004

Beyer, W., Bellan, S., Eberle, G., Ganz, H. H., Getz, W. M., Haumacher, R., Hilss, K. A., Kilian, W., Lazak, J., Turner. W. C. & Turnbull P. C. B. (2012): Distribution and molecular evolution of *Bacillus anthracis* genotypes in Namibia. *PLOS Neglected Tropical Diseases* **6**(3). doi:10.1371/journal.pntd.0001534

Bindernagel, J. A. (1972): Liver fluke *Fasciola gigantica* in African buffalo and antelopes in Uganda, East Africa. *Journal of Wildlife Diseases* **8**(4): 315–317.

Bolortsetseg, S., Enkhtuvshin, S., Nyamsuren, D., Weisman, W., Fine, A., Yang, A. & Joly, D. O. (2012): Serosurveillance for foot-and-mouth disease in Mongolian gazelles (*Procapra gutturosa*) and livestock on the eastern steppe of Mongolia. *Journal of Wildlife Diseases* **48**(1): 33–38.

Boomker J., Horak I. G. & De Vos V. (1986): The helminth parasites of various artiodactylids from some South African nature reserves. *Onderstepoort Journal of Veterinary Research* **53**(2): 93–102.

Boomker, J., Horak, I. G., Gibbons, L. M. & De Vos, V. (1983): *Haemonchus contortus* from the vaal ribbok, *Pelea capreolus*, and the bontebok, *Damaliscus dorcas dorcas*, in the Bontebok National Park. *Onderstepoort Journal of Veterinary Research* **50**(3): 179–181.

Boomker, J., Horak, I. G., De Vos, V. (1986): The helminth parasites of various artiodactylids from some South African nature reserves. *Onderstepoort Journal of Veterinary Research* **53**(2): 93–102.

Boomker, J., Horak, I. G. & MacIvor K. M. (1989): Helminth parasites of grysbok, common duikers and Angora and Boer goats in the Valley Bushveld in the eastern Cape Province. *Onderstepoort Journal of Veterinary Research* **56**(3): 165–172.

Boomker, J., Horak, I. G. & Knight, M. M. (1991): Parasites of South African wildlife. IX. Helminths of kudu, *Tragelaphus strepsiceros*, in the eastern Cape Province. *Onderstepoort Journal of Veterinary Research*, **58**(3): 203–204.

Boomker, J., Horak, I. G. & Flamand, J. R. (1991): Parasites of South African wildlife. X. Helminths of red duikers, *Cephalophus natalensis*, in Natal. *Onderstepoort Journal of Veterinary Research* **58**(3): 205–209.

Boomker, J. & Horak, I. G. (1992): Parasites of South African wildlife. XIII. Helminths of grey rhebuck, *Pelea capreolus*, and of bontebok, *Damaliscus dorcas dorcas*, in the Bontebok National Park. *Onderstepoort Journal of Veterinary Research* **59**(3): 175–182.

Boomker, J., Booyse, D. G., Watermeyer, R., De Villiers, I. L., Horak, I. G. & Flamand, J. R. (1996): Parasites of South African wildlife. XIV. Helminths of nyalas (*Tragelaphus angasii*) in the Mkuzi Game Reserve, KwaZulu-Natal. *Onderstepoort Journal of Veterinary Research* **63**(4): 265–271.

Brosch, R. Gordon, S. V., Marmiesse, M. Brodin, P. Buchrieser, C. Eiglmeier, K. Garnier, T. Gutierrez, C., Hewinson, G. Kremer, K. Parsons, L. M., Pym, A. S., Samper, S. van Soolingen, D. & Cole S. T. (2002): A new evolutionary scenario for the mycobacterium Tuberculosis complex. *PNAS* **99**(6): 3684–3689.

Butt, B. & Turner, M. D. (2012): Clarifying competition: the case of wildlife and pastoral livestock in East Africa. *Pastoralism: Research, Policy and Practice* **2**(1): 9. doi:10.1186/2041-7136-2-9

Butynski, T. M. (2000): Independent Evaluation of the Hirola antelope (*Beatragus hunteri*) conservation status and conservation action in Kenya: 1–180. Kenya Wildlife Service and Hirola Management Committee. Nairobi, Kenya.

Child, B. A., Musengezi, J., Parent, G. D. & Child, G. F. T. (2012): The economics and institutional economics of wildlife on private land in Africa. *Pastoralism: Research, Policy and Practice* **2**(1): 18. doi:10.1186/2041-7136-2-18

Cleaveland, S., Packer, C., Hampson, K., Kaare, M., Kock, R., Craft, M., Lembo, T., Mlengeya, T. & Dobson, A. (2008): The multiple roles of infectious diseases in the Serengeti ecosystem. In In Serengeti III: Human impacts on dynamic ecosystems: 209–240. Sinclair A. R. E., Packer C., Mduma S. & Fryxell, J.M. (Eds.). Chicago: University of Chicago Press.

Clegg, S. B., Turnbull, P. C., Foggin, C. M. & Lindeque, P. M. (2007): Massive outbreak of anthrax in wildlife in the Malilangwe Wildlife Reserve, Zimbabwe. *Veterinary Record* **160**: 113–118.

Crutzen, P. J., Aselmann, I. & Seiler, W. (1986): Methane production by domestic animals, wild ruminants, other herbivorous fauna, and humans. *Tellus B* **38**(3–4): 271–284.

Davies, F. G. (1981): Bovine petechial fever (Ondiri disease). *Veterinary Microbiology* **34**(2): 103–121.

de Garine-Wichatitsky, M., Caron, A., Gomo, A., Foggin, C., Dutlow, K., Pfukenyi, D., Lane, E., Le Bel, S., Hofmeyr, M., Hlokwe, T. & Michel, A. (2010): Bovine tuberculosis in buffaloes, Southern Africa. *Emerging Infectious Diseases*, **16**(5): 884–5.

de Graaff, G. & Penzhorn, B. L. (1976): The reintroduction of springbok Antidorcas marsupialis into South African National Parks-A documentation *Koedoe* **19**, no1 75–82.

Dethlefsen, L., McFall-Ngai, M. & Relman D. A. (2007): Review Article: An ecological and evolutionary perspective on human–microbe mutualism and disease. *Nature* **449**: 811–818. doi:10.1038/nature06245

Driscoll, E. E., Hoffman, J. I., Green, L. E., Medley, G. F. & Amos, W. (2011): A preliminary study of genetic factors that influence susceptibility to bovine tuberculosis in the British cattle herd. *PLoS One* **6**(4): e18806.

Duncan, C., Chauvenet, A. L. M., McRae, L. M. & Pettorelli, N. (2012): Predicting the future impact of droughts on ungulate populations in arid and semi-arid environments. *PloS One* **7**(12): e51490. doi:10.1371/journal.pone.0051490

Elzein, E. M., Housawi, E. A., Bashareek, F. M. T., Gameel, Y., Al-Afaleq, A. A. & Anderson, E. (2004): Severe PPR infection in gazelles kept under semi-free range conditions. *Journal of Veterinary Medicine. Series B, Infectious Diseases and Veterinary Public Health* **51**(2): 68–71. doi:10.1111/j.1439-0450.2004.00731.x

Evans, A., Gakuya, F., Paweska, J. T., Rostal, M., Akoolo, L., Van Vuren, P. J., Manyibe, T., Macharia, J. M., Ksiazek, T. G., Feikin, D. R., Breiman, R. F. & Kariuki Njenga, M. (2008): Prevalence of antibodies against Rift Valley fever virus in Kenyan wildlife. *Epidemiology and Infection* **136**: 1261–1269.

Ezenwa, V. O. & Jolles, A. E. (2011): From host immunity to pathogen invasion: the effects of helminth coinfection on the dynamics of microparasites. *Integrated Comparative Biology* **51**(4): 540–551.

FAO (2012): *Development of a long term roadmap for the progressive control of FMD in Eastern Africa*. Nairobi: Report of a Workshop held in Nairobi, Kenya March 5–6, 2012. Convened by FAO, OIE and AU-IBAR as a joint meeting and workshop. http://www.fao.org/fileadmin/user_upload/eufmd/docs/PCP_nairobi/Report.pdf (as 11 February 2016).

Frankenberg, E. (1992): Management of mountain gazelles in Israel. *Ongulés/Ungulates* **91**: 353–355.

Ferguson, K. J., Cleaveland, S., Haydon, D. T., Kock, R. A., Lembo, T. & Hopcraft, J. G. C. (2013): Evaluating the potential for the environmentally sustainable control of foot and mouth disease in sub-Saharan Africa. *Ecohealth* **10**: 314–322. DOI 10.1007/s10393-013-0850-6

Ferguson, K. & Hanks, J. (2010): Fencing impacts: a review of the environmental, social and economic impacts of game and veterinary fencing in Africa with particular reference to the Great Limpopo and Kavango-Zambezi Transfrontier Conservation Areas. *Pretoria Mammal Research Institute.*

Formenty, P., Domenech, J., Lauginie, F., Ouattara, M., Diawara, S., Raath, J.P., Grobler, D., Leforban, Y. & Angba, A. (1994): Epidemiological study of bluetongue in ovines, bovines and different species of animals in Côte d'Ivoire [in French]. *Review scientifique et technique Office international des Epizooties* **13**(3): 737–751.

Fourie, L. J. & Horak, I. G. (1987): Tick induced paralysis of springbok. *South African Journal of Wildlife Research* **17**(4): 131–133.

Fry, D. M. (1995): Reproductive effects in birds exposed to pesticides and industrial chemicals. *Environ Health Perspect*, **103**(Suppl 7): 165–171.

Funk, S., Nishiura, H., Heesterbeek, H., Edmunds, W. J. & Checchi, F. (2013): Identifying transmission cycles at the human-animal interface: the role of animal reservoirs in maintaining gambiense human African trypanosomiasis. *PloS Computational Biology* **9**(1): e1002855. doi:10.1371/journal.pcbi.1002855

Furley, C. W., Taylor, W. P. & Obi, T. U. (1987): An outbreak of peste des petits ruminants in a zoological collection. *Veterinary Record* **121**(19): 443–447.

Gainer, R. S. (1987): Epizootiology of anthrax and Nyasa wildebeest in the Selous Game Reserve, Tanzania. *Journal of Wildlife Diseases* **23**(1): 175–178.

Gakuya, F., Ombui, J., Heukelbach, J., Maingi, N., Muchemi, G., Ogara, W., Mijele, D. & Alasaad, S. (2012): Knowledge of mange among Masai pastoralists in Kenya. *PLoS One* **7**(8): e43342. doi: 10.1371/journal.pone.0043342.

Glew, L., Hudson, M. D. & Osborne, P. E. (2010): Evaluating the effectiveness of community-based conservation in northern Kenya. A Report to the Nature Conservancy, August 2010. School of Civil Engineering and the Environment, University of Southampton, England.

Gous, T. A. & Williams, M. C. (2009): The pathology of tuberculosis caused by *Mycobacterium tuberculosis* in a herd of free ranging springbok *(Antidorcus marsupialis): Onderstepoort Journal Veterinary Research* **76**: 419–441.

Grant, C. C., Davidson, T., Funston, P. J. & Pienaar, D. J. (2002): Challenges faced in the conservation of rare antelope: a case study on the northern basalt plains of the Kruger National Park. *Koedoe* **45**(2): 45–62.

Green, R. E., Taggart, M. A., Das, D., Pain, D. J., Sashi Kumar, C., Cunningham, A. A. & Cuthbert, R. (2006): Collapse of Asian vulture populations: risk of mortality from residues of the veterinary drug diclofenac in carcasses of treated cattle. *Journal of Applied Ecology* **43**(5): 949–956. doi:10.1111/j.1365-2664.2006.01225.x

Greth, A., Gourreau, J. M., Vassart, M., Vy, N. B., Wyers, M. & Leferve, P. C. (1992): Capripox disease in an Arabian oryx (*Oryx leucoryx*) from Saudi Arabia. *Journal of Wildlife Diseases* **28**(2): 295–300.

Gupta, P. P. & Singh, S. P. (1987): Tuberculosis in an Indian blackbuck. *Journal of Research, Punjab Agricultural University* **24**(2): 315–316.

Hamblin, C., Anderson, E. C., Jago, M., Mlengeya, T. & Hirji, K. (1990): Antibodies to some pathogenic agents in free-living wild species in Tanzania. *Epidemiology and Infection* **105**(3): 585–594.

Hargreaves, S. K., Foggin, C. M., Anderson, E. C., Bastos, A. D. S., Thomson, G. R., Ferris, N. P. & Knowles, N. J. (2004): An investigation into the source and spread of foot and mouth disease virus from a wildlife conservancy in Zimbabwe. *Rev. Sci. Tech.* **23**(3): 783–790.

Harwood, J. & Hall, A. (1990): Mass mortality in marine mammals—its implications for population-dynamics and genetics. *Trends in Ecology & Evolution* **5**: 254–257.

Hedger, R. S. & Hamblin, C. (1983): Neutralising antibodies to lumpy skin disease virus in African wildlife. *Comparative Immunology and Microbiology of Infectious Diseases* **6**(3): 209–213.

Hesse, C. & Cavanna, S. (2010): *Modern and mobile: the future of livestock production in Africa's drylands*. International Institute for Environment and Development and SOS Sahel, UK.

Hethcotea, H. E. W., Wangb, W., Hanc, l. & Zhien Mad, Z. (2004): A predator-prey model with infected prey. *Theoretical Population Biology* **66**(3): 259–268.

Hoffmann, B., Wiesner, H., Maltzan, J., Mustefa, R., Eschbaumer, M., Arif, F. A. & Beer, M. (2012): Fatalities in wild goats in Kurdistan associated with peste des petits ruminants virus. *Transboundary and Emerging Diseases* **59**(2): 173–176. doi:10.1111/j.1865-1682.2011.01270.x

Holdo, R. M., Sinclair, A. R. E., Dobson, A. P., Metzger, K. L., Bolker, B. M., Ritchie, M. E., & Holt, R. D. (2009): A disease-Mediated trophic cascade in the Serengeti and its implications for ecosystem C. *PLoS Biology* **7**(9): e1000210. doi:10.1371/journal.pbio.1000210

Homewood, K. M., Trench, P. & Brockington, D. (2012): Pastoralist livelihoods and wildlife revenues in East Africa: a case for coexistence? *Pastoralism: Research, Policy and Practice* **2**(1): 19. doi:10.1186/2041-7136-2-19

Horak, I. G. (1978): Parasites of domestic and wild animals in South Africa. *X. Helminths in impala. Onderstepoort Journal of Veterinary Research* **45**(4): 221–228.

Horak, I. G. (1979): Parasites of domestic and wild animals in South Africa. XII. Artificial transmission of nematodes from blesbok and impala to sheep, goats and cattle. *Onderstepoort Journal of Veterinary Research* **46**: 27–30.

Horak, I. G. (1981): The seasonal incidence of the major nematode genera recovered from sheep, cattle, impala and blesbok in the Transvaal. *Journal of the South African Veterinary Association* **52**(3): 213–223.

Horak, I. G., Boomker, J. & De Vos, V. (1980): A description of the immature stages of *Kirkioestrus minutus* (Rodhain; Bequaert, 1915) (*Diptera, Oestridae*), and the life cycle and seasonal prevalence of this fly in blue wildebeest. *Onderstepoort Journal of Veterinary Research* **47**(1): 23–30.

Horak, I. G., Brown, M. R., Boomker, J., De Vos, V., Van Zyl, E. A. (1982a): Helminth and arthropod parasites of blesbok, *Damaliscus dorcas phillipsi*, and of bontebok, *Damaliscus dorcas dorcas*. *Onderstepoort Journal of Veterinary Research* **49**(3): 139–146.

Horak, I. G., De Vos, V. & De Klerk, B. D. (1982b): Helminth and arthropod parasites of vaal ribbok, *Pelea capreolus*, in the western Cape Province. *Onderstepoort Journal of Veterinary Research* **49**(3): 147–148.

Horak, I. G., Meltzer, D. G. & De Vos, V. (1982c): Helminth and arthropod parasites of springbok, *Antidorcas marsupialis*, in the Transvaal and Western Cape Province. *Onderstepoort Journal of Veterinary Research* **49**(1): 7–10.

Horak, I. G., Potgieter, F. T., Walker, J. B., De Vos, V. & Boomker, J. (1983): The ixodid tick burdens of various large ruminant species in South African nature reserves. *Onderstepoort Journal of Veterinary Research* **50**(3): 221–228.

Horak, I. G., Sheppey, K., Knight, M. M. & Beuthin, C. L. (1986): Parasites of domestic and wild animals in South Africa. XXI. Arthropod parasites of vaal ribbok, bontebok and scrub hares in the western Cape Province. *Onderstepoort Journal of Veterinary Research* **53**(4): 187–197.

Horak, I. G., Keep, M. E., Flamand, J. R. & Boomker, J. (1988): Arthropod parasites of common reedbuck, *Redunca arundinum*, in Natal. *Onderstepoort Journal of Veterinary Research*, **55**(1): 19–22.

Horak, I. G., Keep, M. E., Spickett, A. M. & Boomker, J. (1989): Parasites of domestic and wild animals in South Africa. XXIV. Arthropod parasites of bushbuck and common duiker in the Weza State Forest, Natal. *Onderstepoort Journal of Veterinary Research* **56**(1): 63–66.

Horak, I. G., Boomker, J., Flamand. J. R. (1991): Ixodid ticks and lice infesting red duikers and bushpigs in north-eastern Natal. *Onderstepoort Journal of Veterinary Research* **58**(4): 281–284.

Horak, I. G., Anthonissen, M., Krecek, R. C. & Boomker, J. (1992a): Arthropod parasites of springbok, gemsbok, kudus, giraffes and Burchell's and Hartmann's zebras in the Etosha and Hardap Nature Reserves, Namibia. *Onderstepoort Journal of Veterinary Research* **59**(4): 253–257.

Horak, I. G., Boomker, J., Spickett, A. M. & De Vos, V. (1992b): Parasites of domestic and wild animals in South Africa. Ectoparasites of kudus in the eastern Transvaal Lowveld and the eastern Cape Province. *Onderstepoort Journal of Veterinary Research* **59**(4): 259–273.

Horak, I. G., Boomker, J., Flamand, J. R. (1995): Parasites of domestic and wild animals in South Africa. XXXIV. Arthropod parasites of nyalas in north-eastern KwaZulu-Natal. *Onderstepoort Journal of Veterinary Research*, **62**(3): 171–179.

Horak, I. G. & Boomker, J. (1998): Parasites of domestic and wild animals in South Africa. XXXV. Ixodid ticks and bot fly larvae in the Bontebok National Park. *Onderstepoort Journal of Veterinary Research* **65**(3): 205–211.

Hove T., Sithole N., Munodzana D. & Masaka S. (1998): Isolation and characterisation of a Babesia species from *Rhipicephalus evertsi* picked off a sable antelope (*Hippotragus niger*) which died from acute babesiosis. *Onderstepoort Journal of Veterinary Research* **65**(2): 75–80.

Howard, G. W. (1977): Prevalence of nasal bots (*Diptera: Oestridae*) of some Zambian Hartebeest. *Journal of Wildlife Diseases* **13**: 400–404.

Hubschle O. J. B. (1988): Rabies in kudu antelope. *Reviews of Infectious Diseases* **10**(Supplement 4): S629–S633.

Huchzermeyer, F. W., Penrith, M. L. & Elkan, P. W. (2001): Multifactorial mortality in bongos and other wild ungulates in the north of the Congo Republic. *Onderstepoort Journal of Veterinary Research* **68**: 263–269.

Hugh-Jones, M. E. & Vos, V. De. (2002): Anthrax and Wildlife. *Revue Scientifique et Technique Office Internationale Epizooties* **21**(2): 359–383.

Jumba, I. O., Kisia, S. M. & Kock, R. (2007): Animal health problems attributed to environmental contamination in Lake Nakuru National Park, Kenya: a case study on heavy metal poisoning in the waterbuck *Kobus ellipsiprymnus defassa* (Ruppel 1835). *Archives of Environmental Contamination and Toxicology* **52**(2): 270–81.

Keet, D. F., Hunter, P., Bengis, R. G., Bastos, A. & Thompson, G. R. (1996): The 1992 foot and mouth disease epizootic in the Kruger national Park. *Journal of the South African Veterinary Association* **67**(2): 83–87.

Kindyakov, V. I., Nagumanov, F. M., Balgan'aev, E. Kh., Zinov'ev, B. S., Pankratov, L. D. & Chufarin, A.M. (1970): Epizootiological significance of wild even-toed ungulates in the case of foot and mouth disease *Moscova Veterinariya* **47**: 52–53.

Kinne, J., Kreutzer, R., Kreutzer, M., Wernery, U. & Wohlsein P. (2010): Peste des petits ruminants in Arabian wildlife. *Epidemiology andInfection* **138**: 1211–1214.

Kirkwood, J. K. & Cunningham, A. A. (1994): Epidemiological observations on spongiform encephalopathies in captive wild animals in the British Isles. *Veterinary Record*, **135**: 296–303.

Kock, R. A. (2005): The Wildlife Domestic Animal Disease Interface – should Africa adopt a hard or soft edge? *Transactions of the Royal Society of South Africa* **59**(1): 10–15.

Kock R. A. (2006): Rinderpest and wildlife. In *Rinderpest and peste des petits ruminants virus*. Plagues of large and small ruminants: 144–162. Barrett, T., Pastoret, P.-P. & Taylor, W. (Eds.). Biology of Animal Infections. London: Elsevier publications Academic Press.

Kock, R. A. (2010): The newly proposed Laikipia disease control fence in Kenya. In Fencing impacts: a review of the environmental, social and economic impacts of game and veterinary fencing in Africa with particular reference to the Great Limpopo and Kavango-Zambezi Transfrontier Conservation Areas: 71–75. Ferguson, K. & Hanks, J. (Eds.). Pretoria: Mammal Research Institute.

Kock, R. A., Wambua, J. M., Mwanzia, J., Wamwayi, H., Ndungu, E. K., Barrett, T., Kock, N. D. & Rossiter P. B. (1999): Rinderpest epidemic in wild ruminants in Kenya 1993–7. *Veterinary Record* **145**: 275–283.

Kock, R. (2012): Mission Report (1) Khartoum, Sudan February 2012 Royal Veterinary College and Republic of Sudan Livestock Epidemio-Surveillance Project-Northern Sub-Project (LESP-NS) of the Directorate-General AH&EDC, Ministry of Livestock, Fisheries and Rangelands Khartoum, Sudan.

Kock R. A., Wamwayi, H. M., Rossiter, P. B., Libeau, G., Wambwa, E., Okori, J., Shiferaw, F. S. & Mlengeya, T. D. (2006): Rinderpest in East Africa: continuing re-infection of wildlife populations on the periphery of the Somali ecosystem. *Preventive Veterinary Medicine*, **75** (1–2): 63–80.

Kock, R., Kock, M., Cleaveland, S. & Thomson, G. (2010): Health and disease in wild rangelands. In *Wild rangelands: conserving wildlife while maintaining livestock in semi-arid ecosystems*: 235–265. du Toit, J., Kock, R. & Deutsch, J. (Eds.). Oxford: Wiley-Blackwell Publishers.

Kock, R., Grachev, Y., Zhakypbayev, A., Usenbayev, A., Zuther, S., Klimanova, O., Dieterich, T., Sapanov, M. & Izimbergenova, G. (2012): A retrospective assessment of saiga antelope *Saiga tatarica* die-off in Western Kazakhstan 2010–2011. *Saiga News* **14**(1): 1–4.

Kock, R., Khomenko, S., Zuther, S. & Orynbayev, M. (2015): Betpak-Dala – Turgai and Tengiz aggregations during calving May 2015. *Report to the Convention on Migratory Species (CMS) 11th June 2015: Disease outbreak investigation & disease ecology*. Bonn, Germany: CMS UNEP internal report.

Laxminarayan, R., Duse, A., Wattal, C., Zaidi, A. K. M., Wertheim, H. F. L., Sumpradit, N., ... Cars, O. (2013): Antibiotic resistance-the need for global solutions. *The Lancet Infectious Diseases*, **13**(12), 1057–98. doi:10.1016/S1473-3099(13)70318-9

Lindeque, P. M. & Turnbull, P. C. B. (1994): Ecology and epidemiology of anthrax in the Etosha National Park, Namibia. *Onderstepoort Journal of Veterinary Research* **61**(1): 71–83.

Lundervold, M., Milner-Gulland, E.J., O'Callaghan, C.J., Hamblin, C. (2003): First evidence of bluetongue virus in Kazakhstan. *Veterinary microbiology*, **92**(3):281–7.

Lundervold, M., Milner-Gulland, E. J., O'Callaghan, C. J., Hamblin, C., Corteyn, A. & Macmillan, A. P. (2004): A serological survey of ruminant livestock in Kazakhstan during post-Soviet transitions in farming and disease control. *Acta Veterinaria Scandinavica* **45**: 211–224.

Luquet, E., Garner, T. W., Léna, J. P., Bruel, C., Joly, P., Lengagne, T., Grolet, O. & Plénet, S. (2012): Genetic erosion in wild populations makes resistance to a pathogen more costly. *Evolution* **66**(6): 1942–1952. doi: 10.1111/j.1558-5646.2011.01570.x.

Mack R. N. & Thompson J. N. (1982): Evolution in Steppe with few large, hooved mammals. *American Naturalist* **119**(6): 757–773.

Macphee, R. D. E. & Greenwood, A. D. (2013): Infectious disease, endangerment, and extinction. *International journal of evolutionary biology*, **2013**: 571939. doi:10.1155/2013/571939

Magwedere, K., Hemberger, M. Y., Hoffman, L. C. & Dziva, F. (2012): Zoonoses: a potential obstacle to the growing wildlife industry of Namibia. *Infection Ecology & Epidemiology* **2**: 1–16. doi:10.3402/iee.v2i0.18365

Mahapatra, M., Murali, S. K. M., Eblate E., Fyumagwa, R., Shilinde, S., Maulid Mdaki, M., Keyyu, J., Parida, S. & Kock, R. (In press): Spill-over of peste des petits ruminants virus from domestic sheep and goats to wild ruminants in the Serengeti ecosystem, Northern Tanzania. *Emerging Infectious Diseases*.

Mansell, J. L.,Tang, K. N., Baldwin, C. A., Styler, E. L. & Liggett A. D. (1995): Disseminated chlamydial infection in antelope. *Journal of Veterinary Diagnostic Investigation* **7**(3): 397–399.

Martin, L. B., Hawley, D. M. & Ardia, D. R. (2011): An introduction to ecological immunology. *Functional Ecology* **25**(1): 1–4. doi:10.1111/j.1365-2435.2010.01820.x

Mbaiwa, J. E. & Mbaiwa, O. I. (2006): The effects of veterinary fences on wildlife populations in Okavango Delta, Botswana. *International Journal of Wildlife* **12**: 17–41.

McDermott, M., Mahanty, S. & Schreckenberg, K. (2012): Examining equity: a multidimensional framework for assessing equity in payments for ecosystem services. *Environmental Science & Policy* **33**: 1–12. doi:10.1016/j.envsci.2012.10.006

Morgan, E. R., Shaikenov, B., Torgerson, P. R., Medley, G. F. & Milner-Gulland, E. J. (2005): Helminths of saiga antelope in Kazakhstan: implications for conservation and livestock production. *Journal of wildlife diseases* **41**(1): 149–162.

Morgan, E. R., Lundervold, M., Medley, G., Shaikenov, B. S., Torgerson, P. R. & Milner-Gulland, E. J. (2006): Assessing risks of disease transmission between wildlife and livestock: the Saiga antelope as a case study. *Biological Conservation* **131** (2): 244–254.

Morozov, A., Best & A. (2012): Predation on infected host promotes evolutionary branching of virulence and pathogens' biodiversity. *Journal of theoretical biology* **307C**: 29–36.

Müller, B., de Klerk-Lorist, L.-M., Henton, M. M., Lane, E., Parsons, S., Gey van Pittius, N. C., Kotze, A., van Helden, P. D. & Tanner, M. (2011): Mixed infections of *Corynebacterium pseudotuberculosis* and non-tuberculous mycobacteria in South African antelopes presenting with tuberculosis-like lesions. *Veterinary microbiology* **147**(3–4): 340–345.

Munir, M. (2013): Role of wild small ruminants in the epidemiology of peste des petits ruminants. *Transboundary and Emerging Diseases* 1–14: doi:10.1111/tbed.12052

Munyeme, M., Muma, J. B., Siamudaala, V. M., Skjerve, E., Munang'andu, H. M. & Tryland, M. (2010): Tuberculosis in Kafue lechwe antelopes (*Kobus leche kafuensis*) of the Kafue Basin in Zambia. *Preventive Veterinary Medicine* **95**(3–4): 305–308. doi:10.4061/2011/918743

Nagy D. W. (2004): *Parelaphostrongylus tenuis* and other parasitic diseases of the ruminant nervous system. *Veterinary Clinics of North America: Food Animal Practice* **20** (2): 393–412.

Ngene, N. S., Ihwagi F., Nzisa, M., Mukeka, J., Njumbi, S. & Omondi P. (2011): Total aerial census of elephants and other large mammals in the Tsavo-Mkomazi ecosystem. Nairobi: Kenya Wildlife Service, 18.

Ngeranwa, J. J., Shompole, S. P., Venter, E. H., Wambugu, A., Crafford, J. E. & Penzhorn, B. L. (2008): Detection of Anaplasma antibodies in wildlife and domestic species in wildlife-livestock interface areas of Kenya by major surface protein 5 competitive inhibition enzyme-linked immunosorbent assay. *Onderstepoort Journal Veterinary Research* **75**(3): 199–205.

Niamir-Fuller, M., Kerven, C., Reid, R. & Milner-Gulland, E. (2012): Co-existence of wildlife and pastoralism on extensive rangelands: competition or compatibility? *Pastoralism: Research, Policy and Practice* **2**(1): 8. doi:10.1186/2041-7136-2-8

Nijhof, A. M., Pillay, V., Steyl, J., Prozesky, L., Stoltsz, W. H., Lawrence, J. A., Penzhorn, B. L., *et al.* (2005): Molecular characterization of Theileria species associated with mortality in four species of African antelopes. *Journal of Clinical Microbiology* **43**(12): 5907–5911. doi:10.1128/JCM.43.12.5907

Nkedianye, D., de Leeuw, J., Ogutu, J. O., Said, M. Y., Saidimu, T. L., Kifugo, S. C., Kaelo, D. S. & Reid, R. S. (2011): Mobility and livestock mortality in communally used pastoral areas: the impact of the 2005–2006 drought on livestock mortality in Maasailand. *Pastoralism*, **1**(1): 1–17. doi:10.1186/2041-7136-1-17

Norton-Griffiths, M. (2007): How many wildebeest do you need? *Growth (Lakeland)* **8**(2): 41–64.

Olive, M.-M., Goodman, S. M. & Reynes J.-M. (2012): The role of wild mammals in the maintenance of Rift Valley Fever Virus. *Journal of Wildlife Diseases* **48**(2): 241–266.

Paterson, S., Wilson, K. & Pemberton, J. M. (1998): Major histocompatibility complex variation associated with juvenile survival and parasite resistance in a large unmanaged ungulate population (*Ovis aries* L.). *Proceedings of the National Academy of Sciences of the United States of America* **95**: 3714–3719.

Peter, T. F., Anderson, E. C., Burridge, M. J. & Mahan, S. M. (1998): Demonstration of a carrier state for *Cowdria ruminantium* in wild animal ruminants from Africa. *Journal of Wildlife Diseases*, **34**(3): 567–575.

Pletcher, J. M., Horak, I. G., de Vos, V. & Boomker, J. (1984): Nodular abomasitis in impala (*Aepyceros melampus*) caused by the nematode *Longistrongylus sabie*. *Journal of Parasitology* **70**(6): 907–910.

Pounds, J. A., Bustamante, M. R., Coloma, L. A., Consuegra, J. A., Fogden, M. P. L, Foster, P. N., La Marca, E., Masters, K. L., Merino-Viteri, A., Puschendorf, R., Ron, S. R., Sánchez-Azofeifa, G. A., Still, C. J. & Young, B. E. (2006): Widespread amphibian extinctions from epidemic disease driven by global warming. *Nature* **439**: 161–167.

Reed, D. E., Daley, C. A. & Shave, H. J. (1976): Reovirus-like agent associated with neonatal diarrhoea in pronghorn antelope. *Journal of Wildlife Diseases* **12**(4): 488–491.

Reed, D. H., O'Grady, J. J., Brook, B. W., Ballou, J. D. & Frankham, R. (2003): Estimates of minimum viable population sizes for vertebrates and factors influencing those estimates. *Biological Conservation* **113**(1): 23–34. doi:10.1016/S0006-3207(02)00346-4

Rementsova, M. M. (1987): *Brucellosis in wild animals*. New Delhi: Amerind Publishing Co. Pvt. Ltd.

Renwick, A. R., White, P. C. L. & Bengis, R. G. (2007): Bovine tuberculosis in southern African wildlife: a multi-species host-pathogen system. *Epidemiology and infection* **135**(4): 529–540. doi:10.1017/S0950268806007205

Riginos, C., Porensky, L. M., Veblen, K. E., Odadi, W. O., Sensenig, R. L., Kimuyu, D., Keesing, F., Wilkerson, M. L. & Young, T. P. (2012): Lessons on the relationship between livestock husbandry and biodiversity from the Kenya Long-term Exclosure Experiment (KLEE). *Pastoralism: Research, Policy and Practice* **2**(1): 10. doi:10.1186/2041-7136-2-10

Rouquet, P., Froment, J.-M., Bermejo, M., Kilbourn, A., Karesh, W., Reed, P., Kumulungui, B., Yaba, P., Délicat, A., Rollin, P. E. & Leroy, E. M. (2005): Wild animal mortality monitoring and human ebola outbreaks, Gabon and Republic of Congo, 2001–2003. *Emerging infectious diseases* **11**(2): 283–90. doi:10.3201/eid1102.040533

Russell, G. C., Stewart, J. P. & Haig, D. M. (2009): Malignant catarrhal fever: a review. *Veterinary Journal* **179**(3): 324–335. doi:10.1016/j.tvjl.2007.11.007

Sayer, J. A. (1982): The pattern of the decline of the korrigum *Damaliscus lunatus* in West Africa. *Biological Conservation* **23**: 95–110.

Schaller, G. B. & Lhagvasuren, B. (1998): A disease outbreak in Mongolian gazelles. *Gnusletter* **17**(2):15.

Shimshony, A. (1988): Foot and mouth disease in mountain gazelle in Israel. *Revue Scientifique et Technique Office International des Epizooties* **7**(4): 917–923.

Siembeda, J. L., Kock, R. A, McCracken, T. A. & Newman, S. H. (2011): The role of wildlife in transboundary animal diseases. *Animal Health Research Reviews* **12**: 95–111. doi:10.1017/S1466252311000041

Smies, M. (1980): Environmental impact of tsetse control/eradication operations. *International Journal Pest Management* **26**: 237–240.

Stafford, K. J. (1991): A review of diseases of parasites of the Kafue lechwe (*Kobus leche kafuensis*). *Journal Wildlife Diseases* **27**(4): 661–667.

Steyl, J. C., Prozesky, L., Stoltsz, W. H. & Lawrence, J. A. (2012): Theileriosis (Cytauxzoonosis) in Roan antelope (Hippotragus equinus): Field exposure to infection and identification of potential vectors. Onderstepoort Journal of Veterinary Research, **79**(1): 1–8. 10.4102/ojvr.v79i1.367 DOI:10.4102/ojvr.v79i1.367#_blank

Thomson, G. R., Vosloo, W. & Bastos, A. D. S. (2003): Foot and mouth disease in wildlife. *Virus Research* **91**(1): 145–161. http://www.ncbi.nlm.nih.gov/pubmed/12527441

Thorne, E. T., Williams, E. S., Spraker, T. R., Helms, W. & Segerstrom T. (1988): Bluetongue in free-ranging pronghorn antelope (*Antilocapra Americana*) in Wyoming: 1976–1984. *Journal of Wildlife Diseases* **24** (1): 113–119.

Van Gils, J. A, Munster, V. J., Radersma, R., Liefhebber, D., Fouchier, R. A. M. & Klaassen, M. (2007): Hampered foraging and migratory performance in swans infected with low-pathogenic avian influenza A virus. *PloS One* **2**(1): e184. doi:10.1371/journal.pone.0000184

Verhulst, A., Van Meirvenne, N., Buscher, P., Pandey, V. S., van Meirvenne, N., Uilenberg, G. & Hamers, R. (1993): Wild animals as reservoirs of animal and human trypanosomosis. Resistance or tolerance of animals to disease. *Veterinary epidemiology and diagnostic methods* **161**: abstract.

Vosloo, W., Thompson, P. N., Botha, B., Bengis, R. G. & Thomson, G. R. (2009): Longitudinal study to investigate the role of impala (*Aepyceros melampus*) in foot-and-mouth disease maintenance in the Kruger National Park, South Africa. *Transboundary and Emerging Disease* **56**(1–2): 18–30.

Waghela, S. & Karstad, L. (1986): Antibodies to Brucella spp. among blue wildebeest and African buffalo in Kenya. *Journal of Wildlife Diseases* **22**(2): 189–192.

Weaver, G. V., Domenech, J., Thiermann, A. R. & Karesh, W. B. (2013): Foot and mouth disease: a look from the wild side. *Journal of Wildlife Diseases* **49**(4): 759–785. doi:10.7589/2012-11-276

Weber, A., van Hoven, W. & van Hoven, W. (1992): Tuberculosis of the parotid salivary gland in a kudu, *Tragelaphus strepsiceros*. *Koedoe* **35**(1): 119–122.

Whitney, J. B. & el Moghraby, A. I. (1982): Dinder National Park, Sudan: development versus preservation. *Environmental Conservation* **9**: 248–250. doi:10.1017/S0376892900020476.

Wright, F. C. & Glaze, R. L. (1988): Blackbuck antelope (*Antilope cervicapra*), a new host *for Psoroptes cuniculi* (Acari: Psoroptidae). *Journal of Wildlife Diseases* **24**(1): 168–169.

Wyatt, K. B., Campos, P. F., Gilbert, M. T. P., Kolokotronis S.-O., Hynes, W. H., DeSalle R., Daszak, P., MacPhee, R. D. E. & Greenwood, A. D. (2008): Historical mammal extinction on Christmas Island (Indian Ocean) correlates with introduced infectious disease, *PLoS ONE* **3** (11): e3602.

Young, E., Basson, P. A. & Weiss, K. E. (1970): Experimental infection of Giraffe (*Giraffa camelopardalis* [Linnaeus 1962]), Impala (Aepyceros melampus [Lichtenstein 1812]) and the Cape buffalo (*Syncerus caffer* [Sparman, 1779]) with lumpy-skin disease virus. *Onderstepoort Journal of Veterinary Research* **37**: 78–88.

Zieger, U., Pandey, G. S., Kriek, N. P. J. & Cauldwell, A. E. (1998): Tuberculosis in Kafue lechwe (*Kobus leche kafuensis*) and in a bush buck (*Tragelaphus scriptus*) on a game ranch in Central Province, Zambia. *Journal of the South African Veterinary Association* **69**(3): 98–101.

Zinsstag, J., Schelling, E., Waltner-Toews, D. & Tanner, M. (2010): From 'one medicine' to 'one health' and systemic approaches to health and well-being. *Preventive Veterinary Medicine* **101**(3–4). doi:10.1016/j.prevetmed.2010.07.003

8

Hands-on Approaches to Managing Antelopes and their Ecosystems: A South African Perspective

Michael H. Knight[1,2], Peter Novellie[1,3], Stephen Holness[1,2], Jacobus du Toit[4], Sam Ferreira[1], Markus Hofmeyr[1], Christina Grant[1,2], Marna Herbst[1], and Angela Gaylard[1]

[1]*Conservation Services, South African National Parks, Republic of South Africa*
[2]*Centre for African Conservation Ecology, Nelson Mandela Metropolitan University, Port Elizabeth, Republic of South Africa*
[3]*Sustainability Research Unit, Nelson Mandela Metropolitan University, George, Republic of South Africa*
[4]*Du Toit Wildlife Services, Pretoria, Republic of South Africa*

Introduction

Protected areas (UNEP-WCMC, 2008) in Africa (Naughton-Treves *et al.*, 2005) remain one of the principal strategies to conserve its unique biodiversity and wildlife assemblages. Since 1970 the total protected area coverage in sub-Saharan Africa has increased nearly twofold to include a total of 3.06 million km^2 of terrestrial and marine habitats (WCMC, 2004). The continent now has more than 1,100 national parks and wildlife reserves. However, despite this the picture for wildlife conservation as a whole in Africa is rather depressing, although there are some promising signs in certain regions. It has been suggested at the continental scale that large (>5 kg) mammalian herbivores population abundance in Africa's protected areas has declined by 59% since 1970 (Craigie *et al.*, 2010), although Scholte (2011) questions its universal validity. They noted that there were regional differences with antelope populations in southern African protected areas appearing to maintain their abundances, while those in eastern and western Africa generally showed moderate and drastic declines, respectively.

Antelope Conservation: From Diagnosis to Action, First Edition. Edited by Jakob Bro-Jørgensen and David P. Mallon.

The declines in wildlife populations are thought to be largely attributable to illegal overexploitation and habitat change, driven by a rapid growth in the human population and increased resource consumption (Baillie *et al.*, 2004).

The better performance of large mammal populations in southern African protected areas is related to particular conservation values and historical and socio-economic circumstances. Southern African protected area managers are generally better resourced than in other regions. This stems from higher realized economic value of wildlife and associated tourism (Norton-Griffiths, 2007; Carruthers, 2008; Craigie *et al.*, 2010). The utilitarian ideological approach to wildlife management in state protected and private/communal lands in southern Africa has been identified as the basis for the success of the region's wildlife populations (Lindsey *et al.*, 2009). This approach takes into account socioeconomic interests of the population by linking conservation to revenue generation, poverty alleviation, social justice, capacity building, and empowerment (Adams & Hulme, 2001; Winterhalder & Lu, 1997) with activities covering recreational hunting, trophy hunting, meat production, game breeding, and ecotourism. It is estimated that there was an annual conversion rate of 2–2.5% of farms from livestock to wildlife production in South Africa from the mid 1980s up to the mid 2000s (Republic of South Africa, 2005), with from 9,000 to 10,000 private wildlife ranches, covering an estimated 16.8% of South Africa, and 10.3% and 8.7% of Namibia and Zimbabwe's (prior to recent political changes) freehold land, respectively (Cumming, 2004). These socio-economic conditions have allowed wildlife managers in southern Africa to use a range of management interventions necessary to maintain large mammal populations. The capacity for strategic intervention has been particularly important to ensure success, given that most protected areas are generally not large enough to support fully functioning ecosystem processes (Sinclair *et al.*, 2007). But with the increasing transformation of lands surrounding protected areas they are becoming ecologically degraded (loss of ecological pattern and process) isolated islands (Newmark, 2008), with few, if any, considered large enough to accommodate the projected effects of climate change (Erasmus *et al.*, 2002).

Mammalian herbivore diversity and biomass are influenced by a combination of rainfall and soil nutrient status through the quantity and quality of food made available (Fritz & Duncan, 1994). Although most African savanna herbivore systems are dominated in terms of metabolic biomass by three principal species, mostly from among the bulk roughage feeders such as elephant (*Loxodonta africana*), hippopotamus (*Hippopotamus amphibius*), buffalo (*Syncerus caffer*), and zebra (*Equus quagga*), but also wildebeest (*Connochaetes taurinus*) (Cumming, 1982), the major collective contribution of antelopes as ecosystems architects deserves specific attention.

In this chapter, we identify a key set of desirable conservation, socio-economic, and political outcomes that are necessary to maintaining viable antelope populations in and out of protected areas, and analyse the potential influence of a suite of management interventions that are designed to achieve these conservation outcomes. Examples, primarily from southern Africa, will be used to illustrate how these interventions have had both positive and negative impacts, particularly on the conservation of antelopes, and what lessons have been learnt.

Conservation management framework, outcomes and strategic interventions

Protected areas are crucial for antelope conservation, and in the following we will consider how the main conservation outcomes we consider to be important in this regard can be achieved by the variety of management applications (Figure 1). The desired features include: (i) being situated within a connected landscape; (ii) comprising as broad an array of habitats as possible (habitat diversity); (iii) representation of the full possible spectrum of species (trophic and ecological structure) that historically occurred in the area in question; (iv) the inclusion of as full diversity of ecological processes that we are aware of

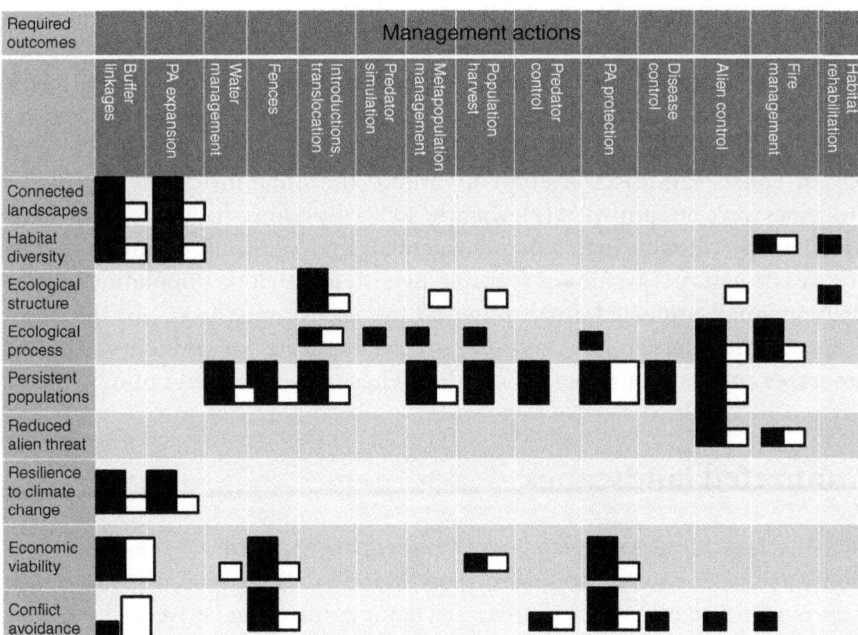

Figure 1 **Management actions required to deliver on conservation outcomes for 'large' (>10,000 km^2, open squares) and 'small' (solid squares) protected areas. The size of the square indicates the extent to which such activities would be required (large square = extensively used or important; small square = infrequently used or less important).**

as possible; (v) the need for species persistence; (vi) the need to reduce threats from invasive alien biota; (vii) resilience to climate change; (viii) socio-economic viability; and (ix) a favourable socio-political environment (conflict avoidance).

We compared the different management applications used by wildlife managers in conserving antelope species in 'large' and 'small' sized wildlife areas respectively, arbitrarily defining large protected areas to be >10,000 km^2 (considered by Frankel & Soulé (1981) to be of sufficient size to support ecological patterns and processes, although this is contested by Western & Sesemakula (1981)). Although the importance of large protected areas in conserving biodiversity is readily accepted, small protected areas predominate worldwide (Chape *et al.*, 2008). The management applications we consider include: the need for conservation-compatible land-use in the buffer areas around protected areas; expansion of protected areas; provisions of supplementary water management supplies, and the erection of fences to control wildlife and livestock movements; repatriation of wildlife through translocations and restocking; simulation of predator impacts on wildlife populations; meta-population management of wildlife; harvesting of wildlife populations for ecological and economic reasons; control of predators; active law enforcement; management of emerging wildlife diseases; control of alien species; use of fire as an ecosystem driver; and habitat rehabilitation.

From Figure 1 it appears that smaller protected areas generally require considerably more (62%) active management interventions to compensate for their more limited habitat diversity and incomplete ecological processes. As small protected areas notably restrict the full functioning of ecological processes, conservationists attempt to mimic the outcomes as if ecological processes were not impeded (Newmark, 2008). By comparison, the larger protected areas still required some management interventions, although mostly of a minor nature. The aim of securing persistent antelope populations, especially in small protected areas, required the use of most (64%) of the listed management interventions, as well as compensating for limited ecological processes and reducing societal conflict. These are expanded upon below.

Connected landscapes

With Africa's protected areas largely inadequate to conserve the continent's biodiversity and wildlife populations on their own, there is an urgent need to increase the size of protected areas, strive for connection between them, and soften their hard boundaries through cooperative and compatible land-use management practices, in turn increasing ecological linkages and reducing fragmentation (Figure 1) (Harris & Eisenberg, 1989; Georgiadis *et al.*, 2007). Although fragmentation plays a major role in the slow attrition of species diversity (Gonzalez, 2013), one cannot ignore the fact that poor management can undermine even the largest, well planned reserves (Cumming, 2004).

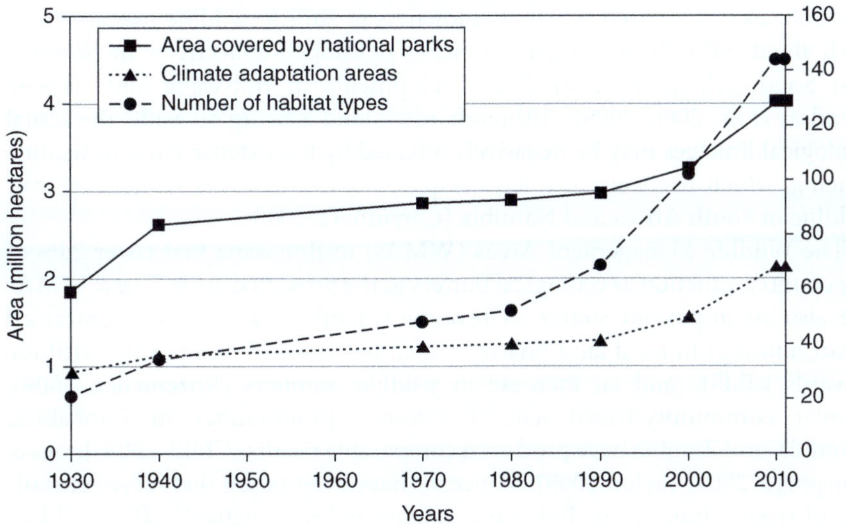

Figure 2 **Trends in the total area (ha), number of vegetation units, and the 'climate adaptation area' (ha) covered by national parks in South Africa since 1930. A 'climate adaptation area' is defined as an area of high biotic and abiotic diversity, including steep gradients in temperature, precipitation, and altitude (Holness & Bradshaw, 2012). Vegetation units refer to a complex of plant communities occupying the landscape (Mucina & Rutherford, 2006).**

The expansion of protected areas has been the focus of iconic examples such as the Serengeti Regional Conservation Strategy (Mbano *et al.*, 1995). In southern Africa, increases in wildlife numbers have resulted from the transboundary conservation initiatives (Whande & Suich, 2009), expansion of protected areas through innovative contractual arrangements (Castley *et al.*, 2009; Pollard *et al.*, 2003), community conservation projects (Jones & Murphree, 2004) and the development of conservancies on private lands (Lindsay *et al.*, 2009). Although most of these cases tend to be wildlife-focused, greater attention will need to be given to looking at further novel ways of providing healthy mixes of wildlife and livestock farming, thus providing economic, social as well as ecological resilience (Georgiadis *et al.*, 2007) by reducing conflict with agriculture and entrenched societal values.

With conservation closely linked to development and job creation, the expansion of the national protected area network is firmly embedded in South African national policies with a commitment to meeting international convention targets (NPAES, 2008). Since 1998, a total of 6,400 km^2 have been added to South African national parks alone through acquisition and contractual arrangements (Figure 2, Knight & Holness, unpublished data). In addition, supportive legislation in South Africa and Namibia (and until recently in

Zimbabwe) incentivised private landowners to shift to wildlife management, with about 295,000 km^2 of game ranches developed collectively in Namibia and South Africa, with smaller developments in Botswana and Zambia (Lindsay *et al.*, 2006, 2009). Although, more land is being set aside, the actual ecological linkages may be negatively affected by the extensive use of wildlife fencing, which ironically is a legal requirement in granting ownership over wildlife in South Africa and Namibia (Carruthers, 2008).

The Wildlife Management Areas (WMAs) in Botswana that cover 24% of the country, function as ecological buffers to the protected areas. These WMAs are also an important source of revenue (mainly sourced from contracted safari hunting) to local land boards, which has resulted in a positive attitude towards wildlife and an increase in wildlife numbers (Rozemeijer, 2009). Similar community-based natural resource programmes in Zimbabwe, Namibia, and Zambia have produced comparable results (Child, 2009; Jones & Murphree, 2004; Taylor, 2009). Concerns have been raised that recent curtailing of trophy hunting in Botswana is expected to negatively affect wildlife populations and the attitude of local people towards them (Boyes, 2013).

In increasingly fragmented landscapes, the need to maintain the use of existing wildlife corridors to facilitate ecological processes has been widely advocated (Newmark, 1993). However, in most cases they are failing due to human-wildlife conflict issues (Davenport *et al.*, 2010). Attempts to encourage the establishment of novel, ecologically relevant corridors have been advanced (Rouget *et al.*, 2006) with some indications of success, notably the use by herbivores such as zebra in Botswana (Bartlam-Brooks *et al.*, 2011). It remains important to assess whether wildlife actually use the planned corridors before strongly advocating their protection (Caro *et al.*, 2009).

Habitat diversity

The loss of habitat and its associated biological diversity, one of Diamond's (1989) famous 'evil quartet', should be mitigated as far as possible. Expanding protected areas and integrating them into the surrounding landscape, as well as using other ecological drivers such as fire, mega-herbivores, or rehabilitation, play important roles in increasing habitat, environmental, and functional heterogeneity (Figure 1; Archibald, 2008; Newmark, 2008; Owen-Smith, 2004; Cromsigt *et al.*, 2009; Fynn, 2012). As heterogeneity, in its abiotic and biotic forms remains the ultimate source of diversity, it should be the focus of ecological and restoration management (Pickett *et al.*, 2003).

The expansion of South Africa's national parks since 1990 has increased the spatial and habitat heterogeneity at the macro scale with the number of vegetation units represented in protected areas increasing from 40 to 145 (Figure 2)

(Knight & Holness, unpublished data). This potentially provides greater ecological resistance and resilience in the face of global climate change (Lawler, 2009) as reflected in the increase in climate adaptation areas in the parks (Figure 2), but also provides opportunities for the conservation of a greater diversity of wildlife species and larger numbers, further diversifying the system. The expansion of 13 South African national parks saw their antelope populations increase by an average of 2.6 times since 2000 to a total of 36,000 head of game in 2010 (Knight *et al.*, unpublished data). Furthermore, the introduction of mega-herbivores, such as white rhinos (*Ceratotherium simum*), as system engineers can create grazing lawns and alter patterns of fire for the benefit of other grazers (Waldram *et al.*, 2007). This emphasises the importance of conserving as much ecological heterogeneity as possible, in its many forms (topographic, edaphic, and hydrologic conditions) at different scales, in ecosystem management.

Trophic and ecological structure

With the close relationship between African herbivore and predator assemblages and their profound collective impact on their environment (Sinclair *et al.*, 2007), any restoration efforts should strive to re-establish the ecological patterns and processes that historically occurred in the area. Small protected areas normally lack the full suite of herbivores and carnivores, and lose species through ecological relaxation (also referred to as extinction debt; Diamond, 1972) and impeded ecological processes, as well as small-population effects (Caughley, 1994). Hence, management of populations in smaller protected areas generally would require a strategy employing reintroductions and periodic restocking (Figure 1).

South Africa has a long history of animal translocations, fulfilling a number of ecological benefits including genetic enhancement, restoring ecological functions, and assisted migration to mitigate for climate change (Spear & Chown, 2009), as well as economic returns. Between 2001 and 2010, South African National Parks (SANParks) alone captured a total of 19,800 ungulates of 16 different species, 53% of which were translocated to other national parks, 9% were relocated within parks, and 37% were either sold or culled (Knight, unpublished data). Thus, extensive resources and skills are required to manage a system based on small protected areas. In re-establishing and restocking wildlife populations, the conservation authorities have moved towards introducing species that historically occurred in the area, provided that suitable habitat existed, sourcing from closest natural populations and introducing reasonable founder numbers (Novellie & Knight, 1994; Alpers *et al.*, 2004). Carnivores have also been reintroduced to many smaller reserves in southern Africa to provide the missing large predator trophic level and their associated ecological processes on wildlife populations, as well as providing greater financial opportunities from ecotourism, hunting, or sales (Bothma *et al.*, 2009; Hayward *et al.*, 2007). Restoring or mimicking ecological structure through re-introductions does carry some pitfalls as it may not necessarily address the threats that led to the species losses in the first place and in most cases lacks theoretical guidance or experimental approaches (Seddon *et al.*, 2007). Hence each antelope re-introduction programme may differ substantially between sites and ideally should be planned as adaptive experiments.

Ecological processes

Following the ecosystem paradigm, protected-area management recognises that ecological processes are of key importance for ecosystem function, including interactions between the larger mammalian herbivores and the plant and predator communities, as well as natural disturbance events (Gordon *et al.*, 2004). These processes act in top-down (through predation) and bottom-up (through nutrient cycles effect on vegetation) ways that regulate herbivore dynamics (Estes *et al.*, 2011). The absence of some such processes, especially in smaller protected areas, poses numerous challenges to conservationists wishing to restore key ecological processes that influence these herbivorous effects.

Establishing or reinforcing missing ecological processes is closely linked to management actions aimed at re-establishment of as full a spectrum of naturally occurring herbivores and predators as described in the preceding section. However, it may also involve the use of drivers such as fire to create habitat heterogeneity, action to allow naturally functioning rivers, acceptance of the role of natural disturbances (such as floods, droughts, disease), and efforts to minimise the impact of alien biota (Figure 1) (Archibald, 2008; Rogers & O'Keefe, 2003; Richardson *et al.*, 1997).These actions all influence the primary hydrological and mineral cycles, which are the basis of ecological processes (Vold & Buffett, 2008). The extent and manner in which any of these management actions are used depend upon the conservation objectives of the reserve in question. The management approach in many smaller formally protected areas has changed over time, with a move away from managing for a stable state to one of accepting a heterogeneity paradigm, with an emphasis on trying to maintain the ecological processes and their natural flux, that provides greater ecological resilience, which is important in the face of expected global climate change (Walker, 1989). By contrast within the private and some communal areas in southern Africa, the objectives are different with more of an emphasis on the agricultural approach of maximising sustainable yield and stability with greater financial returns from wildlife (du Toit, 2002; Krug, 2001). In many cases this requires the removal of perceived threats (such as predators, disease), homogenised spatial distributions of wildlife through water provision, and a focus on individual animal welfare (rather than population-focused management) (Fuhlendorf & Engle, 2001; Casanovas *et al.*, 2012; Gortázar *et al.*, 2010; Houser *et al.*, 2011). As a result these areas may not be considered 'wild' in the conventional sense (Mallon & Stanley Price, 2013). Yet, the extent to which one should and can allow flux in smaller protected areas, especially those with rare and endangered species remains to be investigated.

Small protected areas with their associated smaller wildlife populations set in a fragmented landscape, require active metapopulation management strategies (Nicholls *et al.*, 1996) to maintain their genetic viability, periodic population harvesting, and even in the rare case of predator simulations (selective removal of wildlife to simulate predator impact) (Figure 1). However, even larger parks, such as Kruger National Park, may require management interventions to reinforce ecological processes such as fire management, alien species control, and species introductions (Freitag-Ronaldson & Foxcroft, 2003). The reintroduced scimitar-horned oryx (*Oryx dammah*) population that now exist in a number of fenced reserves in North Africa are planned to be managed as a meta-population (Devillers & Devillers-Terschuren, 2005). A number of locally rare antelopes species such as roan antelope (*Hippotragus equinus*), sable antelope (*H. niger*), and tsessebe (*Damaliscus lunatus lunatus*) are managed as a meta-population in Namibia (Anon., 2003) and South Africa (Grant & van der Walt, 2000). Meta-population management requires considerable expertise and resources to run effectively. Although the extensive movement of wildlife primarily within the private ranching sector in South Africa has witnessed a rapid growth in herbivore populations (Carruthers, 2008), it is suspected to lead to genetic homogenization and the loss of vital evolutionary gene flow processes that would be important in providing necessary ecological resilience (Spear & Chown, 2009) in the face of global climate change. Hence, some argue that wildlife ranching may not be considered true conservation (Cousins *et al.*, 2010).

In the absence of larger carnivores in smaller protected areas, conservation and private wildlife managers have relied upon wildlife harvesting and/or carnivore simulations to control the perceived negative habitat impacts of expanding herbivore populations (Johnson *et al.*, 2009). Such systems are often run following stable-state ecosystem principles, that can be financially lucrative (Krug, 2001), but potentially ecologically detrimental (Fulendorf & Engle, 2001). However, the increasing introduction of apex carnivores into small reserves has largely changed the approach to managing wildlife populations, reducing the need of controlling wildlife populations, and a move to a more hands-off management approach. As large carnivores in confined systems can have profound effects on wildlife populations (more so than in large uncontained areas), often leading to local population crashes, and behavioural responses by certain prey species (Tambling *et al.*, 2012), such introductions need to be closely monitored, again ideally in an adaptive experiment. With apex predators in small wildlife areas traditionally managed following set carnivore stocking rates based upon estimated herbivore carrying capacity models (Hayward *et al.*, 2007), Ferreira and Hofmeyr (2014) have advocated an alternative carnivore management strategy that attempts to mimic social carnivore behaviour and social dynamics. The strategy, which is currently in its experimental phase, will require considerable conservation expertise and management resources to execute, possibly discouraging its use in the many private reserves with large carnivores.

The extirpation of large apex carnivores from many areas has led to situations where the meso-predators, such as black-backed jackals (*Canis mesomelas*) and caracal (*Felis caracal*), play an unnaturally greater influence on prey populations (McKenzie, 1990). Klare *et al.* (2010) noted that jackals could annually consume up to 31% of ungulate populations that hid their newborns, thus having a considerable impact on wildlife ranching activities and revenue generation. This has led to periodic control activities of jackals in private and even some national protected areas, which has had mixed results (Ferreira & Gaylard, unpublished data). The vital point is that such management activities should be framed as experiments to assess and measure the key drivers.

Fire remains one of the most important abiotic ecological processes affecting African savannas and as such is a tool extensively used by conservation managers. Aptly described as a large generalist herbivore (Bond, 1997), it sometimes replaces, competes, or facilitates herbivory, and as such remains a very topical subject. As fire and herbivory interact in very dynamic and integrated ways, the manner in which it is employed in protected areas depends on the conservation objectives, ecological context, and size of area in question. The increase in the blue wildebeest population in the Serengeti from the 1960s to the 1980s coincided with the reduced extent of fires (Sinclair *et al.*, 2007). Increasing white rhino numbers in Imfolozi-Hlulhuwe Park, South Africa, also led to a reduction in the incidence and impact of fires (Waldram *et al.*, 2007). When fire has been suppressed, changes in vegetation structure and composition and associated herbivores have been noted. Creating as heterogeneous a system as possible through the use of fire should remain a focus of any protected area but needs to be implemented using theoretical support and an experimental framework.

Population persistence

In southern Africa, there has been a general shift away from the focus on the small-population paradigm with its traditional emphasis on avoiding the risk of extinction, to one of addressing the drivers causing declining populations, which focuses more on the relationships between spatial patterns and ecological processes at the broader landscape level (Mulder & Coppolillo, 2005). This has implications for issues such as reserve design and connectivity, to name a few.

In an analysis of the persistence of a total of 29 antelope species (inclusive of an extralimital Lichtenstein hartebeest (*Alcelaphus buselaphus lichtensteinii*) population in Kruger National Park) in 17 South African national parks, we found that only an average of 68% of the 160 populations in small parks were either growing or stable in comparison to 79% of the 27 populations in larger parks (Kruger and Kalahari Gemsbok National Parks). This could be a result of a greater resilience of larger populations (Knight *et al.*, unpublished data; Figure 3). Overall, about 11% of the populations were showing some form of population decline in both large and small parks, which may be consequential to increasing competition from increased numbers of other herbivores and carnivores or other factors such as habitat modification. These populations were predominantly of species that were habitat specialists whose populations were marginal (at the extreme of their historical distribution), examples including tsessebe, roan and sable antelope (Owen-Smith *et al.*, 2011). Others species with declining populations, such as mountain reedbuck (*Redunca fulvorufula*), grey rhebok (*Pelea capreolus*), springbok (*Antidorcas marsupialis*), and common duiker (*Sylvicapra grimmia*), may be responding to pressure from introduced carnivores and/or meso-predator release. The relatively high (~20%) unknown population status arises from inadequate survey results of small cryptic species (such as suni *Neotragus moschatus*, klipspringers *Oreotragus oreotragus*, grysboks *Raphicerus* spp. and the duikers) that are notoriously difficult to survey – a universal problem.

With 74% of the 187 antelope populations having population sizes of less than 500 individuals, consideration needs to be given to managing for the possible adverse consequences of small population size (Knight *et al.*, unpublished data), notably using a meta-population approach to population management, and improving integration into the surrounding land-use. The large size of a park may not entirely free the need for such management as 39% of the antelope populations in the Kruger and Kalahari Gemsbok National Parks total less than 500 individuals (Figure 3). The active management of small protected areas and private wildlife ranches has facilitated the recovery and conservation of endemic, threatened, and rare species such as bontebok (*Damaliscus pygargus pygargus*), blesbok (*D. p. phillipsi*), and black wildebeest (*Connochaetes gnou*) (Friedmann & Daly, 2004). The increases of these three species on private land have been primarily driven by the financial incentive of conserving these rarer southern African endemics.

Fencing in southern Africa largely arose from the agricultural requirements of separating livestock from wildlife to prevent the transmission of disease (Carruthers, 2008). Fencing has further proliferated with the transfer

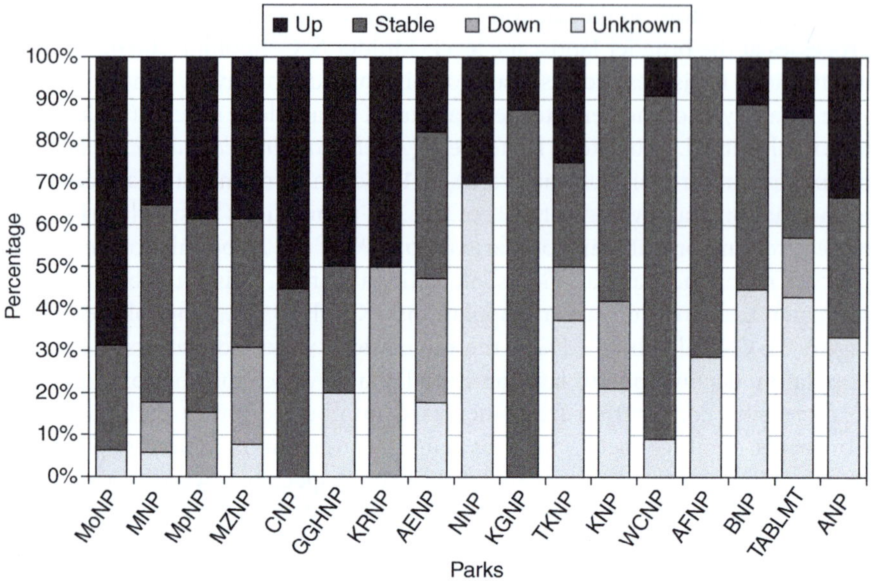

Figure 3 **The population trends of antelope species within 17 South African National Parks (NPs). (MoNP = Mokala, MNP = Marakele, MpNP = Mapungubwe, MZNP = Mountain Zebra, CNP = Camdeboo, GGHNP = Golden Gate Highlands, KRNP = Karoo, AENP = Addo Elephant, NNP = Namqua, KGNP = Kalahari Gemsbok, TKNP = Tankwa Karoo, KNP = Kruger, WCNP = West Coast, AFNP = Augrabies Falls, BNP = Bontebok, TABLMT = Table Mountain, ANP = Agulhas).**

of ownership of wildlife to private and communal landowners in southern Africa. Although fencing has had a role in enhancing the growth of wildlife populations, especially of threatened species, it also led to greater fragmentation of the landscape and the breakdown of ecological linkages (Hayward & Kerley, 2009; Woodroffe *et al.*, 2014), and the demise of migratory antelope populations in southern Africa (Williamson & Williamson, 1985; Whyte & Joubert, 1988). However, further legislative support and financial incentives for land-owners to join into larger conservancies and link to formal protected areas, are leading to larger areas being set aside for wildlife, and the removal of fences (Lindsey *et al.*, 2009), thereby restoring opportunities for key ecological processes to play out across the landscape. The dropping of fences on the western boundary of Kruger National Park and the expansion of Addo Elephant National Park through private contractual relationships are achieving the same result (Castley *et al.*, 2009; Freitag-Ronaldson & Foxcroft, 2003). By contrast, East Africa, once a bastion of unfenced range lands is seeing an increased use of fencing with its negative impacts (Georgiadis *et al.*, 2007) – something southern Africa experienced many years ago.

The management of water resources plays a mixed role in the persistence of antelope species and as such remains a largely emotive and controversial issue (Gaylard *et al.*, 2003). In many cases, the provision of supplementary water supplies in protected areas can lead to a conflict between ecological and tourism objectives. It ranks alongside fire, harvesting of wildlife, and fencing as one of the most invasive management techniques employed by managers. Moreover, the availability and distribution of water can affect ecosystem structure and function at different scales and thus should be used with extreme caution (Gaylard *et al.*, 2003). In southern Africa, smaller protected areas, with incomplete ecosystems, often make extensive use of supplementary water supplies. Larger protected areas, such as Kruger National Park, have reduced the number and distribution of waterholes in an attempt to reintroduce natural seasonal movement patterns and reverse the negative impact that such supplies had on rarer species, such as roan antelope, through increased competition from zebra and the impact of apex predators that followed them (Owen-Smith *et al.*, 2011). The introduction of supplementary water into the Kalahari Gemsbok National Park, South Africa, has caused the once largely nomadic blue wildebeest population to become resident and has led to generally higher local wildlife population densities than would be observed in its absence (Knight, 1991, 1995). Where protected areas are ecologically incomplete, as is often the case, supplementary water supplies can be used to simulate natural ecological processes; however, this should ideally by done in an adaptive manner, with the expectation that it may deliver unexpected outcomes.

The management of wildlife diseases has also become increasingly important in ensuring antelope population persistence (Chapter 7). Ironically, with the focus initially on protecting livestock from wildlife, there has been a shift towards the converse given the decline in wildlife populations over most of Africa (Carruthers, 2008). Wildlife diseases can also be viewed as fulfilling a micro-predator ecological role (Bengis *et al.*, 2003). Their impact can be slow and localized (such as the exotic bovine tuberculosis (BTB) on buffalo), to the rapidly spreading and wide-ranging diseases, such as rinderpest and anthrax, leading to herbivore population reductions and indirect impacts on vegetation and nutrient dynamics (Sinclair *et al.*, 2007). The risk of wildlife being affected by wildlife diseases, either indigenous and exotic, increases with a reduction in herbivore population sizes (given greater exposure to demographic stochasticity and catastrophic events), infrastructural introductions (such as artificial waterholes), greater exposure to domestic animals, and expected global climate change. The rare and endemic antelope species will often require close veterinary monitoring to guarantee their persistence, as they tend to be particularly vulnerable; for instance, the roan antelope is highly vulnerable to anthrax, which has been artificially spread by the introduction of waterholes in Kruger National Park (de Vos & Bryden, 1996). However, the potential impact of disease on more common and more ecologically important species (Gaston & Fuller, 2007), such as wildebeest or buffalo (given their biomass contributions in many ecosystems), should not be ignored. It thus remains crucial to understand the potential ecological role of wildlife diseases, and to be clear when veterinary intervention is needed for the functioning of an ecosystem.

Reduced threats from alien species

Since invasive alien species are regarded as one of the six most important threats to biodiversity (Millennium Ecosystem Assessment, 2005), attention needs to be given to reducing their impact. A total of 350 alien plant species have been recorded in Kruger National Park alone (Spear *et al.*, 2011), with some species, such as the flowering perennial triffid weed *Chromolaena odorata* for example, known to reduce herbivore diversity and biomass (Dumalisile, 2008). Further transformation and fragmentation of the landscape surrounding protected areas, and impacts of global climate change are expected to increase the threats to conservation from invasive alien plant species. As eradication attempts are generally unsuccessful (Rejmánek & Pitcairn, 2002), there is an urgent need to allocate sufficient long-term funding to at least contain and minimise the impact of alien species in protected areas (Freitag-Ronaldson & Foxcroft, 2003). In some cases, mega-herbivores – provided at high enough densities – have facilitated the control of alien plants (Moolman & Cowling, 1994).

Although possibly having less of an impact on biodiversity than exotic plants, the negative role of non-native mammalian species (other than livestock) on African herbivores are still considered to be substantial (Castley *et al.*, 2001) and these should be removed from protected areas as a matter of practice. The removal of Himalayan tahr (*Hemitragus jemlahicus*), from Table Mountain National Park given their extensive negative habitat impact and the later reintroduction of klipspringers (*Oreotragus oreotragus*) is a point in case (Knight, unpublished). Even the introduction of extralimital African species, such as impala (*Aepyceros melampus*) and nyala (*Tragelaphus angasii*) into habitats not historically occupied by such species, has had negative impacts upon indigenous bushbuck (*T. scriptus*) (Coates & Downs, 2005). The introduction of extralimital species has occurred extensively on private land, primarily driven by the financial incentive of diversifying the hunting and game viewing product (Cousins *et al.*, 2010). Attempted policy interventions by the South African environmental authorities to limit this have met considerable resistance from the private wildlife sector (Brink *et al.*, 2011).

Resilience to climate change

The diversity of Africa's iconic large herbivore systems were shaped predominantly during the dramatic Pleistocene climatic changes (Lorenzen *et al.*, 2012). However, the current rate of climatic change far exceeds any climatic change records from the past, and is likely to be too rapid for evolutionary adaptation of most species (Malcolm & Markham, 2000). This creates challenging conservation planning and management problems.

Management strategies that enable antelope species to potentially cope with climate change include the expansion of protected areas (Figure 2) and enhanced linkages. The options for expanding protected areas in turn include acquisition, contractual relationships, land stewardship, and easements, many of which have been widely used in southern Africa and should provide greater ecological resilience to the protected areas system (Cumming, 2004; Castley *et al.*, 2009). Enhancing linkages between protected areas and the surrounding landscapes through conservation-friendly land-use arrangements offers the most cost-effective means of conserving the broader landscape. It also offers further conservation options for wildlife, as well as partially addressing rural poverty. Encouraging such land-use changes, e.g., through community-based natural resource management (CBNRM), is dependent upon providing incentives to land/community owners based on a devolution of decision-making powers and control over natural resources (Armitage, 2005). This is best achieved through a supportive legal framework, as occurs in many southern African countries (Child, 2004).

Other management strategies suggested to enhance adaptability to climate change include reducing non-climate-change-related threats, reconfiguring the current protected areas systems (considered unlikely given the exorbitant costs), assisted colonization (considered controversial and unethical in some quarters (Seddon *et al.*, 2014)), and as a last resort *ex situ* (e.g., zoos) conservation (Lawler, 2009). With limited conservation resources available to conserve everything, systems-based approaches focusing on causal mechanisms may be far more successful than traditional symptomatic issue-based approaches (Hannah *et al.*, 2002). When applying the latter, there is a need to possibly prioritise species for conservation effort, for instance, by focusing on ecologically important species (such a wildebeest and topi *Damaliscus lunatus*); this entails hard and controversial choices in a context of conflicting objectives and diverse interest group involvement.

Economic viability and conflict avoidance

The growth in wildlife populations in southern Africa stems primarily from embracing the sustainable utilization philosophy (Child, 2004). This was encouraged by the development of progressive land-use policies and legislative frameworks that provided greater ownership and control over wildlife resources, local decision making, and an incentive to conserve wildlife (Child & Child, 2015). This in turn led to attractive financial returns from wildlife, which encouraged greater collective cooperation between landowners and communities, and sometimes protected areas, to form conservancies with larger ecological benefits (Jones & Murphree, 2004). With most of Africa having a communal land tenure system, the CBNRM programmes in southern Africa have largely delivered on increasing the value and benefits of wildlife to local communities, together with positive conservation results (Jones & Murphree, 2004). Some CBNRM programmes have had considerable success, such as in Namibia, which has witnessed about 17.6% of the country's landmass and approximately 12% of its population gaining benefit from wildlife-based community programmes (Weaver *et al.*, 2011). This came about through legislative changes that saw a devolution of rights to wildlife and wildlife benefits to resident land stewards. Although CBNRM has been promoted as the panacea to addressing rural poverty, Murphree (2009) cautioned that it normally has benefit in conjunction with other forms of agricultural production. Thus diverse and integrated approaches to land management that have positive conservation outcomes need be further explored and developed.

In the case where large freehold lands exists, such as in South Africa and Namibia, increasing attention needs to be given to land redistribution to reduce potential conflict and threats to wildlife populations and future sustainable options (Cumming, 2004). Financial incentives led to the proliferation of wildlife ranches in southern Africa; this witnessed the conversion of 17.9 % and 18.0 % of private and communal lands in South Africa and Namibia, respectively from livestock farming to wildlife-focused land-use activities from the 1980s, while by comparison the additional contribution to the formal state-protected areas has been relatively minimal (Cumming, 2004; Carruthers, 2008). Yet, the degree to which these enabling policies have been instituted differ across the region. By contrast, the loss of 60–70% of Kenya's wildlife has been ascribed to institutional, policy, and market failures that arise from top-down regulatory approaches, lack of direct ownership of wildlife, and its reduced suite of associated potential income-generating opportunities (Norton-Griffiths, 2007).

The private wildlife industry has played a key role in wildlife conservation, particularly the proliferation of rare species with for example about four times as many sable antelope and black wildebeest on private lands as in formal reserves (du Toit, unpubl.). However, this success has also produced its challenges with conflicting conservation objectives, such as selective breeding for recessive colour morphs, hybridization, and introduction of extra-limital species, to name a few (Cousins, 2010). There is a need for improved policies that further incentivise the wildlife industry to follow positive conventional conservation objectives (Brink *et al.*, 2011). Nonetheless, delivering an economically viable wildlife economy often promotes greater linkages across the landscape, the strategic use of fencing to protect populations, sustainable use of wildlife resources, and the essential requirement of effective and more professional natural resource management (Figure 1).

Conclusion

With the increasing pressures on land-use at the regional scale, protected areas are increasingly being drawn into the developmental arena where social development and agricultural policies have often failed to alleviate poverty. Ultimately, the survival of Africa's iconic antelope populations depends on the persistence of protected areas and the active link to their surrounding informal wildlife areas. To achieve this, we believe there is an urgent need to provide enabling national legislative environments that provide ownership and control of wildlife resources to communal and private land owners, with the right to manage these resources accordingly. Developing financially sustainable partnerships between protected areas and surrounding users should be given urgent attention. These should be directed at encouraging a matrix of different land-use options at the regional scale to deliver on positive conservation outcomes. Without this, wildlife is doomed to be restricted to isolated protected areas with all its ecological and socioeconomic problems discussed above.

Harmonising land-use and access to resources across the protected-area network and the surrounding matrix of other land-uses will provide a diverse landscape required to meet the daunting ecological challenges that are expected to arise from global climate change. This will require a new breed of conservation managers, who are well versed in multidisciplinary subjects including conservation biology, resources economics, social development, and governance issues, along with the required resources to deliver on conservation outcomes (Craigie *et al.*, 2010; Cumming, 2004). As wildlife areas shrink outside of southern Africa, we see a need for more hands-on interventions to manage the increasingly fragmented populations in order to deliver conservation outcomes. This opens up the debate on the definition of 'wilderness' and the concept of wild species. Management initiatives should ideally be framed in a manner that can be used as a learning exercise. With neither national governments nor NGOs expected to substantially increase funding towards conservation in the future, alternative sources of funding have to be sought. Increasing the value of wildlife, maximizing economic opportunities from consumptive and non-consumptive uses, and incentivising the local communal/private sectors may offer the only long-term sustainable future for our wildlife populations. Providing such enabling environments will largely reduce the burden of over-stretched national conservation agencies to deliver on their mandates.

References

Adams, W. & Hulme, D. (2001): Conservation and community: changing narratives, policies and practices in African conservation. In *African wildlife and livelihoods: the promise and performance of community conservation*: 9–23. Hulme, D. & Murphree, M. D. (Eds). Oxford: James Currey.

Alpers, D. L., van Vuuren, B. J., Arctander, P. & Robinson, T. J. (2004): Population genetics of roan antelope (*Hippotragus equinus*) with suggestions for conservation. *Molecular Ecology* **13**: 1771–1784.

Anon. (2003): Species Management Plans: Roan antelope *Hippotragus equinus*, sable antelope *Hippotragus niger niger*, tsessebe *Damaliscus lunatus lunatus*. Windhoek, Namibia: Ministry of Environment and Tourism.

Archibald, S. (2008): African grazing lawns—how fire, rainfall, and grazer numbers interact to affect grass community states. *Journal of Wildlife Management* **72**(2): 492–501.

Armitage, D. (2005): Adapative capacity and community-based natural resource management. *Environmental Management* **35**(6): 703–715.

Baillie, J., Hilton-Taylor, C. & Stuart, S. N. (2004): *IUCN Red List of threatened species: a global species assessment*. Gland, Switzerland & Cambridge, UK: IUCN.

Bartlam-Brooks, H. L. A, Bonyongo, M. C. & Harris, S. (2011): Will reconnecting ecosystems allow long-distance mammal migrations to resume? A case study of a zebra *Equus burchelli* migration in Botswana. *Oryx* **45:** 210–216.

Bengis R. G., Grant, R. & de Vos, V. (2003): Wildlife diseases and veterinary controls: A savanna ecosystem perspective. In *The Kruger experience: ecology and management of savanna heterogeneity*: 349–369. du Toit, J. T., Rogers, K. H. & Biggs, H. C. (Eds). Washington DC: Island Press.

Bond, W. J. (1997): Fire. In *The Vegetation of Southern Africa*: 421–446. Cowling, R. M., Richardson, D. M. & Pierce, S. M. (Eds). Cambridge: Cambridge University Press.

Bothma, J. du P., Suich, H. & Spenceley, A. (2009): Extensive wildlife production on private land in South Africa. In *Evolution and innovation in wildlife conservation: parks and game ranches to transfrontier conservation areas*: 147–162. Suich, H. & Child, B. with Spenceley, A. (Eds). London: SASUSG, IUCN. Earthscan.

Boyes, S. (2013): No trophy hunting in Botswana and Zambia. http://newswatch.nationalgeographic.com/2013/01/22/no-trophy-hunting-in-botswana-and-zambia/

Brink, M., Cameron, M., Coetzee, K., Currie, B., Fabricius, C., Hattingh, S., Schmidt, A. & Watson, L. (2011): Sustainable management through improved governance in the game industry. *South African Journal of Wildlife Research* **41**(1): 110–119.

Caro, T., Jones, T. & Davenport, T. R. B. (2009): Realities of documenting wildlife corridors in tropical countries. *Biological Conservation* **142**: 2807–2811.

Carruthers, J. (2008): Wilding the farm or farming the wild? The evolution of scientific game ranching in South Africa from the 1960s to the present. *Transactions of the Royal Society of South Africa* **63**(2): 160–181.

Casanovas, J. G., Barrull, J., Mate, I., Zorrilla, J. M., Ruiz-Olmo, J., Gosàlbez, J. & Salicrú, M. (2012): Shaping carnivore communities by predator control: competitor release revisited. *Ecological Research* **27**: 603–614.

Castley, J. G., Boshoff, A. F. & Kerley, G. I. H. (2001): Compromising South Africa's natural biodiversity – inappropriate herbivore introductions. *South African Journal of Science* **97**: 344–348.

Castley, J. G., Knight, M. H. & Gordon, J. (2009): Making conservation work: innovative approaches to meeting conservation and socio-economic objectives, an example from the Addo Elephant National Park, South Africa. In *Evolution and innovation in wildlife conservation: parks and game ranches to transfrontier conservation areas*: 307–322. Suich, H. & Child, B. with A. Spenceley, A. (Eds). London: SASUSG, IUCN. Earthscan.

Caughley, G. (1994): Directions in conservation biology. *Journal of Animal Ecology* **63**: 215–244.

Chape, S., Spalding, M. & Jenkins, M. (2008). *The world's protected areas: status, values and prospects in the 21st century*. Berkeley: UNEP World Conservation Monitoring Centre, University of California Press.

Child, B. (2004): Parks in transition: biodiversity, development and the bottom line. In *Parks in transition: biodiversity, development and the bottom line*: 1–27. Childs, B. (Ed.). London: Earthscan.

Child, B. (2009): Conservation in transition. In *Evolution and innovation in wildlife conservation: parks and game ranches to transfrontier conservation areas*: 3–18. Suich, H. & Child, B. with Spenceley, A. (Eds). London: SASUSG, IUCN. Earthscan.

Child, G. & Child, B. (2015): The conservation movement in Zimbabwe: an early experiment in devolved community based regulation. *African Journal of Wildlife Research* **45**(1): 1–16.

Coates, G. D. & Downs, C. T. (2005): Survey of the status and management of sympatric bushbuck and nyala in KwaZulu-Natal, South Africa. *South African Journal of Wildlife Research* **35**(2): 179–190.

Cousins, J. A., Sadler, J. P. & Evans, J. (2010): The challenge of regulating private wildlife ranches for conservation in South Africa. *Ecology and Society* **15**(2): 28. http// www.ecologyandsociety.org/vol15/iss2/art28/(as at 15 February 2016).

Craigie, I. D, Baillie, J. E. M., Balmford, A., Carbone, C., Collen, B., Green, R. E. & Hutton, H. M. (2010): Large mammal population declines in Africa's protected areas. *Biological Conservation* **143**: 2221–2228.

Cromsigt, J. P. G. M., Prins, H. H. T. & Olff, H. (2009): Habitat heterogeneity as a driver of ungulate diversity and distribution patterns: interaction of body mass and digestive strategy. *Diversity and Distributions* **15**: 513–515.

Cumming, D. H. M. (1982). The influence of large herbivores on savannas structure in Africa. In *The ecology of tropical savannas*: 217–245. Huntley, B. J. & Walker, B. H. (Eds.). Berlin: Springer-Verlag.

Cumming, D. (2004): Performance of parks in a century of change. In *Parks in transition: biodiversity, development and the bottom line:* 105–124. Childs, B. (Ed.). London: Earthscan.

Davenport, T. R. B., Jones, T. & Caro, T. (2010): Wildlife corridors in Tanzania. *19 Proceedings of the Tanzania Wildlife Research Institute Scientific Conference* **7**: 20, 22–28.

Devillers, P. & Devillers-Terschuren, J. (2005): *Oryx dammah. In Sahelo-Saharan antelopes. Status and perspectives. Report on the conservation status of the six Sahelo-Saharan antelopes.* In Beudels, R. C., Devillers, P., Lafontaine, R. M., Devillers-Terschuren, J. & Beudels, M. O. (Eds). CMS SSA Concerted Action (2nd edn). CMS Technical Series Publication N°11, 2005. Bonn, Germany: UNEP/CMS Secretariat.

de Vos, V. & Bryden, B. (1996): Anthrax in the Kruger National Park: temporal and spatial patterns of disease occurrence. *Salisbury Medical Bulletin, special supplement* **87**: 26–31.

Diamond, J. (1989): Overview of recent extinctions. In *Conservation for the twenty-first century:* 37–41. Western, D. & Pearl, M. (Eds). Oxford: Oxford University Press.

Diamond, J. M. (1972). Biogeographic kinetics: estimation of relaxation times for avifaunas of southwest pacific islands. *Proceedings of the National Academy of Sciences of the United States of America* **69**(11): 3199–3203.

Dumalisile, L. (2008): Effects of *Chromolaena odorota* on mammalian biodiversity in Hluhluwe-iMfolozi Park, South Africa. MSc (Wildlife Management) thesis, University of Pretoria, Pretoria.

Fritz, H. & Duncan, P. (1994): On the carrying capacity for large ungulates of African savanna ecosystems. *Proceedings of the Royal Society of London, Series B, Biological Sciences* **256**: 77–82.

Du Toit. J. (2002): Wildlife harvesting guidelines for community-based wildlife management: a southern African perspective. *Biodiversity and Conservation* **11**: 1403–1416.

Du Toit J. G. (Unpublished): *Role of the private sector in the wildlife industry.* Tswane: Wildlife Ranchers SA/Du Toit Wilddienste, 2007.

Erasmus, B. F. N., Van Jaarsveld, A. S., Chown, S. L., Kshatriya, M. & Wessels, K. J. (2002): Vulnerability of South African animal taxa to climate change. *Global Change Biology* **8**(7): 599–693.

Estes, J. A., Terborgh, J., Brashares, J. S., Power, M. E., Berger, J., Bond, W. J., Carpenter, S. R., Essington, T. E., Holt, R. D., Jackson, J. B., Marquis, R. J., Oksanen, L., Oksanen, T., Paine, R. T., Pikitch, E. T., Ripple, W. J., Sandin, S. A., Scheffer, M., Schoener, T. W., Shurin, J. B., Sinclair, A. R., Soulé, M. E., Virtanen, R. & Wardle, D. A. (2011): Trophic downgrading of planet earth. *Science* **333**: 301–306.

Ferreira, S. & Gaylard, A. (unpublished). Risks of demographic and component Allee effects to ungulate persistence in small protected areas. South African National Parks.

Ferreira, S. M. & Hofmeyr, M. (2014): Managing charismatic carnivores in spatially restricted areas: large felids in South Africa. *South African Journal of Wildlife Research* **44**(1): 32–42.

Frankel, O. H. & Soulé, M. E. (1981): Conservation and evolution. London: Cambridge University Press.

Freitag-Ronaldson, S. & Foxcroft, L. (2003): Anthropogenic influences at the ecosystem level. In *The Kruger experience: ecology and management of savanna heterogeneity*: 391–421. du Toit, J. T., Rogers, K. H. & Biggs, H. C. (Eds). Washington D. C.: Island Press.

Friedmann, Y. & Daly, B. (2004): *Red data book of the mammals of South Africa: a conservation assessment.* South Africa: Endangered Wildlife Trust, CBSG Southern Africa, Conservation Breeding Specialist Group (SSC/IUCN).

Fuhlendorf, S. D. & Engle, D. M. (2001): Restoring heterogeneity on rangelands: ecosystem management based on evolutionary grazing patterns. *Bioscience* **51**(8): 625–632.

Fynn, R. W. S. (2012): Functional resource heterogeneity increases livestock and rangeland productivity. *Rangeland Ecology and Management* **65**(4): 319–329.

Gaston, K. J. & Fuller, R. A. (2007): Commonness, population depletion and conservation biology. *Trends in Ecology and Evolution* **23**(1):14–19.

Gaylard, A., Owen-Smith, N. & Redfern, J. (2003): Surface water availability: implications for heterogeneity and ecosystem processes. In *The Kruger experience. ecology and management of savanna heterogeneity*: 171–188. du Toit, J. T., Rogers, K. H. & Biggs, H. C. (Eds). Washington D. C.: Island Press.

Georgiadis, N. J., Nasser Olwero, J. G., Ojwang, G. & Romanach, S. S. (2007): Savanna herbivore dynamics in a livestock-dominated landscape: I. Dependence on land use, rainfall, density, and time. *Biological Conservation* **137**: 461–472.

Gonzalez, A. (2013): Biodiversity: The ecological deficit. *Nature* **503**: 206–207. doi:10.1038/503206a

Gordon, I. J., Hester, A. J. & Festa-Bianchet, M. (2004): The management of wild large herbivores to meet economic, conservation and environmental objectives. *Journal of Applied Ecology* **41**: 1021–1031.

Gortázar. C., Ferroglio. E., Lutton, C. E. & Acevedo, P. (2010): Disease-related conflicts in mammal conservation. *Wildlife Research* **37**: 668–675.

Grant, C. C. & van der Walt, J. L. (2000): Towards an adaptive management approach for the conservation of rare antelope in the Kruger National Park – outcome of a workshop held in May 2000. *Koedoe* **43**: 103–112.

Hannah, L., Midgley, G. F. & Millar, D. (2002): Climate change-integrated conservation strategies. *Global Ecology and Biogeography* **11**: 485–495.

Harris, L. & Eisenberg, J. (1989): Enhanced linkages: necessary steps for success in conservation of faunal diversity. In *Conservation for the twenty-first century*: 166–181. Western, D. & Pearl, M. (Eds). Oxford: Oxford University Press.

Hayward, M. W. & Kerley, G. I. H. (2009): Fencing for conservation: restriction of evolutionary potential or riposte to threatening processes? *Biological Conservation* **142**: 1–13.

Hayward, M. W., O'Brien, J. & Kerley, G. I. H. (2007): Carrying capacity of large African predators: predictions and tests. *Biological Conservation* **139**: 219–229.

Holness, S. & Bradshaw, P. (2012): Resilience of biodiversity to climate change. In *National biodiversity assessment 2011: an assessment of South Africa's biodiversity and ecosystems. Synthesis Report*: 109-118. Driver, A., Sink, K. J., Nel, J. L., Holness, S., Van Niekerk, L., Daniels, F., Majiedt, P. A., Jonas, Z. & Maze, K. (Eds). Pretoria: South African National Biodiversity Institute and Department of Environmental Affairs.

Houser, A., Gusset, M., Bragg, C. J., Boast, L. K. & Somers, M. J. (2011): Pre-release hunting training and post-release monitoring are key components in the rehabilitation of orphaned large felids. *South African Journal of Wildlife Research* **41**: 11–20.

Johnson, S. R., Boonzaaier, W., Collinson, R. & Davies, R. (2009): Changing institutions to respond to challenges: North West Parks, South Africa. In *Evolution and innovation in wildlife conservation: parks and game ranches to transfrontier conservation areas*: 289–305. Suich, H. & Child, B. with Spenceley, A. (Eds). London: SASUSG, IUCN, Earthscan.

Jones, B. T. B. & Murphree, M. W. (2004): Community-based natural resource management as a conservation mechanism: lessons and directions. In *Parks in transition: biodiversity, development and the bottom line*: 63–103. Childs, B. (Ed.). London: Earthscan.

Klare, U., Kamler, J. F., Stenkewitz, U. & Macdonald, D. W. (2010): Diet, prey selection, and predation impact of black-backed jackals in South Africa. *Journal of Wildlife Management* **74**(5): 1030–1042.

Knight, M. H. (1991): *Ecology of the gemsbok Oryx gazella gazella (Linnaeus) and blue wildebeest Connochaetes taurinus (Burchell) in the southern Kalahari*. PhD thesis, University of Pretoria, Pretoria.

Knight, M. H. (1995): Drought-related mortality of wildlife in the southern Kalahari and the role of man. *African Journal of Ecology* **33**: 377–394.

Knight, M. H. (Unpublished): Proposed introduction strategy of the Klipspringer *Oreotragus oreotragus* to the Cape Peninsula National Park. Unpublished report to South African National Parks, Pretoria. 1998.

Knight, M. H., Castley, G., Ferreira, S. & Gaylard, A. (Unpublished). Large mammal survey data 2000–2010. Raw data.

Knight, M. H. & Holness, S. (Unpublished). [Has the expansion of South African National Parks resulted in an efficient reserve network or not?], Raw data.

Krug, W. (2001): *Private supply of protected land in southern Africa: a review of markets, approaches, barriers and issues*. Paris: Report to Environment Directorate, OECD.

Lawler, J. J. (2009): Climate change adaptation strategies for resource management and conservation planning. *New York Academy of Sciences* **1162**: 79–98.

Lindsay, P., Alexander, R., Frank, L. & Romañach, S. (2006): The potential of the trophy hunting to create incentives for wildlife conservation in Africa where alternative wildlife based land uses may not be viable. *Animal Conservation* **9**: 283–298.

Lindsey, P. A., Romañach, S. S. & Davies-Mostert, H. T. (2009): The importance of conservancies for enhancing the value of game ranch land for large mammal conservation in southern Africa. *Journal of Zoology* **277**: 99–105.

Lorenzen, E. D., Heller, R. & Siegismund, H. R. (2012): Comparative phylogeography of African savanna ungulates. *Molecular Ecology* **21**: 3656–3671.

Malcolm, J. R. & Markham, A. (2000): *Global warming and terrestrial biodiversity decline*. Washington, DC: WWF.

Mallon, D. P. & Stanley Price, M. R. (2013). The fall of the wild. *Oryx* **47**(4): 467–468.

Mbano, B. N. N., Malpas, R. C. Maige, M. K. S., Symonds, P. A. K. & Thompson, D. M. (1995): The Serengeti Regional Conservation Strategy. In *Serengeti II: dynamics, management and conservation of an ecosystem*: 605–616. Sinclair, A. R. E. & Arcese, P. (Eds). Chicago: University of Chicago Press.

McKenzie, A. A. (1990): *Cooperative hunting in black-backed jackal* Canis mesomelas, *Schreber*. PhD thesis, University of Pretoria, Pretoria, South Africa.

Millennium Ecosystem Assessment (2005): *Ecosystems and human well-being: current state and trends*. Washington, DC: Island Press.

Moolman, H. J. & Cowling, R. M. (1994): The impact of elephant and goat grazing on the endemic flora of South African succulent thicket. *Biological Conservation* **68**: 53–61.

Mucina, L. & Rutherford, M. C. (2006): The vegetation of South Africa, Lesotho and Swaziland. *Strelitzia* 19. Pretoria: South African Biodiversity Institute.

Mulder, M. B & Coppolillo, P. (2005): *Conservation: linking ecology, economics and culture*. Princeton: Princeton University Press.

Murphree, M. (2009): The strategic pillars of communal natural resource management: benefit, empowerment and conservation. *Biodiversity and Conservation* **18**: 2551–2562.

Naughton-Treves, L., Buck Holland, M. & Brandon, K. (2005): The role of protected areas in conserving biodiversity and sustaining local livelihoods. *Annual Review Environmental Resources* **30**: 219–252.

Newmark, W. D. (1993): The role and design of wildlife corridors with examples from Tanzania. *Ambio* **22**(8): 500–504.

Newmark, W. D. (2008): Isolation of African protected areas. *Frontiers in Ecology and the Environment* **6**(6): 321–328.

Nicholls, A. O., Viljoen, P. C., Knight, M. H. & van Jaarsveld, A. S. (1996): Evaluating population persistence of censused and unmanaged herbivore populations from the Kruger National Park, South Africa. *Biological Conservation* **76**(1): 57–67.

Norton-Griffiths, M. (2007): How many wildebeest do you need? *World Economics* **8**(2): 41–64.

Novellie, P. A. & Knight, M. H. (1994): Repatriation and translocation of ungulates into South African national parks: an assessment of past attempts. *Koedoe* **37**(1): 115–119.

NPAES. (2008): *National protected area expansion strategy for South Africa: priorities for expanding the protected area network for ecological sustainability and climate change resilience*. South Africa: Department of Environmental Affairs.

Owen-Smith, N. (2004): Functional heterogeneity in resources within landscape and herbivore population dynamics. *Landscape Ecology* **19**: 761–771.

Owen-Smith, N., Chirima, G. J. & Macandza, V. A. (2011): Shrinking sable antelope numbers in Kruger National Park: what is suppressing population recovery? *Animal Conservation* **15**(2): 195–204.

Pickett, S. T. A, Cadenasso, M. L. & Benning, T. L. (2003): Biotic and abiotic variability as key determinants of savanna heterogeneity at multiple spatiotemporal scales. In *The Kruger experience: ecology and management of savanna heterogeneity*: 22–40. du Toit, J. T., Rogers K. H. & Biggs, H. C. (Eds). Washington DC: Island Press.

Pollard, S., Shackleton, C. & Carruthers, J. (2003): Beyond the fence – people and the lowveld landscape. In *The Kruger experience: ecology and management of savanna heterogeneity*: 59–80. du Toit, J. T., Rogers K. H. & Biggs, H. C. (Eds). Washington, DC: Island Press.

Rejmánek, M. & Pitcairn, M. J. (2002): When is eradication of exotic pest plants a realistic goal? In *Turning the tide: the eradication of invasive species:* 249–253. Veitch, C. R. & Clout, M. N. (Eds). Gland and Cambridge: IUCN.

Republic of South Africa. (2005): *Panel of experts on professional and recreational hunting in South Africa*. Pretoria: Department of Environmental Affairs and Tourism.

Richardson, D. M., MacDonald, I. A. W., Hoffmann, J. H. & Henderson, L. (1997): Alien plant invasions. In *Vegetation of southern Africa*: 534–570. Cowling, R. M., Richardson, D. M. & Pierce, S. M. (Eds). Cambridge: Cambridge University Press.

Rogers, K. H. & O'Keeffe, J. (2003): River heterogeneity: ecosystem structure, function and management. In *The Kruger experience: ecology and management of savanna heterogeneity*: 189–218. du Toit, J. T., Rogers K. H. & Biggs, H. C. (Eds). Washington DC: Island Press.

Rouget, M., Cowling, R. M., Lombard, A. T., Knight, A. T. & Kerley, G. I. H. (2006): Designing large-scale conservation corridors for pattern and process. *Conservation Biology* **20**(1): 549–561.

Rozemeijer, N. (2009): CBNRM in Botswana. In *Evolution and innovation in wildlife conservation: parks and game ranches to transfrontier conservation areas*: 243–256. Suich, H. & Child, B. with Spenceley, A. (Eds). London: SASUSG, IUCN. Earthscan.

Scholte, P. (2011): Towards understanding large mammal population declines in Africa's protected areas: A West- Central African perspective. *Tropical Conservation Science* **4**(1): 1–11.

Seddon, P. J., Armstrong, D. P. & Maloney, R. E. (2007): Developing the science of reintroduction biology. *Conservation Biology* **21**: 303–312.

Seddon, P. J., Griffiths, C. J., Soorae, P. S. & Armstrong, D. P. (2014): Reversing defaunation: restoring species in a changing world. *Science* **345**(6195): 406–412.

Sinclair, A. R., Mduma, S. A., Hopcraft, J. G., Fryxell, J. M., Hilborn, R. & Thirgood, S. (2007): Long-term ecosystem dynamics in the Serengeti: lessons for conservation. *Conservation Biology* **21**(3): 580–590.

Spear, D. & Chown, S. L. (2009): The extent and impacts of ungulate translocations: South Africa in a global context. *Biological Conservation* **142**: 353–363.

Spear, D., McGeoch, M. A. Foxcroft, L. C. & Bezuidenhout, H. (2011): Alien species in South Africa's National Parks. *Koedoe* **53**(1): 4 pp. doi:10.4102/koedoe.v53i1.1032

Tambling, C. J., Druce, D. J., Hayward, M. W., Castley, J. G., Adendorf, J. & Kerley, G. I. H. (2012): Spatial and temporal changes in group dynamics and range use enable anti-predator responses in African buffalo. *Ecology* **93**(6): 1297–1304.

Taylor, R. (2009): The performance of CAMPFIRE in Zimbabwe: 1989–2006. In *Evolution and innovation in wildlife conservation: parks and game ranches to transfrontier conservation areas*: 201–222. Suich, H. & Child, B. with Spenceley, A. (Eds). London: SASUSG, IUCN. Earthscan.

UNEP-WCMC. (2008): *About protected areas: guidelines for applying protected areas management categories*. Dudley, N. (Ed.). Switzerland: IUCN.

Vold, T. & Buffett. D. A. (2008): Ecological concepts, principles and applications to conservation. British Columbia. http://www.biodiversitybc.org (as at 15 February).

Waldram, M. S., Bond, W. J. & Stock, W. D. (2007): Ecological engineering by a megagrazer: white rhino impacts on a South African savanna. *Ecosystems* **11**(1): 101–112.

Walker, B. (1989): Diversity and stability in ecosystem conservation. In *Conservation for the twenty-first century*: 121–130. Western, D. & Pearl, M. (Eds). Oxford: Oxford University Press.

Weaver, L. C., Hamunyela, E., Diggle, R., Matongo, G. & Pietersenet, T. (2011): The catalytic role and contributions of sustainable wildlife use to the Namibia CBNRM programme. In *The relevance of CBNRM to the conservation and sustainable use of CITES-listed species in exporting countries*: 59–70. Abensperg-Traun, M., Roe, D. & O'Criodain, C., (Eds). Gland, Switzerland: IUCN; London, UK: IIED.

Western, D. & Sesemakula, J. (1981): The future of savanna ecosystems: ecological islands or faunal enclaves. *African Journal of Ecology* **19**: 7–19.

Whyte, I. J. & Joubert, S. C. J. (1988): Blue wildebeest population trends in the Kruger National Park and the effects of fencing. *South African Journal of Wildlife Research* **18**: 78–88.

Whande, W. & Suich, H. (2009): Transfrontier conservation initiatives in southern Africa: observations from the Great Limpopo Transfrontier Conservation Area. In *Evolution and innovation in wildlife conservation: parks and game ranches to transfrontier conservation areas*: 373–391. Suich, H. & Child, B. with Spenceley, A. (Eds). London: SASUSG, IUCN. Earthscan.

WCMC. (2004): World Conservation Monitoring Centre, UK: http://www.unep-wcmc.org/(as at 24 February 2016).

Williamson, D. & Williamson, J. (1985). Botswana fences and the depletion of Kalahari wildlife. *Parks* **10**: 5–7.

Winterhalder, B. & Lu, F. (1997): A forager-resource population ecology model and implications for indigenous conservation. *Conservation Biology* **11**: 1354–1364.

Woodroffe, R., Hedges, S. & Durant, S. M. (2014). To fence or not to fence. *Science* **344**: 46–48.

9

DNA in the Conservation and Management of African Antelope

Eline D. Lorenzen

Natural History Museum of Denmark, University of Copenhagen, Denmark

Species lists have become the main currency of conservation practices (Mace, 2004). The number, distribution, and abundance of species are estimated through biodiversity assessments. In this regard, genetic analysis provides an invaluable resource. By providing information on patterns and levels of genetic diversity not apparent at the morphological level, DNA from individuals and populations can reveal cryptic variation and aid in the identification of distinct populations worthy of conservation.

During the past two decades, genetic studies have been published on many populations of antelope across the world. In particular, many studies have focused on sub-Saharan Africa, reflecting in part the sheer diversity and abundance of antelope species in the region, but also in part the presence of many wildlife parks in East and southern Africa, which provide the infrastructure necessary for sample collection. Although the focus of the work has often been demographic history and population genetics (reviewed in Lorenzen *et al.*, 2012), the data have provided much-needed information on patterns of biodiversity essential for informed species management and conservation efforts; only by understanding patterns of diversity within and among species will we be able to prioritize conservation efforts.

Several important topics in species conservation have been addressed using population genetics, including delimiting species and subspecies, revealing cryptic diversity, and assessing hybridization between distinct populations. The molecular information from these studies is publicly available and can be found in online databases, and represents a valuable reference database of antelope DNA diversity.

Antelope Conservation: From Diagnosis to Action, First Edition. Edited by Jakob Bro-Jørgensen and David P. Mallon.

Patterns of evolution in sub-Saharan Africa

For the evolution of antelope, sub-Saharan Africa is a region of particular intrigue. DNA provides a rich source of information, which, when coupled with insights from the fossil record and long-term climate records spanning tens of thousands of years (e.g., pollen records from lake sediments), can elucidate what factors shaped the incredible diversity of antelope alive today.

Although habitats and vegetation vary enormously, sub-Saharan Africa can be loosely divided into two major vegetation zones: savanna and tropical forests. The savanna biome harbours the highest diversity of antelope on earth; more species are found here, and in higher abundance, than anywhere else (Figure 1a). The close association between the antelope biomass and the distribution of savanna habitat reflects their interdependence over evolutionary timescales.

The distributional balance between savanna habitat and tropical forests is governed by precipitation. Long-term climate data indicate that the past ~ 3 million years have encompassed at least five significant shifts between wet, warm periods, where savannas were displaced by tropical forests, and dry, cold periods, where savannas expanded. We know from the fossil record that this period spans the emergence of most of today's

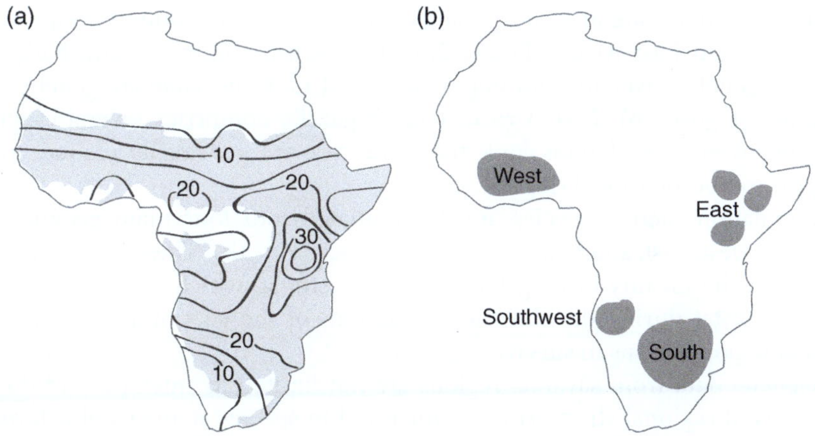

Figure 1 **Distribution, diversity, and regional biogeography of sub-Saharan African antelope species. (a) Diversity isoclines of ungulates (hoofed mammals) across sub-Saharan Africa. Bovids, a taxonomic group dominated by antelopes, make up 80% of ungulate species. Distribution of savanna in grey shading, redrawn from Toit & Cumming (1999). (b) The four main biogeographic regions elucidated from comparative phylogeographic data of antelope and other savanna-adapted species. In many species, East Africa is the only region that harbours several endemic, distinct genetic lineages. Figure adapted from Lorenzen *et al.* (2012).**

antelope as distinct species, and their evolution has therefore been heavily influenced by these known climatic shifts and corresponding environmental changes (Vrba, 1995).

Savannas repeatedly expanded and contracted during long-lasting climatic states, impacting herbivore species reliant on grassland habitat for survival. Although tropical forests expanded at the cost of savannas during humid, warm periods, we believe that core refuge areas of stable savanna habitat were maintained. Refugia enabled the continued survival of savanna-adapted taxa, including many antelope (even if we know from a rich fossil record that many species did not survive into the present). The expansion of savanna habitat during arid periods allowed long-isolated populations to come into contact again. Over time, populations isolated in refugia diverged due to the powerful forces of natural selection and genetic drift, ultimately shaping differences in ecology, morphology, and behaviour, and leading to the astounding number of antelope species that inhabit Earth.

Regional structuring. All widely distributed antelope species are split into regional populations that are genetically distinct from each other. Populations within species may be distinct to the human observer, although this is not always the case. For example, genetically distinct black-faced impala in southwest Africa are also morphologically distinct from common impala in southern and East Africa (Figure 2a). In contrast, common eland harbour cryptic genetic diversity, and populations that look the same are genetically distinct (Figure 2b). East African Grant's gazelle comprises several cryptic species; some populations look the same, yet are genetically distinct and appear to not interbreed (Figure 2c).

Populations across species are differentiated into four main geographic regions: west, east, south, and southwest Africa (Fig. 1b). These regions likely pinpoint the locality of long-term savanna refugia, which provided appropriate habitat during the climatic fluctuations of the past millions of years, enabling populations to survive.

Genetic data from savanna regions are not limited to antelope, and these patterns of regional structuring are mirrored in species across trophic levels; from the grasses and acacias that provide food and shelter for antelope species, to the apex carnivores that hunt them as prey (Lorenzen *et al.*, 2012). Intriguingly, the evidence from grasses is indirect and comes from a pathogen of related domesticated plants. The maize stalk borer (*Busseola fusca*) is a parasite of domesticated crops and shows the same correlation between genetic lineages and geographic regions as we see in antelope. Although the maize stalk borer is currently associated with cultivated plants, its biogeographic history mirrors that of its wild C_4 plant host(s) prior to domestication (Sezonlin *et al.*, 2006).

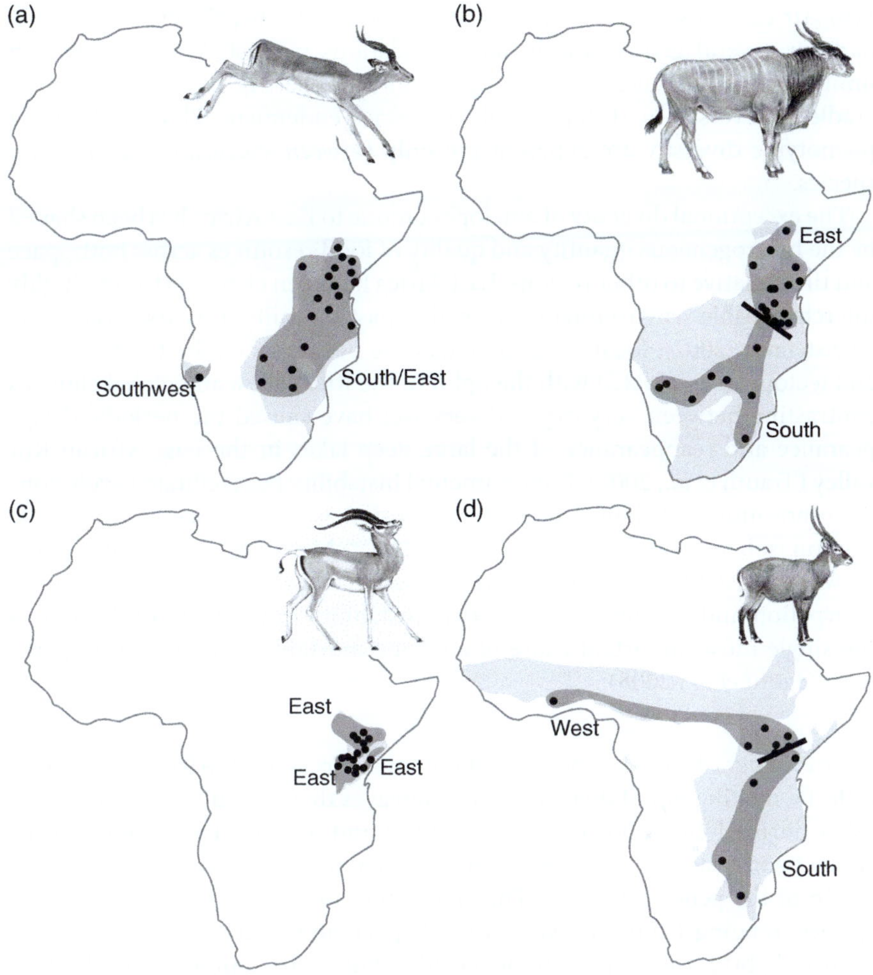

Figure 2 **Four examples of regional genetic structuring inferred from genetic data. Populations that have been genotyped are indicated by black dots. Species distributions are in light grey, and biogeographic regions inferred from genetic data are in dark grey. The species and genetic studies are (a) impala (*Aepyceros melampus*) (Lorenzen, Arctander, *et al.*, 2006a), (b) eland (*Tragelaphus oryx*) (Lorenzen *et al.*, 2010), (c) Grant's gazelle (*Nanger granti*) (Lorenzen *et al.*, 2007) and (d) waterbuck (*Kobus ellipsiprymnus*) (Lorenzen, Simonsen, *et al.*, 2006b). Solid black lines in (b) and (d) indicate hybrid zones. Antelope illustrations by Jonathan Kingdon.**

The presence of similar regional patterns of genetic variation and differentiation across co-distributed species and trophic levels adapted to the same ecosystem provides strong evidence of ecologically driven evolution (Lorenzen *et al.*, 2012).

East Africa. Most antelope species studied so far are surprisingly diverse and harbour several genetically distinct populations in East Africa, often with limited geographic distributions (e.g., Grant's gazelle, Fig. 2c). The region is a cradle of diversity, with high levels of species endemism. Elevated levels of phenotypic diversity are apparent not only *between* species, but also *within* species.

The exceptional diversity of antelopes unique to East Africa has been shaped by the heterogeneous quantity and quality of food resources across both space and time relative to other regions. East Africa has been characterized by highly interchangeable environments over the past 2 million years (Maslin & Christensen, 2007); local environments have been affected by the volcanism and tectonism associated with the uplift of the Rift Valley, and global climates contrasting between very dry and very wet have caused the periodic disappearance and reappearance of the large deep lakes in the East African Rift valley (Trauth *et al.*, 2007). Environmental instability has facilitated evolutionary innovation and change, as natural selection acted in concert with genetic drift on isolated populations (Demenocal, 2004; Maslin & Christensen, 2007). However, unstable environments are also associated with a greater risk of local extirpation and species loss; ecological instability in the form of drought is the single most important cause of antelope mortality in East African parks (e.g., Ogutu *et al.*, 2008).

Suture zones. Hybrid zones are ubiquitous across East Africa. The zones reflect a melting pot of diversity, where lineages that have diverged from each other during long periods of isolation meet and overlap. The location of the contact zone varies from Uganda to Tanzania, depending on the geographic origin of the genetic lineages. The extent and form of contact varies among species, ranging from admixture to full speciation, and is exemplified by a narrow hybrid zone in common eland (Figure 2b), limited hybridization across a tight boundary in waterbuck (Figure 2d), pronounced one-way gene flow in kob, and reproductive isolation in Grant's gazelle (Figure 2c).

Unique regional histories. The geographic regions of sub-Saharan Africa (Figure 1b) represent unique evolutionary scenarios. This is reflected by common idiosyncrasies in the genetic patterns of the antelope populations inhabiting them. In many cases, we know from the fossil record that species have been present in East Africa for hundreds of thousands of years; however, the genetic data indicate recent colonization from southern Africa, suggesting that some original eastern populations went extinct and were recolonized. Such patterns of extinction/replacement are well known from northern hemisphere ungulates (e.g., Campos, 2010).

Across taxa, higher levels of genetic variation and less pronounced structuring of populations in southern Africa suggest that habitats and faunal populations have remained more homogeneous, in contrast to the mosaic of populations present in the east. This is supported by palaeo-climate data that indicate stable levels of moisture and greater environmental stability in the south throughout the Pleistocene. A continuous southern African refuge area acted as a museum, conserving populations and species over evolutionary timescales. However, local loss of populations in southern Africa and subsequent colonization from the east is apparent in some species, suggesting the biogeographic history of the south has also been dynamic.

Conservation and management

Molecular data have become an increasingly important tool in informed species conservation and sustainable wildlife management. The movement of antelope through translocations, reintroductions, and population augmentations is common practice in wildlife management. But before moving animals from one area to another, managers can use DNA information to avoid mixing genetically distinct populations (termed evolutionary significant units, or ESUs). DNA sequencing is a valuable tool for assessing rare or nocturnal species, which can be difficult to monitor by other means. DNA data can also be used to assess levels of inbreeding in small populations (or the reverse by estimating levels of outbreeding when genetically highly distinct populations mix), or to investigate hybridization between introduced and endemic populations of the same species (which may be an unfortunate downstream outcome of translocating non-native populations).

The sample needed for genetic analysis is minute (~1 ug), and can be any biological material, including bone (in case of a deceased animal), antler, keratin, tissue, hair, and faecal pellets. The latter provides an especially promising tool in conservation, as sampling does not affect the animal (unlike a tissue biopsy). A recent example of the applicability of DNA sequencing of dung was the rediscovery of the endemic giant sable (*Hippotragus niger variani*) in Angola. There had been no confirmed observations of the species during 25 years of civil war. By comparing DNA sequenced from fresh dung sampled in Angola with DNA from giant sable museum specimens, researchers were able to provide definitive evidence that the species was indeed still present (Pitra *et al.*, 2006), which was subsequently confirmed by sightings.

The limitation of faecal material is the rapid degradation of DNA and low endogenous DNA content of the sample, which makes reliable genotyping challenging. However, methods to extract and sequence degraded DNA are rapidly improving, in part due to the surge of interest in ancient DNA research, where sub-fossil material tens to hundreds of thousands of years old can be genotyped (Lorenzen *et al.*, 2011). Indeed, the future of truly non-invasive genotyping as a viable tool in conservation and management holds great promise.

Wildlife forensics

DNA-led species identification using genetic barcoding is an effective use of genetic data within forensics. DNA barcoding is a taxonomic method that uses a short genetic marker in an organism's DNA to identify it as belonging to a particular species (Bitanyi *et al.*, 2011). It is a simple but effective approach to large-scale monitoring of, for instance, the bushmeat trade, to confirm the origin of raw and processed meat and importantly, to uncover illegally harvested species. For example, a recent study showed that estimates of bushmeat utilization using genetic barcoding differs markedly from the information provided by local providers and traders (Bitanyi *et al.*, 2012). Barcoding may also be used in less serious offences; a study of South African biltong (air-dried meat) demonstrated that the game species such as eland, impala, and kudu (*Tragelaphus strepsiceros*) advertised on packets rarely match what is inside (Cawthorn *et al.*, 2013).

Another potent way in which to combat the illegal harvest and trade of threatened and protected species is the use of genetic assignment methods. In this age of globalization, international wildlife crime is rife, and the most effective way to contain illegal trade is to reliably determine from where wildlife is being removed. This allows authorities to (i) direct law enforcement to poaching hot spots, (ii) prevent countries from denying their poaching issues, (iii) thwart illegal trade before it enters into an increasingly complex web of international criminal activity, and (iv) importantly, increase the rate of criminal prosecution.

When the species identity of a sample is already known, assignment methods use the genotype of an individual to assign it to the population from which it most probably originated. This is done using DNA information from a reference panel of populations that have already been genotyped.

An illustrative example of the powerful application of DNA in combating wildlife crime was the seizure of over 6.5 tons of contraband ivory in Malawi in 2002, the largest ivory seizure since the 1989 ivory trade ban and the second-largest seizure ever recorded. The shipment was destined for the Far East. Wildlife authorities initially suspected that this ivory came from several forest and savanna locations across Africa. However, DNA analysis determined that the ivory originated exclusively from savanna elephants (*Loxodonta africana*). The most probable origin was a narrow east-to-west band centred on Zambia. The findings enabled law enforcement to focus their investigation on a smaller area and fewer trade routes, and led to changes within the Zambian government to improve anti-poaching efforts (Wasser *et al.*, 2007).

Perspectives

Molecular data have become an important tool in conservation, and provide crucial information about species diversity and population processes that are not apparent at the morphological level, including cryptic diversity and hybridization. Much of the publically available DNA data from antelopes originates from academic studies whose original focus was to understand the evolutionary histories of the world's antelope species. The data now serve a multitude of valuable downstream purposes for species conservation, informed management, and wildlife forensics. A limitation to the application of DNA data for conservation is the inability to sample rare or threatened species for DNA analysis. However, recent improvements in DNA extraction methods have revolutionized our ability to carry out non-invasive sampling of threatened wild and captive populations, by enabling the retrieval of DNA from tiny amounts of biological material.

References

Bitanyi, S., Bjørnstad, G., Nesje M., Ernest, E. M., Mdegela, R. H. & Røed, K. H. (2012): Molecular identification versus local people's information for accurate estimates of bushmeat utilization from the Serengeti ecosystem, Tanzania. *African Journal of Biotechnology* **11**(1): 243–252.

Bitanyi, S., Bitantyi, S., Bjørnstad, G., Ernest, E. M., Nesie, M., Kusiluka, L. J., Keyyu, J. D., Mdegela, R. H. & Røed, K. H. (2011): Species identification of Tanzanian antelopes using DNA barcoding. *Molecular Ecology Resources* **11**(3):442–449.

Campos, P. F., Willerslev, E., Sher, A., Orlando, L., Axelsson, E., Tikhonov, A., Aaris-Sørensen, K., Greenwood, A. D., Kahlke, R. D., Kosintsev, P., Krakhmalnaya, T., Kuznetsova, T., Lemey, P., MacPhee, R., Norris, C. A., Shepherd, K., Suchard, M. A., Zazula, G. D., Shapiro, B. & Gilbert, M. T. P. (2010): Ancient DNA analyses exclude humans as the driving force behind late Pleistocene musk ox (*Ovibos moschatus*) population dynamics. *Proceedings of the National Academy of Sciences U.S.A.* **107**: 5675–5680.

Cawthorn, D.-M., Steinman, H. A. & Hoffman, L. C. (2013): A high incidence of species substitution and mislabelling detected in meat products sold in South Africa. *Food Control* **32**(2):440–449.

Demenocal, P. B. (2004): African climate change and faunal evolution during the Pliocene–Pleistocene. *Earth and Planetary Science Letters* **220**(1–2): 3–24.

Lorenzen, E. D., Masembe, C., Arctander, P. & Siegismund, H. R. (2010): A long-standing Pleistocene refugium in southern Africa and a mosaic of refugia in East Africa: insights from mtDNA and the common eland antelope. *Journal of Biogeography* **37**(3):571–581.

Lorenzen, E. D., Nogués-Bravo, D., Orlando, L., Weinstock, J., Binladen, J., Marske, K. A., Ugan, A., Borregaard, M. K., Gilbert, M. T. P., Nielsen, R., Ho, S. Y. W., Goebel, T., Graf, K. E., Byers, D., Stenderup, J. T., Rasmussen, M., Campos, P. F., Leonard, J. A., Koepfli, K. P., Froese, D., Zazula, G., Stafford, T. W., Aaris-Sørensen, K., Batra, P., Haywood, A. M., Singarayer, J. S., Valdes, P. J., Boeskorov, G., Burns, J. A., Davydov, S. P., Haile, J., Jenkins, D. L., Kosintsev, P., Kuznetsova, T., Lai, X., Martin, L. D., McDonald, H. G., Mol, D., Meldgaard, M., Munch, K., Stephan, E., Sablin, M., Sommer, R. S., Sipko, T., Scott, E., Suchard, M. A., Tikhonov, A., Willerslev, R., Wayne, R. K., Cooper, A., Hofreiter, M., Sher, A., Shapiro, B., Rahbek, C. & Willerslev, E. (2011): Species-specific responses of Late Quaternary megafauna to climate and humans. *Nature* **479**(7373):359–364.

Lorenzen, E. D., Arctander, P. & Siegismund, H. R. (2006a): Regional genetic structuring and evolutionary history of the impala Aepyceros melampus. *Journal of Heredity* **97**(2):119–132.

Lorenzen, E. D., Arctander, P. & Siegismund, H. R. (2007): Three reciprocally monophyletic mtDNA lineages elucidate the taxonomic status of Grant's gazelles. *Conservation Genetics* **9**(3): 593–601.

Lorenzen, E. D., Heller, R. & Siegismund, H. R. (2012): Comparative phylogeography of African savanna ungulates. *Molecular Ecology* **21**(15): 3656–3670.

Lorenzen, E. D., Simonsen, B. T., Kat, P. W., Arctander, P. & Siegismund, H. R. (2006b): Hybridization between subspecies of waterbuck (*Kobus ellipsiprymnus*) in zones of overlap with limited introgression. *Molecular Ecology* **15**(12): 3787–3799.

Mace G. (2004): The role of taxonomy in species conservation. *Philosophical Transactions of the Royal Society B Biological Sciences* **359**: 711–719.

Maslin, M. A. & Christensen, B. (2007): Tectonics, orbital forcing, global climate change, and human evolution in Africa: introduction to the African paleoclimate special volume. *Journal of Human Evolution* **53**(5): 443–464.

Ogutu, J. O., Piepho, H. P., Dublin, H. T., Bhola, N. & Reid, R. S. (2008): Rainfall influences on ungulate population abundance in the Mara-Serengeti ecosystem. *Journal of Animal Ecology* **77**(4): 814–829.

Pitra, C., Vaz Pinto, P., O'Keeffe, B. W. J., Willows-Munro, S., van Vuuren, B. J. & Robinson, T. J. (2006): DNA-led rediscovery of the giant sable antelope in Angola. *European Journal of Wildlife Research* **52**(3):145–152.

Sezonlin, M., Dupas, S., le Rü, B., le Gall, P., Moyal, P., Calatayud, P. A., Giffard, I., Faure, N. & Silvain, J. F. (2006): Phylogeography and population genetics of the maize stalk borer *Busseola fusca* (Lepidoptera, Noctuidae) in sub-Saharan Africa. *Molecular Ecology*, **15**(2): 407–420.

Toit, Du, J. T. & Cumming, D. (1999): Functional significance of ungulate diversity in African savannas and the ecological implications of the spread of pastoralism. *Biodiversity and Conservation* **8**(12):1643–1661.

Trauth, M., Maslin, M. & Deino, A. (2007): High- and low-latitude forcing of Plio-Pleistocene East African climate and human evolution. *Journal of Human Evolution* **53**: 475–486.

Vrba, E. S. (1995): The fossil record of African antelopes (Mammalia, Bovidae) in relation to human evolution and paleoclimate. In Paleoclimate and evolution, with emphasis on human origins. Vrba, E. S. *et al.* (Eds.). 385–424. New Haven: Yale University Press.

Wasser, S. K., Mailand, C., Booth, R., Mutayoba, B., Kisamo, E., Clark, B. & Stephens, M. (2007): Using DNA to track the origin of the largest ivory seizure since the 1989 trade ban. *Proceedings of the National Academy of Sciences U. S. A.* **104**(10): 4228–4233.

10

Biological Conservation Founded on Landscape Genetics: The Case of the Endangered Mountain Nyala in the Southern Highlands of Ethiopia

Anagaw Atickem[1,2], Eli K. Rueness[1], Leif E. Loe[3], and Nils C. Stenseth[1,2]

[1]*Centre for Ecological and Evolutionary Synthesis (CEES), Department of Biosciences, University of Oslo, Norway*
[2]*Department of Zoological Science, Addis Ababa University, Addis Ababa, Ethiopia*
[3]*Norwegian University of Life Sciences, Ås, Norway*

Habitat fragmentation and its effects on antelope populations

The abundance of African large mammals has declined substantially during the last few decades (Nelson, 2008; Craigie *et al.*, 2010; Scholte, 2011), and nearly a third of antelope species are now threatened with extinction (IUCN, 2015). In particular, many long-distance migrants have either disappeared or are at risk of disappearing due to disruption of suitable migration corridors (Berger, 2004; Wilcove, 2007; Berger *et al.*, 2008; Wilcove & Wikelski, 2008; Ogutu *et al.*, 2013). The rapid increase in the human population and the concomitant growth in human settlement, agriculture and infrastructures (Balmford *et al.*, 2001) causes habitat loss and fragmentation (Fahrig, 2003; Laurance *et al.*, 2007), which are two of the most important reasons for the decline in mammals across African countries (Harris *et al.*, 2009).

Habitat fragmentation can affect the survival of wildlife species negatively through reduction of habitat diversity (Carlson & Hartman, 2001), degradation

Antelope Conservation: From Diagnosis to Action, First Edition. Edited by Jakob Bro-Jørgensen and David P. Mallon.

of habitat quality in the remaining patches (Battin, 2004), demographic stochastic effects in small populations (such as uneven birth sex ratio; Johannesen *et al.*, 2003; Henle *et al.*, 2004) and edge effects (including increased predation/disturbance by humans; Ries *et al.*, 2004). Moreover, habitat fragmentation leads to increased isolation between populations and thereby loss of genetic diversity (Bálint *et al.*, 2011; Hill *et al.*, 2011; Cobben *et al.*, 2012; Rubidge *et al.*, 2012), randomly by genetic drift and through inbreeding (Hedrick & Kalinowski, 2000; Ito *et al.*, 2013), and there is compelling evidence for the importance of genetic variation for the long-term survival of a given population (Willis, 1993; Buza *et al.*, 2000; Reed & Frankham, 2003; Spielman *et al.*, 2004). Additionally, isolation of populations may disrupt evolutionary and adaptive processes by hindering the spread of beneficial genes (Chapter 9; Templeton *et al.*, 2001).

While protected areas cover 11.5% of the planet's land surface (Jenkins & Joppa, 2009), the protected-area approach has not provided the ultimate solution to sustain wildlife populations (Mora & Sale, 2011). Large mammal populations, including antelopes, are declining also within Africa's protected areas (Craigie *et al.*, 2010). For instance, between 1977 and 1997, the total number of individuals of all non-migratory wildlife species in the Masai Mara ecosystem declined by 58% (Ottichilo *et al.*, 2000). The decline seemed to be particularly steep in many antelope species, and a decline as high as 72% was recorded for the populations of medium-sized brown antelopes (Ottichilo *et al.*, 2000). Within six protected parks in Tanzania, Newmark (1996) found that five antelope species (roan antelope *Hippotragus equinus*, klipspringer *Oreotragus oreotragus*, mountain reedbuck *Redunca fulvorufula*, steenbok *Raphicerus campestris*, hartebeest *Alcelaphus buselaphus*) became locally extinct over an 83-year period in the last century. The small sizes of the parks and the lack of sufficient connectivity between neighbouring protected areas are considered important reasons for the decline (Newmark, 1996; Ervin, 2003; Newmark, 2008; Western *et al.*, 2009). Due to the increasing scale of habitat fragmentation caused by human development and climate change, current nature reserves may thus be insufficient for the future protection of all of the species they were designed to conserve (Heller & Zavaleta, 2009; Mora & Sale, 2011; Bellard *et al.*, 2012; Pettorelli, 2012). Hence, when planning the future management of wildlife species one should put an effort into maintaining population connectivity in conservation networks (Fahrig, 2003; Ogutu *et al.*, 2013; Reed, 2004; McLachlan *et al.*, 2007). In this context, the landscape-genetic toolbox has become important in conservation management planning (Manel *et al.*, 2003; Segelbacher *et al.*, 2010; Sork & Waits, 2010; Yang *et al.*, 2011; Sommer *et al.*, 2013) by providing essential information for designing dispersal corridors between fragmented localities to facilitate the long-term survival of highly mobile wildlife species (Scott *et al.*, 2002; Rodrigues *et al.*, 2004; Sharma *et al.*, 2013). Landscape genetics combines population genetics, landscape ecology, and spatial statistics to quantify the relative impact that landscape and environmental features have on gene flow between populations (Holderegger & Wagner, 2008; Balkenhol *et al.*, 2009).

Landscape genetics as a tool in wildlife conservation

The concept of landscape genetics is tightly linked to landscape connectivity, originally defined as 'the degree to which the landscape facilitates or impedes movement among resource patches' (Taylor *et al.*, 1993). Landscape connectivity is a function of structural and functional connectivity (Brooks, 2003). Structural connectivity describes physical relationships among habitat patches, such as habitat corridors or inter-patch distances, while the functional connectivity explicitly considers the behavioural responses of an organism to various landscape elements (Taylor *et al.*, 2006). Quantifying the two components and relating both to measures of genetic variation is central in landscape genetics (Manel *et al.*, 2003). Studies of landscape connectivity may be based on expert opinions; that is, a panel of experts rank habitat types in terms of suitability and/or provide educated guesses on functional connectivity between populations. This approach has been criticized because the result is strongly dependent on the experts' background and research experience (Johnson & Gillingham, 2004; Seoane *et al.*, 2005). The assignment of habitat suitability and functional connectivity in the expert opinion method will thus be largely subjective (Compton *et al.*, 2007; Spear *et al.*, 2010; Zeller *et al.*, 2012). In general, it has been documented that empirical data outperforms expert opinions (Clevenger *et al.*, 2002; Yamada *et al.*, 2003; Johnson & Gillingham, 2004; Seoane *et al.*, 2005). In the following, we describe the much more common method of using objective data and statistical models to evaluate structural as well as functional connectivity.

When evaluating structural connectivity, the effects of different land-cover types or other environmental gradients on animal movement are generally difficult to quantify and remain the biggest challenge in landscape genetics studies (Spear *et al.*, 2010; Zeller *et al.*, 2012). The simplest and most frequently used method is based on the assumption that population connectivity is a function of the habitat preferences of the species; preferred habitat has low resistance and vice versa. This approach requires knowledge of where the species is present. Many types of methodology have been used to obtain information of space use of a focal species. When analyzing changes in species distribution occurring over centuries and large spatial scales, the reference is commonly museum specimens collected by past generations of researchers. On smaller scales, information of space use may be direct observations of individuals in field or signs of presence (such as faeces). Today, the most common method is to mark individuals with biotelemetry equipment. For medium to large-sized species of birds and mammals able to carry a global positioning system (GPS) device, the amount and precision of spatial information per marked individual is extremely high. In addition to spatial data on individuals, environmental data is needed. The limiting factor is that the environmental data needs to be continuous in space. Because it is rarely realistic to do continuous habitat field surveys over a sufficiently large spatial scale, aerial or satellite images are the most used source of information on land use classes and locations of human infrastructure. In addition digital terrain models are used to extract information of elevation, slope, and aspect. The animal locations together with the environmental map layers are used to produce a habitat suitability map, which is a descriptive representation of the relationship between a wildlife population and its preferred and avoided habitat (Hirzel & Lay, 2008). When developing the habitat suitability map, environmental factors at locations with animal presence are contrasted with what is available to the animal using resource selection functions (Dellinger *et al.*, 2013; Erfanian *et al.*, 2013). Finally a resistance map can be calculated as the inverse of the habitat suitability map. Based on the resistance maps, a number of different models can then be applied to predict connectivity between different locations. In applied studies, least-cost path modelling has become the most common choice for modelling of dispersal corridors (Li *et al.*, 2010). This method assumes that the organisms move along the path with the lowest total cumulative resistance (Rosenberg *et al.*, 1997; Adriaensen *et al.*, 2003) and culminates in a map of the most likely dispersal corridors between neighbouring populations.

Functional connectivity is the second integral part of landscape genetics and is studied using data from radio telemetry and population genetic analyses. Recording of movement rates, travel distances, and dispersal routes of animals by radio telemetry allows researchers to quantitatively describe species-specific responses to environmental conditions and landscape structure (Cushman & Lewis, 2010; Ito *et al.*, 2013). Similar to the development in telemetry studies, molecular techniques used for quantifying functional connectivity are improving with technological advances. A range of genetic markers (including allozymes, amplified fragment length polymorphism (AFLPs), random amplified polymorphic DNAs (RAPDs), micro-satellites and mini-satellites, SNPs and mitochondrial DNA (mtDNA)) have been applied in landscape genetic studies (Goldstein & Schlotterer, 1999; Storfer *et al.*, 2010), the goal being to provide fine-scale information on genetic structuring and contemporary gene flow between populations (Wan *et al.*, 2004). Non-invasive methods, where DNA is obtained from sources of cellular material like hair, faeces, and urine can be obtained without physical capturing or other human interference and enable large sample sizes (Chapter 9; Flagstad *et al.*, 1999; Bellemain & Taberlet, 2004; Hedmark *et al.*, 2004). This is particularly useful when studying rare and elusive animals (Maudet *et al.*, 2004; Rodgers & Janečka, 2013). Currently, landscape genetics is getting to a new era as genomic data generated by next-generation sequencing techniques drastically increase the amount of data per study and include both neutral and adaptive genetic variation (Allendorf *et al.*, 2010). This innovation may provide better understanding of the interaction between landscape ecology and evolutionary processes (Manel & Holderegger, 2013).

Data from radio-telemetry and genetic analysis provide complementary pictures of the connectivity between populations and can in some cases give opposing results. In some cases, capturing and fitting collars on a large enough number of individuals to get good measures of a population's dispersal patterns from telemetry data is not possible. For instance, while a 2.5-year study of radio-telemetry data showed no connectivity between two moose populations, genetic data revealed that the two populations were linked (Finnegan *et al.*, 2012). Yet, while the application of molecular techniques may counter the limitations of low sample sizes in telemetry (Epps *et al.*, 2005; Finnegan *et al.*, 2012), gene flow estimates also have shortcomings as direct dispersal cannot be measured. In a study of bobcats (*Lynx rufus*) and coyotes (*Canis latrans*) radio telemetry data revealed that 5–32% of the nearly 200 collared individuals migrated across a border, which is a much larger proportion than what is required to maintain genetic connectivity (ca 0.5%). Yet the populations on either side of the border (a large road) were genetically differentiated (Riley *et al.*, 2006) indicating low reproductive success of the dispersing animals. The difference in the temporal aspect between direct measures of dispersal rates and gene flow estimates can explain contradictory outcomes. In situations with recently separated populations, genetic data may indicate that migration occurs even though studies of habitat resistance and dispersal show no connectivity. Hence, combining ecological and genetic data provides a much more in-depth understanding of migration and dispersal patterns (Sacks *et al.*, 2004; Sacks *et al.*, 2005).

Once the structural and functional connectivity of the landscape has been evaluated, the ultimate task in landscape genetics is to understand how these elements interact in shaping the genetic structure of neighbouring populations. One of the most commonly used methods is the Mantel test that relates a matrix of genetic distances between individuals or populations to matrices of geographical or cost-weighted distances (Mantel, 1967; Balkenhol *et al.*, 2009; Manel & Holderegger, 2013).

The mountain nyala – a case study

The mountain nyala (*Tragelaphus buxtoni*) is an endangered spiral-horned antelope endemic to the south-eastern highlands of Ethiopia (Hillman, 1993; Figure 1 & Plate 1). The largest mountain nyala metapopulation is found within fragmented patches of the Bale Mountains (Atickem *et al.*, 2011) and has been estimated to be about 4000 individuals. The only other smaller remnant populations occur in the Arussi Mountains north of the Bale Massif (Brown, 1969; Yalden & Largen, 1992; Sillero-Zubiri, 2008; Atickem *et al.*, 2011; Figure 2). The majority of the populations exist outside the core area for mountain nyala conservation, the Bale Mountains national park (BMNP; Atickem *et al.*, 2011), and human settlements and livestock grazing occurs also within the BMNP (Atickem and Loe, 2013). Our landscape genetic study (Atickem *et al.*, 2013) included mountain nyala from five localities within the Bale Mountains that comprise about 86% of the total mountain nyala population of the area (Atickem *et al.*, 2011; Figure 2). In addition one locality in the Arussi Mountains was sampled (Figure 2).

We determined the structural connectivity between the different mountain nyala populations across the Bale Mountains by developing habitat-suitability maps using a resource selection function (Manly, 2002) based on generalized linear models (Atickem *et al.*, 2011). We used mountain nyala faecal pellet counts as our indicator of the animals' presence/absence and altogether 1515 plots, each of 20 m^2, within mountain nyala core areas were censused. Our model quantified the relative probability of mountain nyala presence as a function of a set of environmental variables (Johnson & Gillingham, 2008), encompassing six land cover categories (forest, *Erica* shrub, grassland, bushland, agricultural area, and human settlement), slope, and elevation (Atickem *et al.*, 2011).

Figure 1 **Female mountain nyala.**

From the estimates of the resource selection function, we calculated reciprocal pixel values to create a cost-raster that associated favorable landscape with lower pixel values (i.e., lower cost of movement). We then calculated low-resistance dispersal corridors using least-cost path models. Details of the method are given in Atickem *et al.* (2011).

We examined the functional connectivity by analyzing 12 microsatellite markers for a total of 294 individuals sampled in the six localities described above (Figure 2). The DNA was obtained from faecal pellets and was extracted using a protocol particularly developed for such samples (Flagstad *et al.*, 1999). We estimated the levels of genetic variability and gene flow, calculated genetic distances between the populations, and tested for effects of inbreeding and significant population differentiation using a range of population genetic tools (e.g., Goudet, 2002; Rousset, 2008). Finally, we analyzed the correlation of genetic distance to three different measures for dispersal distance/habitat resistance: 1) linear Euclidian distance between populations (km), 2) the least-cost path distance (km), and 3) 'cumulative resistance' (i.e., the sum of resistance values along the least-cost path).

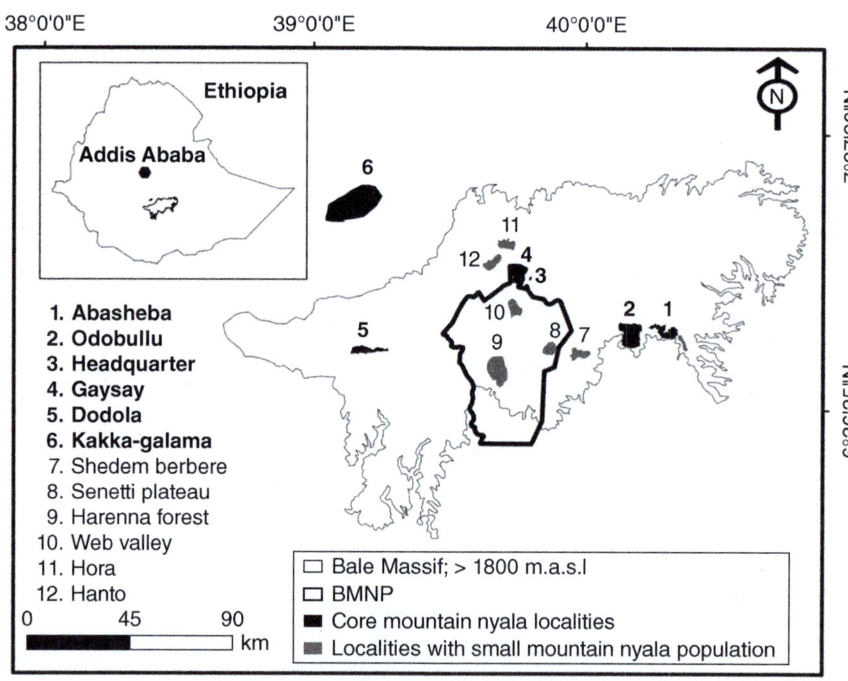

Figure 2 **Mountain nyala distribution in the Bale mountains. As can be seen from this map, most of the populations are found in localities outside the conservation area within the Bale Mountains National Park (BMNP).**

Our study revealed significant genetic population differentiation between all of the mountain nyala populations, with the exception of two locations that are separated by only 5 km (Headquarter and Gaysay; Figure 2). Inbreeding was detected in all of the mountain nyala populations except one (Dodola) and could indicate diversity loss due to population differentiation. Reduced fecundity and survival have been demonstrated in inbred individuals of several antelope species (Ralls *et al.*, 1979; Hedrick & Kalinowski, 2000; Slate *et al.*, 2000; Walling *et al.*, 2011; Ruiz-Lopez *et al.*, 2012). Such effects of inbreeding could be caused by low genetic variability and expression of deleterious genes (Saccheri *et al.*, 1998; Reed *et al.*, 2002; Thevenon & Couvet, 2002). It is unknown whether or not any of the mountain nyala populations experience such negative effects of inbreeding.

We did not find any environmental factors with detectable impact on the population genetic structure. However, possible effects of poaching and other human disturbance on the movement patterns were not evaluated. The correlation between genetic distance and least-cost distance was virtually indistinguishable from the correlation with Euclidian distance (isolation-by-distance). Hence, intrinsic limitations in dispersal distances of mountain nyala may be the reason for the observed differentiation, possibly connected to social and other behavioural constraints for dispersal (Painter *et al.*, 2000; Temple *et al.*, 2006; Clark *et al.*, 2008). Because dispersal is a complex biological phenomenon, it can be difficult to determine why it is limited in some species and common in others, even in the absence of dispersal barriers (Lidicker & Stenseth, 1992; Lion & van Baalen, 2008; Hawkes, 2009). Incorporating knowledge of behaviours that influence gene flow into landscape genetic analyses may be equally important as to study the effects of environmental features (Clark *et al.*, 2008; Bálint *et al.*, 2009). Preliminary results from radio telemetry data from seven GPS-marked mountain nyala females and nine VHF-marked males show that all of the females remained within a small and stable home range (7.04 km 2 ± 4.05 SE based on 95% minimum convex polygon) while two of the males dispersed out of the monitored range. While these results are based on a small sample size and were collected over a relatively short time (two years), it suggests that males may be the dispersing sex, as in most other mammals (Greenwood, 1980). However, a more detailed study on the behavior of mountain nyala movement patterns, also including juvenile individuals, will be required to understand which factors determine dispersal.

Even if we did not find evidence for habitat resistance on gene flow, it is likely that the future connectivity between the subpopulations may decrease as human land use, including settlement and agriculture, increases across the mountain nyala range (Stephens *et al.*, 2001). Migration corridors with low habitat resistance should therefore be preserved to facilitate the future dispersal of mountain nyala. We proposed three corridors where we suggest that the authorities should prevent human influence (Figure 3). These corridors may have the additional benefit of preserving other key species of the study area such as the endemic Bale monkey (*Chlorocebus djamdjamensis*) (Mekonnen *et al.*, 2010), the newly identified long-eared bat (*Plecotus* spp.) as well as many endemic amphibians (Largen, 2001).

Storfer *et al.* (2010) emphasize that the current major challenge in landscape genetics is to obtain spatially explicit information about the habitat. The recent advances in satellite imagery have made the technology faster and more affordable and have opened the opportunity of getting high-quality land cover images of little-studied parts of the world (Nichol & Wong, 2005; Pettorelli, 2013). Studies from Africa, however, represented less than 2% of the landscape genetic literature between 1998 and 2008 (Storfer *et al.*, 2010). Our study demonstrates the application of high-resolution satellite images in landscape genetics in an Afromontane area.

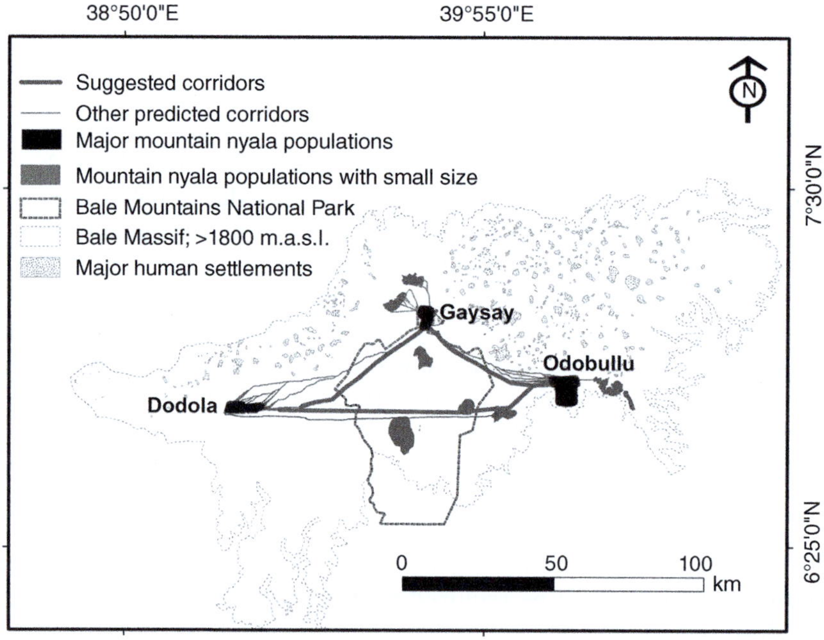

Figure 3 **Predicted dispersal corridors between the mountain nyala populations in the Bale Massif, as derived from least-cost path analysis of habitat suitability maps. The three most recommended conservation corridors are indicated by thick lines (Atickem *et al.*, 2013).**

Conclusion: landscape genetics and antelope conservation

Habitat fragmentation continues at an alarming rate, inhibiting long-distance dispersal of wildlife species. Therefore the conservation strategy of protecting areas as separate entities (national parks) often falls short of protecting large mammals (Newmark, 2008; Craigie *et al.*, 2010) due to the limited size of the conservation areas and their isolation from each other (Woodroofe & Ginsberg, 1998; Brashares *et al.*, 2001). Even the largest parks, like the Serengeti-Mara Ecosystem (SME), which covers 25,000 km^2 on the border of Tanzania and Kenya, are not large enough to contain the entire range of populations of migratory mammals such as the wildebeest (Thirgood *et al.*, 2004). We suggest that future wildlife conservation management should be based on alternative strategies where protected areas are networked with adjacent landscape units to keep populations viable in the form of meta-populations through demographic and genetic rescue (Bennett, 2003; Rouget *et al.*, 2006; Phillips *et al.*, 2008; Pressey & Bottrill, 2009). To monitor this change, it is important to simultaneously investigate changes in landscape connectivity and movement behaviour of the species and monitor the genetic diversity of subpopulations. In the study of the mountain nyala, we measured genetic variability and gene flow within and among the main subpopulations of the species and suggested potential corridors that may preserve dispersal of individuals. We used a wide range of tools including remote sensing, telemetry, habitat suitability modelling, and genetics. The method presented here could be a useful approach for also guiding the conservation of other antelope species.

References

Adriaensen, F., Chardron, J. P., DeBlust, G., Swinnend, E., Villalbad, S., Gulinckd, H. & Matthysena, E. (2003): The application of 'least-cost' modelling as a functional landscape model. *Landscape and Urban Planning* **64**: 233–247.

Allendorf, F. W., Hohenlohe, P. A. & Luikart, G. (2010): Genomics and the future of conservation genetics. *Nature Reviews Genetics* **11**: 697–709.

Atickem, A., & Loe, L. E. (2013): Livestock-wildlife conflicts in the Ethiopian highlands: assessing the dietary and spatial overlap between mountain nyala and cattle. *African Journal of Ecology* **52**: 343–351.

Atickem, A., Loe, L. E., Langangen, Ø., Rueness, E. K., Bekele, A. & Stenseth, N. C. (2011): Estimating population size and habitat suitability for mountain nyala in areas with different protection status. *Animal Conservation* **14**: 409–418.

Atickem, A., Rueness, E. K., Loe, L. E., Serbezov, D., Bekele, A. & Stenseth, N. C. (2013): Population genetic structure and connectivity in the endangered Ethiopian mountain nyala (*Tragelaphus buxtoni*): recommending dispersal corridors for future conservation. *Conservation Genetics* **14**: 427–438.

Bálint, M., Domisch, S., Engelhardt, C. H. M., Haase, P., Lehrian, S., Sauer J., Theissinger, K., Balkenhol, N., Gugerli, F., Cushman, S. A., Waits, L. P., Coulon, A., Arntzen, J. W., Holderegger, R. & Wagner, H. H. (2009): Identifying future research needs inland scape genetics: where to from here? *Landscape Ecology* **24**: 455–463.

Balkenhol, N., Waits, L. P. & Dezzani, R. J. (2009): Statistical approaches in landscape genetics: an evaluation of methods for linking landscape and genetic data. *Ecography* **32**: 818–830.

Balmford, A., Moore, J. L., Brooks, T., Burgess, N., Hansen, L. A., Williams, P. & Rahbek, C. (2001): Conservation conflicts across Africa. *Science* **291**: 2616–2619.

Battin, J. (2004): When good animals love bad habitats: ecological traps and the conservation of animal populations. *Conservation Biology* **18**: 1482–1491.

Bellard, C., Bertelsmeier, C., Leadley, P., Thuiller, W. & Courchamp, F. (2012): Impacts of climate change on the future of biodiversity. *Ecology Letters* **15**: 365–377.

Bellemain, E. & Taberlet, P. (2004): Improved non-invasive genotyping method: application to brown bear (*Ursus arctos*) feces. *Molecular Ecology Notes* **4**: 519–522.

Bennett, A. (2003): *Linkages in the landscape: the role of corridors and connectivity in wildlife conservation*. Gland, Switzerland and Cambridge, UK: IUCN.

Berger, J. (2004): The last mile: how to sustain long-distance migration in mammals. *Conservation Biology* **18**: 320–331.

Berger, J., Young, J. K. & Berger, K. M. (2008): Protecting migration corridors: challenges and optimism for Mongolian saiga. *PLOS Biology* **6**: e165.

Brashares, J. S., Arcese, P., & Sam, M. K. (2001): Human demography and reserve size predict wildlife extinction in West Africa. *Proceedings of the Royal Society B* **268**: 2473–2478.

Brooks, C. P. (2003): A scalar analysis of landscape connectivity. *Oikos* **102**: 433–439.

Brown, L. (1969): Observations on the status, habitat, and behavior of the mountain nyala (*Tragelaphus buxoni*) in Ethiopia. *Mammalia* **33**: 545–597.

Buza, L., Young, A. & Thrall, P. (2000): Genetic erosion, inbreeding, and reduced fitness in fragmented populations of the endangered tetraploid pea *Swainsona recta*. *Biological Conservation* **93**: 177–186.

Carlson, A. & Hartman, G. (2001): Tropical forest fragmentation and nest predation—an experimental study in an Eastern Arc montane forest, Tanzania. *Biological Conservation* **10**: 1077–1085.

Carroll, C., McRae, B. H. & Brookes, A. (2012): Use of linkage mapping and centrality analysis across habitat gradients to conserve connectivity of Gray wolf populations in Western North America. *Conservation Biology* **26**: 78–87.

Clark, R. W., Brown, W. S., Stechert, R. & Zamudio, K. R. (2008): Integrating individual behaviour and landscape genetics: the population structure of timber rattlesnake hibernacula. *Molecular Ecology* **17**: 719–730.

Clevenger, A. P., Wierzchowski, J., Chruszcz, B. & Gunson, K. (2002): GIS-generated, expert-based models for identifying wildlife habitat linkages and planning mitigation passages. *Conservation Biology* **16**: 503–514.

Clobert, J., Ims, R. A. & Rousset, F. (2004): Causes, mechanisms and consequences of dispersal. In *Ecology, genetics and evolution of metapopulations*: 307–335. Hanski, I. & Gaggiotti O. E. (Eds). Amsterdam: Academic.

Cobben, M. M. P., Verboom, J., Opdam, P. F. M., Hoekstra, R. F., Jochem R. & Smulders, M. J. M. (2012): Wrong place, wrong time: climate change-induced range shift across fragmented habitat causes maladaptation and declined population size in a modelled bird species. *Global Change Biology* **18**: 2419–2428.

Compton, B., McGarigal, K., Cushman, S. A. & Gamble, L. (2007): A resistant kernel model of connectivity for vernal pool breeding amphibians. *Conservation Biology* **21**: 788–799.

Craigie, I. D., Baillie, J. E. M., Balmford, A., Carbone, C., Collen, B., Green, R. E. & Hutton, J. M. (2010): Large mammal population declines in Africa's protected areas. *Biological Conservation* **143**: 2221–2228.

Cushman, S. A. & Lewis, J. (2010): Movement behavior explains genetic differentiation in American black bear. *Landscape Ecology* **25**: 1613–1625.

Cushman, S. A., McRae, B., Adriaensen, F., Beier, P., Shirley, M. & Zeller K. (2013): Biological corridors and connectivity. In *Key topics in conservation biology 2*: 384–404. MacDonald, D. & Willis, K. (Eds.). Oxford: Wiley-Blackwell.

Dellinger, J. A., Proctor, C., Steury, T. D., Kelly, M. J. & Vaughan, M. R. (2013): Habitat selection of a large carnivore, the red wolf, in a human-altered landscape. *Biological Conservation* **157**: 324–330.

Epps, C. W., Palsboll, P. J., Wehausen, J. D., Roderick, G. K., Ramey, R. R. I. I. & McCullough, D. R. (2005): Highways block gene flow and cause a rapid decline in genetic diversity of desert bighorn sheep. *Ecology Letters* **8**: 1029–1038.

Erfanian, B., Mirkarimi, S. H., Mahini, A. S. & Rezaei, H. R. (2013): A presence-only habitat suitability model for Persian leopard *Panthera pardus saxicolor* in Golestan National Park, Iran. *Wildlife Biology* **19**: 170–178.

Ervin, J. (2003): Protected area assessments in perspective. *BioScience* **53**: 819–822.

Fahrig, L. (2003): Effects of habitat fragmentation on biodiversity. *Annual Review of Ecology and Systematics* **34**: 487–515.

Finnegan, L. A., Wilson, P. J., Price, G. N., Lowe, S. J., Patterson, B. R., Fortin, M. J. & Murray, D. L. (2012): The complimentary role of genetic and ecological data in understanding population structure: a case study using moose (*Alces alces*). *European Journal of Wildlife Research* **58**: 415–423.

Flagstad, O., Roed, K., Stacy, J. E. & Jakobsen, K. S. (1999): A reliable non-invasive genotyping based on excremental PCR of nuclear DNA purified with a magnetic bead protocol. *Molecular Ecology* **8**: 879–883.

Goldstein, D. B. & Schlotterer, C. (1999): *Microsatellites: evolution and applications*. Oxford: Oxford University Press.

Goudet, J. (2002): *FSTAT: a program to estimate and test gene diversities and fixation indices*. Version 2932.

Greenwood, P. J. (1980): Mating systems, philopatry and dispersal in birds and mammals. *Animal Behaviour* **28**: 1140–1162.

Guisan, A. & Zimmermann, N. E. (2000): Predictive habitat distribution models in ecology. *Ecological Modelling* **135**: 147–186.

Harris, G., Thirgood, S., Hopcraft, J., Cromsigt, J. & Berger, J. (2009): Global decline in aggregated migrations of large terrestrial mammals. *Endangered Species Research* **7**: 55–76.

Hawkes, C. (2009): Linking movement behaviour, dispersal and population processes: is individual variation a key? *Journal of Animal Ecology* **78**: 894–906.

Hedmark, E., Flagstad, Ø., Segerstro, P., Persson, J., Landa, A. & Ellegren, H. (2004): DNA-based individual and sex identification from wolverine (*Gulo gulo*) faeces and urine. *Conservation Genetics* **5**: 405–410.

Hedrick, P. W. & Kalinowski, S. T. (2000): Inbreeding depression in conservation biology. *Annual Review of Ecology and Systematics* **31**: 139–162.

Heller, N. E. & Zavaleta, E. A. (2009): Biodiversity management in the face of climate change: a review of 22 years of recommendations. *Biological Conservation* **142**: 14–32.

Henle K., Davies, K. F., Kleyer, M., Margules, C. & Settele, J. (2004): Predictors of species sensitivity to fragmentation. *Biodiversity and Conservation* **13**: 207–251.

Hill, J. K., Griffiths, H. M. & Thomas, C. D. (2011): Climate change and evolutionary adaptations at species' range margins. *Annual Review of Entomology* **56**: 143–159.

Hillman, J. C. (1993): *Ethiopia: compendium of wildlife conservation information*. Vol. 1: Wildlife conservation in Ethiopia. Addis Ababa: Ethiopian Wildlife Conservation Organization.

Hirzel, A. H. & Lay, G. L. (2008): Habitat suitability modelling and niche theory. *Journal of Applied Ecology* **45**: 1372–1381.

Holderegger, R. & Wagner, H. H. (2008): Landscape genetics. *Bioscience* **58**: 199–207.

IUCN. (2015): Red List, www.iucnredlist.org.

Ito, T. Y., Lhagvasuren, B., Tsunekawa, A., Shinoda, M, Takatsuki S., *et al.* (2013): Fragmentation of the Habitat of Wild Ungulates by Anthropogenic Barriers in Mongolia. *PLoS ONE* **8**: e56995.

Jenkins, C. N. & Joppa, N. (2009): Expansion of the global terrestrial protected area system. *Biological Conservation* **142**: 2166–2174.

Johannesen, E., Aars, J., Andreassen, H. P. & Ims, R. A. (2003): A demographic analysis of vole population responses to fragmentation and destruction of habitat. *Population Ecology* **45**: 47–58.

Johnson, C. J. & Gillingham, M. P. (2004): Mapping uncertainty: sensitivity of wildlife habitat ratings to expert opinion. *Journal of Applied Ecology* **41**: 1032–1041.

Johnson, C. J. & Gillingham, M. P. (2008): Sensitivity of species-distribution models to error, bias, and model design: an application to resource selection functions for woodland caribou. *Ecological Modelling* **213**: 143–155.

Largen, M. J. (2001): Catalogue of the amphibians of Ethiopia, including a key for their identification. *Tropical Zoology* **14**: 307–402.

Laurance, W. F., Nascimento, H. E. M., Laurance, S. G., Andrade, A., Ewers R. M., Harms, K. E., Luizao, R. C. C. & Ribeiro, J. E. (2007): Habitat fragmentation, variable edge effects and the landscape-divergence hypothesis. *PLoS ONE* **2**: e1017.

Li, H., Li, D., Li, T., Qiao, Q., Yang, J. & Zhang, H. (2010): Application of least-cost path model to identify a giant panda dispersal corridor network after the Wenchuan earthquake case study of Wolong Nature Reserve in China. *Ecological Modelling* **221**: 944–952.

Lidicker W. Z. & Stenseth N. C. (1992): To disperse or not to disperse: who does it and why? In *Animal dispersal: small mammals as a model: 21–36*. Stenseth N. C. & Lidicker W. Z. (Eds). London: Chapman and Hall.

Lion, S., & van Baalen, M. (2008): Self-structuring in spatial evolutionary ecology. *Ecology Letters* **11**: 227–295.

Manel, S. & Holderegger, R. (2013): Ten years of landscape genetics. *Trends in Ecology and Evolution* **28**: 614–621.

Manel, S., Schwartz, M. K., Luikart, G. & Taberlet, P. (2003): Landscape genetics: combining landscape ecology and population genetics. *Trends in Ecology and Evolution* **18**: 189–197.

Manly, B. F. J. (2002): *Resource selection by animals: statistical design and analysis for field studies*. Boston: Kluwer Academic Publishers.

Mantel, N. A. (1967): The detection of disease clustering and a generalized regression approach. *Cancer Research* **27**: 209–220.

Maudet, C., Luikart, G., Dubray D., Hardenberg A. V. & Taberlet, P. (2004): Low genotyping error rates in wild ungulate faeces sampled in winter. *Molecular Ecology Notes* **4**: 772–775.

McLachlan, J. S., Hellmann, J. J. & Schwartz, M. W. (2007): A framework for debate of assisted migration in an era of climate change. *Conservation Biology* **21**: 297–302.

McRae, B. H., Dickson, B. G., Keitt, T. H. & Shah, V. B. (2008): Using circuit theory to model connectivity in ecology and conservation. *Ecology* **10**: 2712–2724.

Mekonnen, A., Bekele, A., Hemson, G., Teshome, E. & Atickem, A. (2010): Population size and habitat preference of the Vulnerable Bale monkey *Chlorocebus djamdjamensis* in Odobullu Forest and its distribution across the Bale Mountains Ethiopia. *Oryx* **44**: 558–563.

Miller, C. S., Hebblewhite, M., Goodrich, J. M. & Miquell, D. G. (2010): Review of research methodologies for tigers: telemetry. *Integrative Zoology* **5**: 378–389.

Mora, C. & Sale, P. F. (2011): Ongoing global biodiversity loss and the need to move beyond protected areas: a review of the technical and practical shortcomings of protected areas on land and sea. *Marine Ecology Progress Series* **434**: 251.

Nelson, F. (2008): Are large mammal declines in Africa inevitable? *African Journal of Ecology* **46**: 3–4.

Newmark, W. D. (1996): Insularization of Tanzanian parks and the local extinction of large mammals. *Conservation Biology* **10**: 1549–1556.

Newmark, W. D. (2008): Isolation of African protected areas. *Frontiers in Ecology and the Environment* **6**: 321–328.

Nichol, J. & Wong, M. S. (2005): Satellite remote sensing for detailed landslide inventories using change detection and image fusion. *International Journal of Remote Sensing* **26**: 1913–1926.

Ogutu, J., Owen-Smith, N., Piepho, H., Said, M., Kifugo, S., Reid, R., Gichohi, H., Kahumbu, P. & Andanje, S. (2013): Changing wildlife populations in Nairobi National Park and adjoining Athi-Kaputiei Plains: collapse of the migratory wildebeest. *Open Conservation Biology Journal* **7**: 11–26.

Ottichilo, W. K., Leeuw, J. D., Skidmore, A. K., Prins, H. H. T. & Said M. Y. (2000): Population trends of large non-migratory wild herbivores and livestock in the Masai Mara ecosystem, Kenya, between1977 and1997. *African Journal of Ecology* **38**: 202–216.

Painter, J. N., Crozier, R. H., Poiani, A., Robertson, R. J. & Clarke, M. F. (2000): Complex social organization reflects genetic structure and relatedness in the cooperatively breeding bell miner, *Manorina melanophrys*. *Molecular Ecology* **9**: 1339–1347.

Pauls, S. U. & Nowak, C. (2011): Cryptic biodiversity loss linked to global climate change. *Nature Climate Change* **1**: 313–317.

Pelletier, D., Clark, M., Anderson, M. G., Rayfield, B., Wulder, M. A. & Cardille, J. A. (2014): Applying circuit theory for corridor expansion and management at regional scales: tiling, pinch points, and omnidirectional connectivity. *PLoS ONE* **9**: e84135.

Pettorelli, N. (2012): Climate change as a main driver of ecological research. *Journal of Applied Ecology* **49**: 542–545.

Pettorelli, N. (2013): *The normalized difference vegetation index*. Oxford: Oxford University Press.

Phillips, S. J., Williams, P., Midgley, G. & Archer, A. (2008): Optimizing dispersal corridors for the Cape Proteaceae using network flow. *Ecological Applications* **18**: 1200–1211.

Pressey, R. L. & Bottrill, M. C. (2009): Approaches to landscape and seascape-scale conservation planning: convergence, contrasts, and challenges. *Oryx* **43**: 464–475.

Ralls, K., Brugger, K. & Ballou, J. (1979): Inbreeding and juvenile mortality in small populations of ungulates. *Science* **206**: 1101–1103.

Reed, D. H. (2004): Extinction risk in fragmented habitats. *Animal Conservation* **7**: 181–191.

Reed, D. H. & Frankham, R. (2003): Population fitness is correlated with genetic diversity. *Conservation Biology* **17**: 230–237.

Reed, D. H., Briscoe, D. A. & Frankham, R. (2002): Inbreeding and extinction: effects of environmental stress and lineages. *Conservation Genetics* **3**: 301–307.

Ries, L., Fletcher, R. J., Battin, J., & Sisk, T. D. (2004): Ecological responses to habitat edges: mechanisms, models, and variability explained. *Annual Review of Ecology and Systematics* **35**: 491–522.

Riley, S. P. D., Pollinger, J. P., Sauvajot, R. M., *et al.* (2006) A southern California freeway is a physical and social barrier to gene flow in carnivores. *Molecular Ecology* **15**: 1733–1741.

Rodgers, T. W. & Janečka, J. E. (2013): Applications and techniques for non-invasive faecal genetics research in felid conservation. *European Journal of Wildlife Research* **59**: 1–16.

Rodrigues, A. S. L., Andelman, S. J., Bakarr, M. I., *et al.*, (2004): Effectiveness of the global protected area network in representing species diversity. *Nature* **428**: 640–643.

Rosenberg, D. K., Noon, B. R. & Meslow, E. C. (1997): Biological corridors: form, function, and efficacy. *BioScience* **47**: 677–687.

Rouget, M., Cowling, R. M., Lombard, A. T., Knight, A. T., & Kerley, G. I. H. (2006): Designing large-scale conservation corridors for pattern and process. *Conservation Biology* **20**: 549–561.

Rousset, F. (2008): GENEPOP' 007: a complete re-implementation of the GENEPOP software for Windows and Linux. *Molecular Ecology Resources* **8**: 103–106.

Rubidge, E. M., Patton, J. L., Lim, M., Burton, A. C., Brashares, J. S. & Moritz, C. (2012): Climate-induced range contraction drives genetic erosion in an alpine mammal. *Nature Climate Change* **2**: 285–288.

Ruiz-Lopez, M. J., Ganan, N., Godoy, J. A, Olmo, A. D., Garde, J., Espeso, G., Vargas, A., Martinez, F., Roldan,E. R. S. & Gomendio, M. (2012): Heterozygosity-fitness correlations and inbreeding depression in two critically endangered mammals. *Conservation Biology* **26**: 1121–1129.

Saccheri, I., Kuussaari, M., Kankare, M., Vikman, P., Fortelius, W. & Hanski, I. (1998): Inbreeding and extinction in a butterfly metapopulation. *Nature* **392**: 491–494.

Sacks, B. N., Brown, S. K. & Ernest, H. B. (2004): Population structure of California coyotes corresponds to habitat specific breaks and illuminates species history. *Molecular Ecology* **13**: 1265–1275.

Sacks, B. N., Mitchell, B. R., Williams, C. L. & Ernest, H. B. (2005): Coyote movements and social structure along a cryptic population genetic subdivision. *Molecular Ecology* **14**: 1241–1249.

Scholte, P. (2011): Towards understanding large mammal population declines in Africa's protected areas: a West-Central African perspective. *Tropical Conservation Science* **4**: 1–11.

Schwartz, M. K., Copeland, J. P., Anderson, N. J., Squires, J. R., Inman, R. M., McKelvey, K. S., Pilgrim, K. L., Waits, L. P. & Cushman, S. A. (2009): Wolverine gene flow across a narrow climatic niche. *Ecology* **90**: 3222–3232.

Scott, D., Malcom, J. & Lemieux, C. J. (2002): Climate change and biome representation in Canada's National Park system: implications for system planning and park mandates. *Global Ecology and Biogeography* **11**: 475–484.

Segelbacher, G., Cushman, S. A., Epperson, B. K., Fortin, M. J., Francois, O., Hardy, O., Holderegger, R., Taberlet, P., Waits, L. P. & Manel, S. (2010): Applications of landscape genetics in conservation biology: concepts and challenges. *Conservation Genetics* **11**: 375–385.

Seoane, J., Bustamante, J. & Diaz-Delgado, R. (2005): Effect of expert opinion on the predictive ability of environmental models of bird distribution. *Conservation Biology* **19**: 512–522.

Sharma, S., Dutta, T., Maldonado, J. E., Wood, T. C., Panwar, H. S. & Seidensticker, J. (2013): Forest corridors maintain historical gene flow in a tiger metapopulation in the highlands of central India. *Proceedings of the Royal Society B:* **280**: 1506.

Sillero-Zubiri, C. (2008): *Tragelaphus buxtoni*. In *IUCN 2013. IUCN Red List of Threatened Species*. Version 2013.1. www.iucnredlist.org (as as 26 September 2013).

Slate, J., Kruuk, L. E. B., Marshall, T. C., Pemberton, J. M. & Clutton-Brock, T. H. (2000): Inbreeding depression influences lifetime breeding success in a wild population of red deer (*Cervus elaphus*). *Proceedings of the Royal Society of London, Series B* **267**: 1657–1662.

Sommer, S., McDevitt, A. D. & Balkenho, N. (2013): Landscape genetic approaches in conservation biology and management. *Conservation Genetics* **14**: 249–251.

Sork, V. L. & Waits, L. (2010): Contributions of landscape genetics – approaches, insights, and future potential. *Molecular Ecology* **19**: 3489–3495.

Spear, S. F., Balkenhol, N., Fortin, M. J., McRae, B. H. & Scribner, K. (2010): Use of resistance surfaces for landscape genetic studies: considerations for parameterization and analysis. *Molecular Ecology* **19**: 3576–3591.

Spielman, D., Brook, B. W. & Frankham, R. (2004): Most species are not driven to extinction before genetic factors impact them. *Proceedings of the National Academy of Sciences of USA* **101**: 15261-15264.

Stephens, P. A., Sillero-Zubiri, C. & Leader-Williams, N. (2001): Impact of livestock and settlement on the large mammalian wildlife of the Bale Mountains National Park, Southern Ethiopia. *Biological Conservation* **100**: 307–322.

Storfer, A., Murphy, M., Spear, S., Holderegger, R. & Waits, L. (2010): Landscape genetics: where are we now? *Molecular Ecology* **19**: 3496–3514.

Taylor, P., Fahrig, L. & With, K. A. (2006): Landscape connectivity: back to the basics. In *Connectivity Conservation*: 29–33. Crooks, K. & Sanjayan, M. A. (Eds). Cambridge: Cambridge University Press.

Taylor, P. D., Fahrig, L., Henein, K. & Merriam, G. (1993): Connectivity is a vital element of landscape structure. *Oikos* **68**: 571–573.

Temple, H. J., Hoffman, J. I. & Amos, W. (2006): Dispersal, philopatry and intergroup relatedness: fine-scale genetic structure in the white breasted thrasher, *Ramphocinclus brachyurus*. *Molecular Ecology* **15**: 3449–3458.

Templeton, A. R., Robertson, R. J., Brisson, J. & Strasberg, J. (2001): Disrupting evolutionary processes: the effect of habitat fragmentation on collared lizards in the Missouri Ozarks. *Proceedings of the National Academy of Sciences of the United States of America* **98**: 5426–5432.

Thevenon, S. & Couvet, D. (2002): The impact of inbreeding depression on population survival depending on demographic parameters. *Animal Conservation* **5**: 53–60.

Thirgood, S., Mosser, A., Tham, S., Hopcraft, G., *et al.* (2004): Can parks protect migratory ungulates? The case of the Serengeti wildebeest. *Animal Conservation* **7**: 113–120.

Walling, C. A., Nussey, D. H., Morris, A., Clutton-Brock, T. H., Kruuk, L. E. B. & Pemberton, J. M. (2011): Inbreeding depression in red deer calves. *BMC Evolutionary Biology* **11**: 318.

Wan, Q. H., Wu, H., Fujihara, T. & Fang S. G. (2004): Which genetic marker for which conservation genetics issue? *Electrophoresis* **25**: 2165–2176.

Western, D., Russell, S. & Cuthill, I. (2009): The status of wildlife in protected areas compared to non-protected areas of Kenya. *PLoS ONE* **4**: e6140.

Wilcove, D. S. (2007): No way home: the decline of the world's great animal migrations. Washington, D. C.: Shearwater Books.

Wilcove, D. S. & Wikelski, M. (2008): Going, going, gone: is animal migration disappearing? *PLoS Biol* **6**: e188.

Willis, J. H. (1993): Effects of different levels of inbreeding on fitness components in *Mimulus guttatus*. *Evolution* **47**: 864–876.

Woodroffe, R. & Ginsberg, J. R. (1998): Edge effects and the extinction of populations inside protected areas. *Science* **280**: 2126–2128.

Yalden, D. W. & Largen, M. J. (1992): The endemic mammals of Ethiopia. *Mammal Review* **22**: 115–150.

Yamada, K., Elith, J., McCarthy, M. & Zerger, A. (2003): Eliciting and integrating expert knowledge for wildlife habitat modelling. *Ecological Modelling* **165**: 251–264.

Yang, J., Jiang, Z., Zeng, Y., Turghan, M., Fang, H. & Li, C. (2011): Effect of Anthropogenic landscape features on population genetic differentiation of Przewalski's gazelle: main role of human settlement. *PLoS ONE* **6**: e20144.

Zeller, K. A., McGarigal, K. & Whitely, A. R. (2012): Estimating landscape resistance to movement: a review. *Landscape Ecology* **27**: 777–797.

11

The Use of Camera-Traps to Monitor Forest Antelope Species

Rajan Amin[1], Andrew E. Bowkett[2], and Tim Wacher[1]

[1]*Conservation Programmes, Zoological Society of London, United Kingdom*

[2]*Field Conservation and Research Department, Whitley Wildlife Conservation Trust, Paignton Zoo, United Kingdom*

Introduction

Antelopes and other artiodactyl species constitute a significant component of forest and woodland ecosystems both in terms of biomass (White, 1994) and ecological services (Feer, 1995). Many species are increasingly threatened by habitat loss and hunting for bushmeat (East, 1999). Forest antelopes are primary targets for the trade in bushmeat (Wilkie & Carpenter, 1999; Fa *et al.*, 2005) and have undergone major local and regional declines as a result (e.g., van Vliet *et al.*, 2007). Therefore, understanding the ecology of forest antelopes and monitoring their populations are critical conservation needs. However, forest antelopes are difficult to monitor using traditional methods for forest mammals as many species are solitary and inactive during the day, while all species are shy and spend long periods concealed in dense vegetation. In addressing these challenges, biologists have started to employ modern population sampling methods.

In this chapter, we provide a brief summary of conventional methods, and their limitations, for studying forest antelopes. We then discuss the use of remote photographic sampling techniques that are increasingly being employed in studies of forest antelopes. We conclude with two case studies involving camera-traps. The first case study involves the critically endangered Aders' duiker (*Cephalophus adersi*) where camera-trapping has led directly to further research and advocacy for improved conservation of their habitats while also providing a tool for long-term population monitoring. The second

Antelope Conservation: From Diagnosis to Action, First Edition. Edited by Jakob Bro-Jørgensen and David P. Mallon.

case study is in a montane forest ear-marked for mining development in West Africa. It uses camera-traps to establish a baseline on medium-to-large mammal diversity, of which antelopes form a significant component.

Sampling forest antelopes using conventional methods

Standard methods for monitoring antelope populations in savanna habitats, such as aerial or road counts, are clearly not applicable in dense forests. Much useful information on forest antelope taxonomy, physiology, and feeding ecology has been obtained through examination of hunted or culled specimens (e.g., Hofman & Roth, 2003), and in cases where the number of animals removed by hunting is known, removal methods to monitor population size can be used, such as catch-per-unit-effort (Lancia & Bishir, 1996). However, such information requires close cooperation with hunters, which may not be feasible or legal, and such sampling is not suitable for threatened species. Behavioural sampling of forest antelope under natural conditions is also difficult. Whereas captive studies have helped provide insights into behaviour (Bowman & Plowman, 2002) and diet (Plowman, 2002), such studies are limited by the constraints imposed by captivity and few forest antelope species are currently kept in viable numbers for long-term management (Rohr & Pfefferkorn, 2003).

Accurate species identification is a critical requirement for surveys, and yet even this can be difficult when sightings are rare, or typically brief, as with forest antelope. Field signs, such as spoor and faecal deposits, can provide an indirect measure of presence or relative abundance, but identification to species still needs to be reliable. Dung counts are the most commonly employed indirect measure of forest antelope abundance (e.g., Koster & Hart, 1988; Plumptre & Harris, 1995; Nielsen, 2006), but identification of dung to species has proven to be inaccurate when tested genetically (van Vliet *et al.*, 2008; Bowkett *et al.*, 2009; Faria *et al.*, 2011), even when based on statistical analysis of pellet measurements (Bowkett *et al.*, 2013).

Extrapolating absolute abundance estimates from records/counts of animals or field signs requires a statistical framework to incorporate the probability of detection (Mills, 2013). Such frameworks typically require animals to be identifiable as individuals, for instance, capture-recapture methods (Nichols & Dickman, 1996), or be reliably detected when close to a sampling point, for instance, distance-sampling (Buckland *et al.*, 2001). As noted above, neither of these conditions is easily met for forest antelope. Documented approaches to estimating forest antelope abundance include line transect surveys (Struhsaker, 1997; Rovero & Marshall, 2004; Viquerat *et al.*, 2012), drive counts (Bowland, 1990; Noss, 1999), territory mapping (Bowland & Perrin, 1994), and extrapolation from dung counts (Schmidt, 1983; Plumptre & Harris, 1995).

All these survey methods have limitations for long-term monitoring. Line transect methods require sufficient sightings for accurate density estimates and are difficult to conduct at night when many forest antelope are most active (although see Waltert *et al.*, 2006). Forest antelope are also likely to violate the distance-sampling assumption that animals located at zero distance from the transect line are always detected before being disturbed (Buckland *et al.*, 2001). Drive counts risk injury to the animals and may be expensive to conduct, as they require multiple skilled people to mind the nets. Territory mapping usually involves radio-collaring and is therefore more applicable to research than monitoring. Extrapolation from dung counts requires accurate faecal deposition and decomposition rates amongst other assumptions (Lunt *et al.*, 2007).

An innovative potential survey tool is the use of fake alarm calls to attract duikers, as used by traditional hunters in some areas (van Vliet *et al.*, 2009). Validation of this method requires data on how far individuals travel to investigate such calls and how consistent is their response, as has been investigated in playback survey methods for birds (e.g., Ratcliffe *et al.*, 1998).

The modern development of non-invasive genetics, the analysis of DNA from materials such as saliva or dung (Taberlet *et al.*, 1999), has numerous potential applications to forest antelope monitoring (see also Chapter 9). Amplification of species-specific mitochondrial DNA fragments allows species identification from dung or bushmeat samples (Eaton *et al.*, 2010; Ntie *et al.*, 2010a). Such an approach is analogous to DNA barcoding (Hebert *et al.*, 2003), although the standard universal barcoding gene (COX1) does not differentiate some duikers from their sister species (Johnston *et al.*, 2011), so an alternative genetic marker has to be used (e.g., Ntie *et al.*, 2010a). Genetic identification of dung allows accurate distribution mapping and validation of field identification used for abundance estimates (van Vliet *et al.*, 2008; Bowkett *et al.*, 2013).

Identification of individual animals from faecal DNA is possible using nuclear microsatellite markers (Ntie *et al.*, 2010b; Akomo-Okoue *et al.*, 2013). For forest antelope, this would allow abundance to be estimated by territory mapping or through a capture-recapture framework wherein samples from the same individual collected in different sessions are treated as recaptures (e.g., Zhan *et al.*, 2006). Variable genetic markers can also be used to monitor effective population size (Schwartz *et al.*, 1998) as well as to study many aspects of forest antelope biology including genetic diversity, parentage, and pedigree (Beja-Pereira *et al.*, 2009).

Despite their many advantages, very few wildlife authorities or land managers have the resources to employ non-invasive genetics as part of routine monitoring programs. Even though the cost of genetic studies continues to fall overall (Hall, 2013) and there is increasing expertise in conservation genetics in Africa (Anthony *et al.*, 2012), conservation managers are often vastly under-resourced (Bruner *et al.*, 2004). Therefore, while non-invasive genetic research on forest antelope will hopefully expand rapidly to provide new records and answer key questions, regular genetic monitoring does not appear feasible in the short term.

Sampling forest antelopes using camera-traps

The use of camera-trapping as a survey tool for medium-to-large terrestrial mammals has become increasingly common over recent years (Ahumada *et al.*, 2011; O'Connell *et al.*, 2011) and is a particularly suitable technique for longer-term monitoring in forest habitats (Wacher & Amin, 2015; Gompper *et al.*, 2006; Lyra-Jorge *et al.*, 2008; Roberts, 2011; Silveira *et al.*, 2003). The survey method consists of remotely stationed heat-and-motion sensitive cameras that take photographs or videos of passing animals. Digital cameras operating 24-hours a day are the standard (O'Connell *et al.*, 2011). At night, infrared flash photography that is not visible to animals can be used to minimise disturbance to the animals. The strengths and also limitations of camera-traps for sampling forest antelopes are summarised in Table 1.

Table 1 **Strengths and limitations of camera-traps for sampling forest antelopes.**

Strengths	Limitations
Cost-effective in consideration of effort vs. data generation	Initial high cost of purchasing equipment (although this cost is offset over time)
Operates day and night in range of habitats for length of battery-life	Possible theft from the field (although can come in camouflage packaging and be secured with a lock) or damage by animals
Detects species that are otherwise hard to monitor	Can be time consuming to identify some species and individuals, depending on quality of photograph, and to process images
Non-invasive monitoring method with low bias	Vulnerable to false triggers, particularly in response to vegetation movement and plant growth in front of camera.
Digital photographs provide permanent record, date and time stamps for easy archiving and analysis	Vulnerable to equipment failure, though quality and reliability are improving.
Standardisation allows comparison across survey sites and over time	
Photographs have value for educational and promotional purposes	

Survey designs

Standardized methodologies for systematic camera-trap-based monitoring are now available (Tobler *et al.*, 2008; Ahumada *et al.*, 2011; Amin & Wacher, 2014). For monitoring forest antelopes, we focus on the following survey objectives discussing aspects relevant to camera survey design: namely, camera placement, number of camera sites, and duration of survey. Camera features necessary for obtaining reliable data are also discussed.

Establishing forest antelope species presence. For assessing areas for the presence of forest antelope species, one needs to maximise the chances of capturing clear images of the target species. Camera-traps can be placed opportunistically overlooking natural game trails or natural attractants covering all relevant habitats. Cameras are typically set at a height of 30–45 cm positioned at right angles to game trails at a distance of c. 4–8 m (i.e., within the sensor and flash operation range but not too close to miss part of the animal) with the aim of obtaining full body lateral images for accurate identification. Cameras are programmed to take multiple pictures per trigger with fast trigger speed and high sensitivity. Cameras with infrared flash activated in low light conditions result in monochrome images, which in some cases can be difficult to distinguish species (Figure 1). Behavioural features such as differences in tail movement (e.g., flicked laterally in suni *Nesotragus moschatus*, vertically in blue duiker *Philantomba monticola*; Foley, 2008) can be useful in such cases. Cameras with white flash take colour images, with Xenon flash providing the clearest images for identification. However, white flash can disturb forest antelopes and other species and modify their behaviour.

Figure 1 **Pictures of the white rump-band and leg-spots (left image), key distinguishing features, of Aders' duiker were sometimes partially obscured in infra-red illuminated black and white images (right image).**

Cameras can be deployed as long as needed to ensure survey effort is sufficient for confirming presence/absence of target species.

Estimating population density and indices of abundance. The appropriate methods for density and abundance estimation differ, depending on whether or not animals can be individually identified. Where individual identification is possible, capture-recapture or sight-resight approaches developed for closed populations (see Foster & Harmsen, 2011, for a general discussion) provide a more robust approach to density estimation. However, this approach is seldom appropriate for forest antelope species because reliable individual recognition is difficult. If capture-recapture methods are employed, then it is important to ensure that all individuals in the study population have an equal chance of 'capture'. Capture probabilities greater than or equal to 0.1 are needed for reliable population estimates (White *et al.*, 1982). Thus, at least 2–3 cameras need to be placed per home range of target forest antelope species (Foster & Harmsen, 2011). The area surveyed should be sufficiently large to sample a good proportion of the population; sample size of >20 individuals is recommended (White *et al.*, 1982). Because natural markings on animals are asymmetric, unambiguous individual identifications necessitate photographs of both flanks requiring the deployment of two or more cameras at optimal sites to maximise rate of identification. High-quality photos are necessary, especially for species which are not easily distinguishable. Sampling should be carried out over a sufficiently short time frame during which population closure violations are not expected to occur. Spatially explicit capture-recapture modelling is now the method of choice for density estimates (Gopalaswany *et al.*, 2012).

Methods for obtaining population estimates for unmarked animals from camera-trap data are being developed (Royle, 2004; Rowcliffe *et al.*, 2008). In some situations, under standardised conditions, it may be reasonable to use the mean number of independent photographic events per trap day (event rate) as a relative abundance index for comparing species status at different time periods or locations. But such inferences should only be drawn if it is reasonable to assume that the detection probabilities are similar for the surveys being compared.

Occupancy is defined as the proportion of area, patches, or sites occupied by a species (Mackenzie *et al.*, 2006). It can be used as a surrogate for forest antelope species abundance as they have relatively small, well-defined home ranges. Documented forest antelope home range sizes vary between suni (0.5–4 ha; Kingdon & Hoffman, 2013) and white-bellied duiker (*Cephalophus leucogaster*) (63 ha; Hart, 2013) and with camera-traps sufficiently spaced one can assume that each individual can only appear in one camera-trap.

Occupancy modelling is well suited to camera-trap survey data because detection/non-detection data can be collected more efficiently over a large number of sampling occasions than through other methods such as sign or direct sighting surveys. Camera-trap data are divided into equal time intervals (single or multiple-day occasions) to construct a species detection/non-detection history at each camera survey site. These data are then used to obtain species occupancy and detection probability estimates in Presence software (MacKenzie *et al.*, 2006) or in *R* statistical software package 'unmarked' (Fiske & Chandler 2011; R Development Core Team, 2015). Cameras should be set out systematically in a grid, and the sampled area should cover a representative portion of the population across all relevant habitats. Stratified sampling design should be used to ensure sampling effort is proportional to habitat extent. Camera placement at each grid centroid should be random (i.e., it should not favour locations such as feeding sites where animal presence may be higher than average).

Estimation of occupancy requires that the survey area be closed to changes in occupancy over the sampling period (i.e., animals do not move into the area and become established or move permanently out), so the survey time period should be sufficiently short to reasonably meet this assumption.

The survey effort largely depends on the detection probability and occupancy of the species. Generally, for rare species with low detection probability effort should be devoted to both surveying more sites and more repeated surveys, while for more common species with low detection probability more effort should be devoted to repeated surveys and for species with high detection probability and low occupancy more effort should be devoted to surveying more sites (MacKenzie & Royle, 2005). Studies have shown that occupancy estimates are not accurate for species with estimated detection probability less than 0.1 or recorded in less than 15–20% of the cameras (Rovero *et al.*, 2014).

Camera site and occasion level variables that influence occupancy and detectability can be incorporated into occupancy modelling. Camera site variables such as habitat type can be helpful in investigating their influence on occupancy (e.g., studying habitat preference). Sample occasions can be used to account for variation in detection probability due to seasonal changes etc.

Significant differences in species occupancy over time or between areas surveyed using standardized protocols can be statistically tested (Amin *et al.*, 2014). If the same area is surveyed over time, then multi-season (robust design) occupancy models can also be used to estimate local extinction and colonization (MacKenzie *et al.*, 2006). Multi-species occupancy models can also be used to assess species co-occurrence or avoidance.

Studying forest antelope behaviour and activity patterns. Camera-traps offer a number of advantages compared to more traditional techniques, such as radio telemetry and observation methods, which are not appropriate in many cases for studying forest antelope behaviour due to the nature of their habitat. Camera-traps are particularly effective for studying circadian activity patterns, since they sample continuously over 24-hour cycles at a population level, dealing with multiple species simultaneously in the same survey. Circadian activity patterns are constructed from species time-of-detection data. With multiple or repeated surveys, circadian activity patterns can be compared between different locations or times: for example, to detect environment and disturbance constraints on activity and behaviour. In addition, evaluation of niche partitioning, overlap and competition within similar species pairs or guilds (as frequently occurs with small forest antelopes) can be undertaken (Amin *et al.*, 2014). Information on specific behaviour such as scent marking frequency, reproductive biology, and health can also be obtained.

Traditional statistical methods such as chi-square tests have been used to examine differences in behaviour and activity (Amin *et al.*, 2014). Also, new analytical methods are being developed to quantify levels of activity patterns from camera-trap data (Rowcliffe *et al.*, 2014).

Information on forest antelope activity and behaviour is frequently generated as part of a population status study, and systematic sampling with camera-traps placed in grids is often a suitable design for multiple surveys objectives.

Environmental and social factors to consider

In addition to the ecology and behaviour of antelope species, several environmental factors affect the deployment of camera-traps in forest systems. Rain and humidity can affect functioning of cameras and restrict deployment and servicing of cameras in certain seasons. In some cases, cameras are at risk from animal damage. Insect intrusion can be a problem. Elephants can damage cameras. More often, human vandalism and theft can be an issue. Cameras need to be robust, and camera protective cases, use of dessicant sacs, security locking devices, cryptic positioning, and camouflaging of cameras are possible countermeasures against these problems. Agreement and support from local communities is often required to conduct surveys; in some circumstances, opportunities for direct involvement of local communities may arise.

Photographic sampling of forest antelopes is often required over difficult terrain and large areas. Deploying cameras can demand travel to remote locations and often involve setting up, servicing, and retrieving cameras on foot in difficult conditions. Appropriately trained local staff are therefore essential.

Data management and image processing

Several software packages are available to extract image metadata (e.g., ExifTool by Phil Harvey). Key information is date and time of photo taken and image filename. Additional data such as camera serial number, trigger image sequence, temperature and moon phase, and any image labelling information added by the user may also be available depending on the camera-trap model in use. All photographs are then classified to species. Identification of certain antelope species can sometimes be difficult in some images, particularly in poor angles of view or where a small part of the animal has been captured, and sometimes due to the effects of infrared illumination. Where species identity remains uncertain, genus or family information can be recorded where known, and care must be taken to exclude or include these images as appropriate in subsequent analysis.

Software packages for camera-trap data analysis

A number of software packages are available for camera-trap data management and analysis. For occupancy modelling, standard programs are Presence (Hines, 2006) and package "unmarked" in R (Fiske & Chandler, 2011). Both software packages require the species detection/non-detection matrix as input along with any associated covariates. Outputs include modelled occupancy and detection probability estimates and their standard errors for each combination of covariates. For species with recognizable individuals, software programs that have commonly been used for population density and abundance estimation using capture-recapture models include CAPTURE (White *et al.*, 1982) and MARK (White & Burnham, 1999). Spatially Explicit Capture-Recapture modelling is becoming the method of choice and is available in two versions: "secr" R package implementing likelihood-based inferences (Efford, 2015) and program SPACECAP implementing Bayesian approach (Gopalaswany *et al.*, 2012). Input data include the individual animal capture data, the camera-trap position/location data file, and the associated habitat mask. Outputs include density estimates with standard errors. Graphical presentation of circadian activity patterns can be undertaken in software programs such as excel or R. Several statistical packages are also available to analyse activity data (Kovach, 2011; Rowcliffe *et al.*, 2014). EstimateS (Colwell, 2013) and "BiodiversityR" package in R (Kindt & Coe, 2005) are relevant programs for species richness estimation. Both programs also provide diversity indices.

Integrated camera-trap data management and analysis programs are being developed. The camera-trap analysis package (CTAP) developed by the Zoological Society of London performs a range of data manipulation and analyses providing image display, filtering and editing, survey effort summaries, species list and richness estimates, species event rates, species occupancy estimates, circadian activity patterns and species site distribution maps in Google Earth or in ArcGIS maps (Amin *et al.*, in prep.). Version 2 of the tool will be extended to provide standard report generation, occupancy modelling and mapping with covariates, spatially explicit capture-recapture species density estimation, and cross-survey analysis. The tool will also have an image metadata module to read and update image metadata from jpeg images and allow the data processor to add species information into the image metadata. An image renamer tool for photo file management will also be part of this program.

Camera-trap forest antelope sampling case studies

Case Study 1 Use of camera-traps to study the status of the critically endangered Aders' duiker

Introduction. The chance observation of an Aders' duiker (*Cephalophus adersi*), classified as Critically Endangered by IUCN and recognised as one of the world's rarest and least known antelopes (Wilson, 2001; Finnie, 2008), in the Boni-Dodori forest area in 2004 led to follow-up camera-trap studies as the method of choice to investigate the status of this elusive antelope in the coastal forests of Kenya. This case study shows how a pilot study (Andanje *et al.*, 2011) was used to verify the original observations at Boni-Dodori (Gwynne & Smith 1974; Andanje & Wacher, 2004) and provided justification for a larger-scale survey using formal camera grids. The study highlights the new and detailed information that can accrue about rare forest species when using camera-traps.

Field methods. Following a casual sighting of Aders' duiker in the Boni-Dodori forest during wildlife surveys (2004), a pilot camera-trap survey deployed 19 cameras of three different makes (film and digital) using an opportunistic layout across 27 different locations along convenient tracks and other relatively accessible locations in 2008. Aders' duiker were photographed at 12 sites with an event rate more than 20 times that which had been obtained using a similar approach in Arabuko-Sokoke National Reserve, traditionally believed to be the species' mainland stronghold (Andanje *et al.*, 2011). In the follow-up, a formal camera-trap survey was conducted in 2010–2011 using 21 cameras (all Reconyx RM45) in a 2 x 2 km survey grid, intended to obtain a representative coverage of coastal forest habitat. Camera spacing was selected to achieve comparative independence between camera sites for the majority of forest-adapted mammal species expected to be present with the objective of providing baseline estimates of event rate and occupancy for a full array of coastal forest species in addition to Aders' duiker. Camera grids were set up consecutively for a minimum of 50 days each in order to achieve at least 1000 camera-trap days of sampling effort with 20 fully functioning cameras. Three grids were located within forested sectors of the Boni-Dodori grassland/coastal forest mosaic. A fourth replicate grid was set up in the well-known *Cyanometra* habitat of Aders' duiker in Arabuko-Sokoke National Reserve. New events were scored after an interval of ≥60 minutes since the previous photograph of a particular species at a camera.

Results. Aders' duiker emerged as not only the most frequently recorded antelope species in the Boni-Dodori forest system but also the most frequently recorded mammal, with 3021 events recorded across the three northern grids (Table 2). By contrast only two events, at separate camera sites, were recorded at Arabuko-Sokoke forest. Because camera-trapping also revealed that nearly all forest antelope species detected showed a similar pattern of relative abundance and occupancy, it is possible to speculate that the reason for this large difference is more likely to be due to a common factor, such as the presence of a very large human population immediately adjacent to Arabuko-Sokoke forest, which is not the case in the northern forests. All three Boni-Dodori camera-trap grids reported 95–100% occupancy for Aders' duiker (Table 3). The very high detection probabilities and occupancies obtained for Aders' duikers in the northern forests also indicate that future monitoring focused on this species could involve fewer sampling occasions and camera spacing extended to cover a larger area. Naïve occupancy for Aders' duiker in Arabuko-Sokoke forest was 0.11% and there were too few data to derive meaningful detection probability and model occupancy estimates. The larger number of encounters of Aders' duiker in the northern coastal forest allowed

Table 2 **Aders' duiker event rates in the Boni-Dodori and Arabuko-Sokoke camera sampling grids. Event rate was calculated as the mean of the number of independent photographic events per trap day times 100.**

Camera sampling grid	Number of events	Number of camera-trap days	Event rate (±SE)
Boni NR	1766	1670	106.16 (±3.91)
Dodori NR	643	1087	56.79 (±2.34)
Boni forest	612	1026	60.65 (±2.10)
Arabuko-Sokoke NR	2	2049	0.11 (±0.08)

Table 3 **Modelled occupancy estimates and detection probability estimates for the Aders' duiker in the Boni-Dodori and Arabuko-Sokoke camera sampling grids. Naïve occupancy is presented where encounters were too few to model occupancy.**

Camera sampling grid	Naïve occupancy	Occupancy estimate Ψ (±SE)	Probability of detection P (±SE)
Boni NR	-	1.00 (0.00)	0.97 (0.02)
Boni forest	-	1.00 (0.00)	0.89 (0.03)
Dodori NR	-	0.95 (0.05)	0.85 (0.03)
Arabuko-Sokoke NR	0.11	N/A	N/A

Figure 2 **Aders' duiker grooming a calf, Boni National Reserve, Kenya, May 2010.**

population density estimate of 7.4 Aders' duiker/km^2 (95% CI 4.5–10.1 duikers/km^2) to be obtained using the N-mixture Model (Hines, 2006) from spatially replicated counts (Amin *et al.*, 2014).

With respect to studying small and elusive forest antelopes, new behavioural observations for Aders' duiker deduced from the camera imagery included the following. The majority of the 3021 events indicated solitary individuals, 8.5% comprised at least two individuals (mostly two adults but also mother and young, Figure 2) and 1% comprised trios. At two cameras,

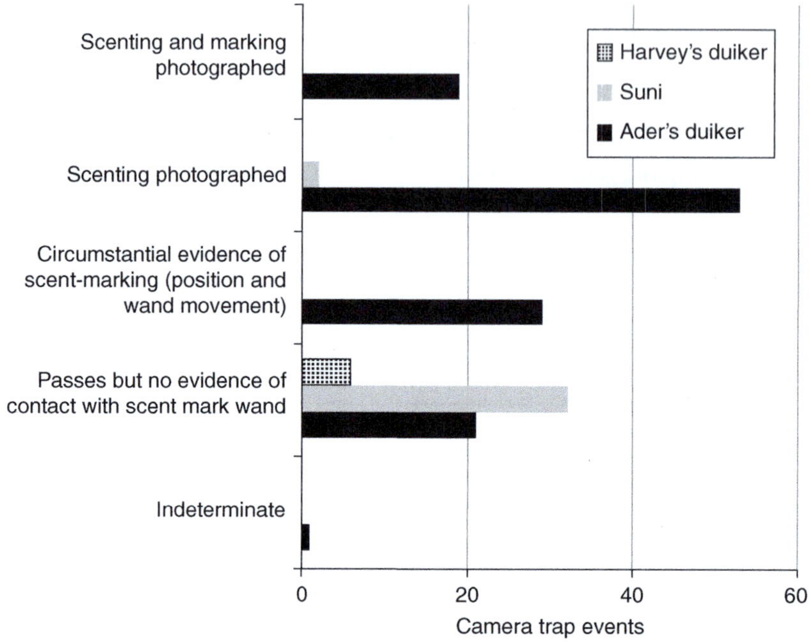

Figure 3 **Camera-trap evidence of forest antelope behaviour at a pre-orbital gland scent-marking wand, Boni National Reserve, Kenya, May 2010.**

scent marking stations were located by chance in the field of view. An analysis of behaviour at a camera station where 123 events involving Aders' duiker were recorded, as well as 34 events involving suni and 6 events involving Harvey's duiker (*Cephalophus harveyi*), is summarised in Figure 3. Photographs showed confirmed or probable scenting and scent marking behaviour by Aders' duiker, that is, sniffing a secretion on a thin plant stem ('wand') then rubbing the pre-orbital gland onto the accretion afresh, in over 80% of appearances in the field of view. A suni was photographed scenting at the same point on one occasion out of the 34 visits, but in a small sample, Harvey's duikers showed no evidence of interacting with the two scent marking sites used by Aders' duiker (note that in the Pic de Fon case study referred to below, bay duiker *Cephalophus dorsalis* and Maxwell's duiker *Philantomba maxwelli* were recorded scenting the same marking station).

Camera-traps also allowed study of frequency of visits to scent-marking stations and behaviour in relation to time elapsed since a previous visit. Photo sequences showed that scent-marking stations were frequently visited at intervals of less than 6 hours, with over 90% of re-visits occurring

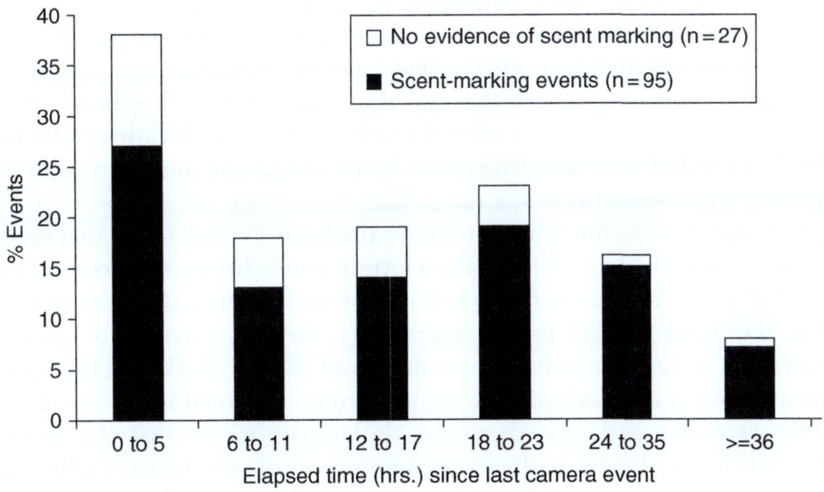

Figure 4 **Camera-trap monitoring of Aders' duiker scent marking behaviour; outcome of n=122 visits to a scent-marking station on a naturally growing wand, Boni National Reserve, Kenya, May 2010.**

within 24 hours. The scent marking station was re-anointed on most visits but with evidence of an increased probability of scent marking on occasions occurring after longer time lapses since the previous visit (Figure 4). Details of nursing behaviour in Aders' duiker were also recorded. In a sequence of 40 images over 10 minutes from 06:09 in May 2010 a female moved into the field of view and stood still on leaf litter below low canopy in an open clearing. A medium-sized calf (shoulder just above level of female's underside) was observed to initiate at least 4 separate bouts of suckling, which appeared to last between 3 minutes (initially) and 16 seconds (final bout recorded). The female interrupted each of the first three bouts by walking forward a few steps, before the final and shortest bout, when she walked off out of camera view, followed by the calf.

Potential inter-specific competition with other forest antelopes can also be examined. Simple correlation analysis indicated no evidence of spatial avoidance between Aders' duiker and suni, which is also present at high occupancy levels in the Boni-Dodori forest system. However, a more detailed examination of local temporal patterns needs to be completed to investigate interactions at a finer scale.

Discussion. Results from this case study have been used selectively here to emphasise the power of camera-trapping to reveal new and detailed information about the behaviour, as well as distribution and abundance, of scarce and elusive forest antelopes. We have not attempted to cover all the outputs that have arisen from the relatively large sample of new information about Aders' duiker that has been obtained. For example, we have also developed a method to adapt camera-trap data for use in a N-mixture model to analyse animal numbers at camera locations (Royle, 2004). Coupled with an estimate of home range size and with camera spacing larger than the home range size, this allows a direct measure of population density for Aders' duiker in the Boni-Dodori forest that does not rely on individual identification (Amin *et al.*, 2014). With respect to applied outcomes, the technology has provided the first strong evidence that the Boni-Dodori forest is probably the main centre of global distribution for the critically endangered Aders' duiker. This new information is coupled with the simultaneous capture of information on other rare and endangered species. It has confirmed that the Boni-Dodori forest hosts a diverse community of predators including endangered African wild dog (*Lycaon pictus*). The poorly known Sokoke bushy-tailed mongoose (*Bdeogale omnivora*) was found to be the most abundant small predator. A new form of giant sengi was documented, and its occupancy and event rates were similar to those recorded for the golden-rumped sengi (*Rhynchocyon chrysopygus*) in the survey grid at Arabuko-Sokoke forest. The camera-trap results are being used to underline the importance of the Boni-Dodori system to forest antelope conservation and national conservation goals (Amin *et al.*, 2014). Thus, besides providing firm evidence of a very significant population of Aders' duiker that would have been difficult to obtain reliably using more traditional methods, the camera-trap surveys have provided extensive data on the entire medium-to-large mammal community.

Case Study 2 Comparison of forest antelope relative abundance, distribution and activity in a West African montane forest

Introduction. At Pic de Fon in south-western Guinea camera-trapping was used to investigate the large mammal fauna of the forest habitats cloaking the flanks of a steep mountain ridge during 2008–2009. The mountain is under development for iron ore mining. A baseline measure of species richness and relative abundance against which to measure future environmental impacts was required under the conditions of the development funding. Prior to camera-trapping, the area had been subject to rapid assessment surveys, which had produced species lists, including possible presence of locally rare species based on dung and track evidence only (e.g., royal antelope *Neotragus pygmaeus* and bongo *Tragelaphus eurycerus*), but without any further insights into relative abundance or habitat use. In this example, the added value of camera-trapping is examined for the study of forest antelopes inhabiting a steep and physically demanding forest environment.

Field methods. The camera-trap sample design at Pic de Fon followed a standard 2 km grid layout (Tobler *et al.*, 2008; O'Brien, 2008) at 42 sites, using Reconyx RM45 and RC60 cameras. The grid was organised into northern and southern sectors covering a 1000 m difference in altitude, with the northern sector representing relatively undisturbed montane forest on the steepest relief and the southern sector representing lower relief and a relatively high proportion of secondary and disturbed forest. The spacing was intended to achieve a standard level of independence in results between cameras relative to the range of movements and home ranges that can be expected among the suite of major target species. The sampling regime was designed to achieve a total of at least 1000 camera trap days for each grid. This was achieved in the northern sector in the wet season grid (1035 camera-trap days), but there was insufficient time to complete 1000 camera-trap days in either the dry season repetition of the northern sector (434 camera-trap days) or the southern grid (834 camera-trap days), owing to an externally imposed need to finish camera-trapping operations by a fixed date. In this study, event rates were calculated by scoring new events after a 30-minute interval between photographs of a given species.

Results. Summary statistics on camera event rate and modelled occupancy for forest antelopes were used to investigate relative abundance and establish baseline metrics for future comparisons (Table 4 and Figures 5 and 6). Maxwell's duiker (*Philantomba maxwellii*) was the most commonly recorded species, followed by black duiker (*C. niger*) and red-flanked duiker (*C. rufilatus*). The two largest duiker species, yellow-backed (*C. sylvicultor*) and bay (*C. dorsalis*), were the least frequently recorded duikers while bushbuck (*Tragelaphus scriptus*), not strictly a forest-adapted species, was the least frequently recorded antelope.

Results also showed that although Maxwell's duiker was photographed more than twice as often as red-flanked duiker the two species report similar occupancy values after correcting for detectability (Table 4; Figure 6). Occupancy modelling provides a robust assessment of species abundance

Table 4 **Event rate, naïve and modelled occupancy for six antelope species photographed at 42 camera stations, Pic de Fon, Guinea, 2009.**

Species	Event rate (events/100 camera-trap days)	Naïve occupancy	Modelled occupancy (Ψ (±SE))
Yellow-backed duiker	0.86	0.41	0.75 (±0.33)
Red-flanked duiker	1.61	0.5	0.51 (±0.11)
Maxwell's duiker	6.95	0.5	0.55 (±0.12)
Black duiker	2.73	0.27	0.35 (±0.14)
Bay duiker	1.03	0.14	0.25 (±0.20)
Bushbuck	0.53	0.23	0.37 (±0.21)

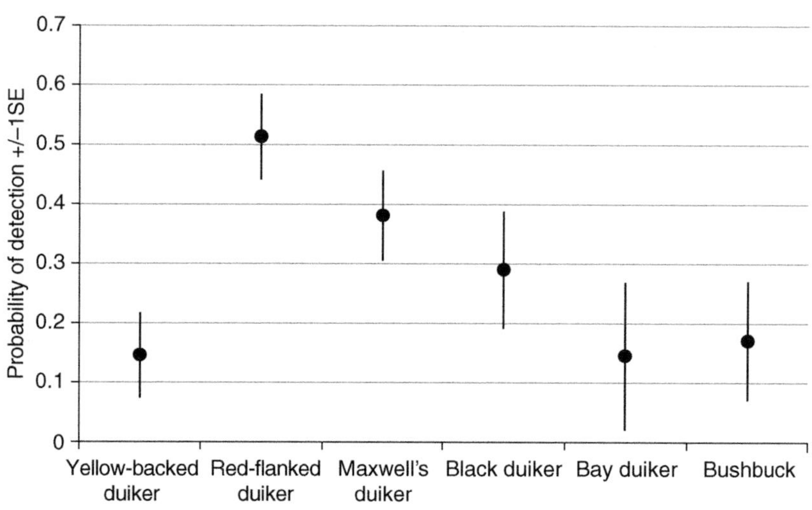

Figure 5 **Probability of detection for six forest antelope species, Pic de Fon, Guinea, 2009.**

over the longer term, given that forest antelope species have relatively small well defined home ranges, allowing camera-traps spaced sufficiently such that each individual only appears in one camera-trap.

Circadian activity patterns were estimated by using number of camera events per hour of the day as a proxy, plotted for six sympatric forest-living antelope species in Figure 7. The results for the five duiker species show that the three more commonly recorded and smaller species exhibited a clearly diurnal (or diurnal/crepuscular in the case of Maxwell's duiker) pattern. In

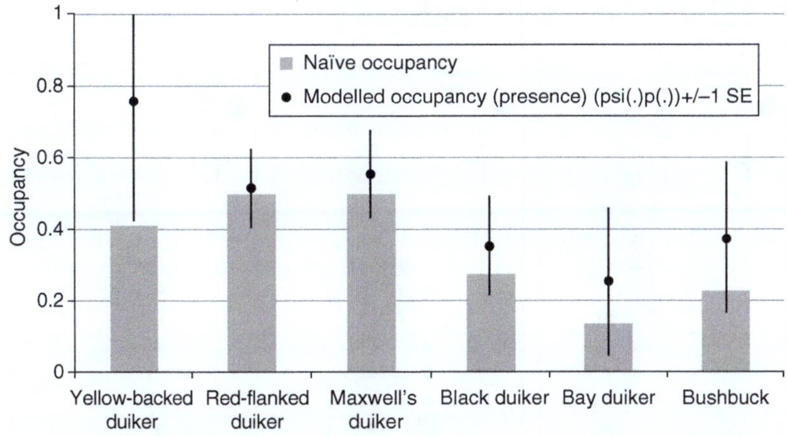

Figure 6 **Comparative occupancy rates.**

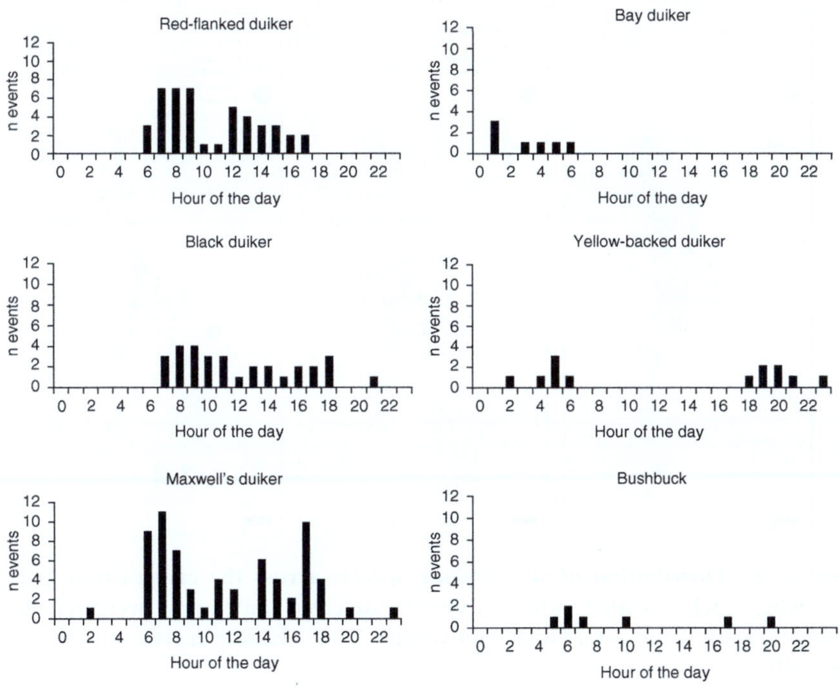

Figure 7 **24-hour distribution of camera-trap events for six antelope species living sympatrically at Pic de Fon, Guinea 2009.**

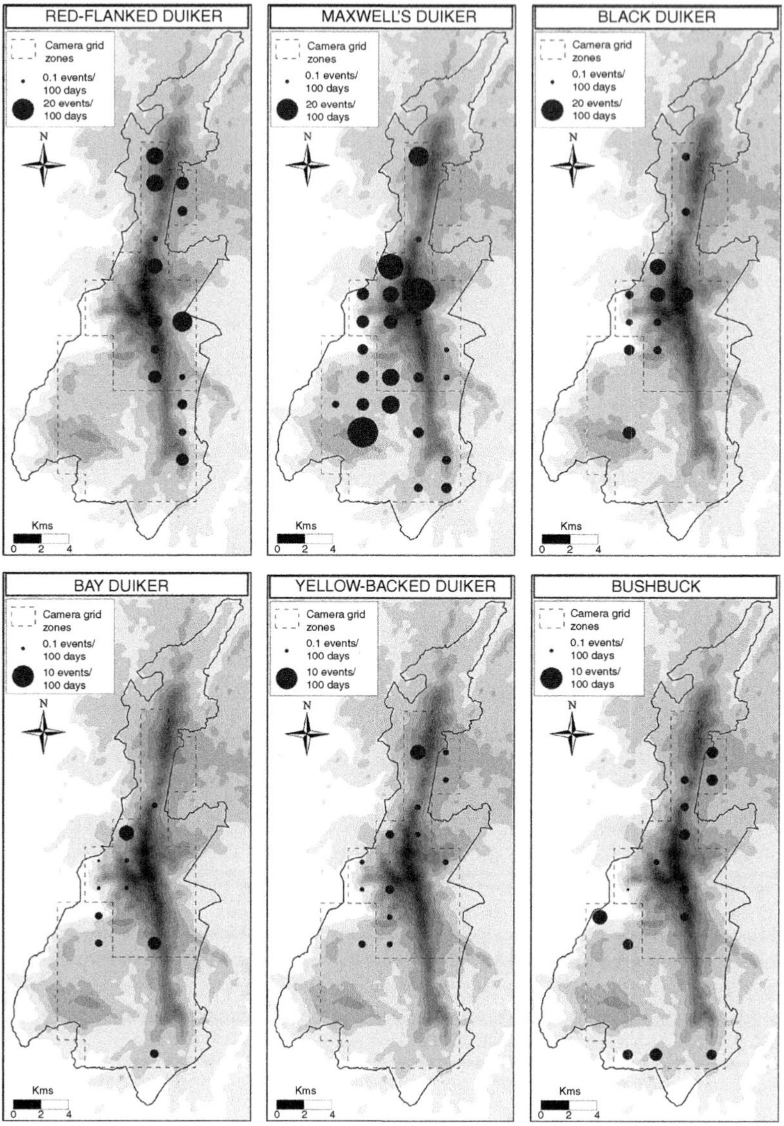

Figure 8 **Distribution of six antelope species across the camera-trap grid, showing each location where cameras detected antelopes, weighted by photographic event rate, Pic de Fon Guinea 2009 [shading represents contour relief].**

contrast, the two less frequently recorded larger duiker species were photographed primarily at night.

Finally, the physical distribution of photographic events for these six antelopes indicates that while all six species were well distributed in the less disturbed central and northern part of the survey area, only Maxwell's duiker was widely distributed in the southern sector, with other species showing much more restricted ranges in this region (Figure 8). This result is of interest in the light of the difference in habitats in the two sectors. In this case, the occupancy analysis could be remodelled employing habitat factors, notably habitat sector and perhaps altitude as a continuous covariate, allowing development of occupancy mapping.

Discussion. The montane forest habitats at Pic de Fon are a steep and immensely challenging environment in which to conduct conventional field work. The data generated in this case study were obtained by deploying a team of three trained field workers spending 10–15 days deploying cameras and 10–15 days recovering them. The results have provided insights into the community ecology of these sympatric antelopes in a way that would not be possible using traditional field methods, and introduce analytical techniques to cater for the wide variations in detectability of forest antelopes, arising from differences in ecology and behaviour. Besides revealing patterns of relative abundance, distribution, and activity, photographs have recorded additional details such as foraging associations between duikers and other forest species (Ahanta francolin *Francolinus ahantensis* and Campbell's monkey *Cercopithecus campbelli*), timing of breeding and social behaviour (e.g., intra-and inter-specific scent marking). It is also possible to relate results to a wider picture: for example, the congruence between observed day-time/night-time activity patterns, body size, and phylogenetic relatedness (van Vuuren, 2003) seen between the two larger and three smaller duiker species. These insights from camera-trapping come in addition to the fulfilment of the primary project goal, which was to establish a baseline reference of the conservation status of this forest mammal community prior to mining development. The forest antelopes are one of the most important components of this system, being significant agents of seed dispersal and economically important as major targets of the bushmeat trade. Despite the challenging physical conditions, the camera-trap survey can be replicated in the future to provide a systematic means to monitor the effectiveness of efforts to mitigate habitat damage while developing the mine site on the mountain.

Conclusion

The widespread adoption of remotely operated, heat-and-motion triggered cameras in wildlife research and monitoring has led to great improvements in our knowledge of cryptic species such as forest antelopes. Although the most immediate application may be in simply documenting the presence of rare species, data from camera-trapping is also being used, in an increasingly sophisticated statistical framework, to estimate abundance, behaviour patterns, and interactions between species over time. Such information is of great value to conservation managers and decision makers as we have shown in the case studies presented in this chapter. In both these case studies, the areas surveyed are under threat from major development projects, and camera-traps have provided baseline population data for ecologically important medium- to large-sized mammal species on which future impacts can be measured. In the Kenyan case study, regional infrastructure developments are likely to have widespread impacts on forests and other coastal habitats, but the importance of the area to globally threatened species (documented with camera-traps) has highlighted the previously unrecognised biodiversity of the area and given conservationists a voice in mitigation efforts. In Guinea, the effects of mining at Pic de Fon may also be severe, but knowledge of forest mammal communities could be used to restrict impacts to those core areas needed for mining access. Camera-traps are therefore proving to be a valuable tool in antelope conservation, and we anticipate that with further advances in technology including remote data access, analytical methods, and reduced cost, camera-trapping will become more widely used as a long-term monitoring method for forest antelopes and other species.

References

Ahumada J. A., Silva C. E., Gajapersad K., Hallam C., Hurtado J., Martin E., McWilliam A., Mugerwa, A., O'Brien T., Rovero F., Sheil D., Spironello W. R., Winarni N. & Andelman S. J. (2011): Community structure and diversity of tropical forest mammals: data from a global camera trap network. *Philosophical Transactions of the Royal Society Series B* **366**: 2701–2711.

Akomo-Okoue E. F., Inoue E., Adenyo C., Hayano A. & Inoue-Murayama, M. I. (2013): Development of microsatellite markers for blue duiker (*Cephalophus monticola*) using next-generation sequencing and cross-amplification in other duikers. *Conservation Genetics Resources* **5**: 779–781.

Amin R., Davey K., Fitzmaurice A. & Wacher, T. (in prep.): A standard toolkit for the management and analysis of camera trap data, Zoological Society of London.

Amin, R. & Wacher, T. (2014): Kenya forest camera trap field setup protocol, Zoological Society of London.

Amin, R., Andanje, S., Ogwonka, B., Ali, H., Bowkett, A. E., Omar, M. & Wacher, T. (2014): The northern coastal forests of Kenya are nationally and globally important for the conservation of Aders' duiker *Cephalophus adersi* and other antelope species. *Biodiversity and Conservation.* **24**: 641–658.

Andanje, S. A., Bowkett, A. E., Agwanda, B. R., Ngaruiya, G., Plowman, A., Wacher, T. & Amin R. (2011): A new population of the critically endangered Aders' duiker *Cephalophus adersi* confirmed from northern coastal Kenya. *Oryx* **45**: 444–447.

Andanje, S. A. & Wacher, T. (2004): Aders' duiker in Dodori forest reserve, North Kenya coast. *Gnusletter* **23**: 4–5.

Anthony, N. M., Mickala, P. & Abernethy, K.A. (2012): Biodiversity and conservation genetics research in Central Africa: new approaches and avenues for international collaboration. *Conservation Genetics Resources* **4**: 523–525.

Beja-Pereira, A., Oliveira R., Alves, P. C., Schwartz, M. K. & Luikart G. (2009): Advancing ecological understandings through technological transformations in noninvasive genetics. *Molecular Ecology Resources* **9**: 1279–1301.

Bowkett, A. E., Jones, T., Laizzer, R. L., Plowman, A. B. & Stevens J. R. (2013): Can molecular data validate morphometric identification of faecal pellets in Tanzanian forest antelope species? *Conservation Genetics Resources* **5**: 1095–1100. DOI 10.1007/s12686-013-9935-3.

Bowkett, A. E., Plowman, A. B., Stevens, J. R., Davenport, T. R. B. & Jansen van Vuuren, B. (2009): Genetic testing of dung identification for antelope surveys in the Udzungwa Mountains, Tanzania. *Conservation Genetics* **10**: 251–255.

Bowland, A. E. (1990): The response of red duikers *Cephalophus natalensis* to drive counts. *Koedoe* **33**: 47–53.

Bowland, A. E. & Perrin, M. R. (1994): Density estimate methods for blue duikers *Philantomba monticola* and red duikers *Caphalophus natalensis* in Natal, South Africa. *Journal of African Zoology* **108**: 505–519.

Bowman, V. & Plowman, A. (2002): Captive duiker management at the duiker and mini-antelope breeding and research institute (Dambari), Bulawayo, Zimbabwe. *Zoo Biology* **21**: 161–170.

Bruner, A. G., Gullison, R. E. & Balmford, A. (2004): Financial costs and shortfalls of managing and expanding protected-area systems in developing countries. *Bioscience* **54**: 1119–1126.

Buckland, S. T., Anderson, D. R. & Burnham, K. P. (2001): *Introduction to Distance Sampling: Estimating Abundance of Biological Populations.* Oxford, Oxford University Press.

Colwell, R. K. (2013): EstimateS: Statistical estimation of species richness and shared species from samples. Version 9. User's Guide and application, http://purl.oclc.org/estimates.

Eaton, M. J., Meyers, G. L., Kolokotronis, S. O., Leslie, M. S., Martin, A. P. & Amato, G. (2010): Barcoding bushmeat: molecular identification of Central

African and South American harvested vertebrates. *Conservation Genetics* **11**: 1389–1404.

East, R. (1999): *African Antelope Database 1998*. IUCN/SSC Antelope Specialist Group. IUCN Gland, Cambridge.

Efford, M. G. (2015): secr: Spatially explicit capture-recapture models, R package version 2.9.3. http://CRAN.R-project.org/package=secr (as at 16 February 2016).

Fa, J. E., Ryan, S. F. & Bell,D. J. (2005): Hunting vulnerability, ecological characteristics and harvest rates of bushmeat species in afro-tropical forests. *Biological Conservation* **121**: 167–176.

Faria, P. J., Kavembe, G. D., Jung'a, J. O., Kimwele, C. N., Estes, L. D., Reillo, P. R., Mwangi, A. G. & Bruford, M. W. (2011): The use of non-invasive molecular techniques to confirm the presence of mountain bongo *Tragelaphus eurycerus isaaci* populations in Kenya and preliminary inference of their mitochondrial genetic variation. *Conservation Genetics* **12**: 745–751.

Feer, F. (1995): Seed dispersal in African forest ruminants. *Journal of Tropical Ecology* **11**: 683–689.

Finnie, D. (2008): *Cephalophus adersi*. The IUCN Red List of Threatened Species. Version 2014.3. www.iucnredlist.org (as at 16 December 2014).

Fiske, I. & Chandler, R. (2011): Unmarked: an R package for fitting hierarchical models of wildlife occurrence and abundance. *Journal of Statistical Software* **43**: 1–23.

Foley, C. (2008): Blue duiker and Suni identikit. *Mammal News Bites* **4**: 3–4.

Foster, R. J. & Harmsen, B. J. (2011): A critique of density estimation from camera trap data. *Journal of Wildlife Management* **76**: 224–236.

Gompper, M. E., Kays, R. W., Ray, J. C., Lapoint, S. D., Bogan, D. A. & Cryan, J. R. (2006): A comparison of noninvasive techniques to survey carnivore communities in northeastern North America. *Wildlife Society Bulletin*, **34**(4): 1142–1151.

Gopalaswany, A. M., Royle, J. A., Hilnes, J. E., Singh, P., Jathanna, D., Kumar, S. & Karanth, K. U. (2012): Program SPACECAP: software for estimating animal density using spatially explicit capture-recapture models. *Methods in Ecology and Evolution*, **3**: 1067–1072.

Gwynne, M. D. & Smith, K. (1974): Garissa county council Boni forest game reserve. Unpublished establishment and development proposals. In *Antelopes in southern Somalia, 1974–1975* (summarised by Antelope Specialist Group from reports of N.A.O. Abel and M.E. Kille). *Gnusletter* **30**: 5–7.

Hall, N. (2013): After the gold rush. *Genome Biology* **14**: 115.

Hart, J. A. (2013): *Cephalophus leucogaster* White-bellied duiker. In *Mammals of Africa. 7. Pigs, hippopotamuses, chevrotain, giraffe, deer and bovids*: 255–258. Kingdon, J. & Hoffmann M. (Eds). London: Bloomsbury Publishing.

Hebert, P. D. N., Cywinska A., Ball S.L. & DeWaard, J. R. (2003): Biological identifications through DNA barcodes. *Proceedings of the Royal Society B Biological Sciences* **270**: 313–321.

Hines, J. E. (2006): Presence. Software to estimate patch occupancy and related parameters. USGS-PWRC.

Hofmann, T. & Roth, H. (2003): Feeding preferences of duiker (*Cephalophus maxwelli*, *C. rufilatus*, and *C. niger*) in Ivory Coast and Ghana. *Mammalian Biology* **68**: 65–77.

Johnston, A. R., Morikawa, M. K., Ntie, S. & Anthony, N. M. (2011): Evaluating DNA barcoding criteria using African duiker antelope (Cephalophinae) as a test case. *Conservation Genetics* **12**: 1173–1182.

Kindt, R. & Coe, R. (2005): Tree diversity analysis: A manual and software for common statistical methods for ecological and biodiversity studies. http://www.worldagroforestry.org/resources/databases/tree-diversity-analysis (as at 16 February 20160.

Kingdon, J. & Hoffmann, M. (2013): *Nesotragus moschatus* Suni. In *Mammals of Africa* 6. *Pigs, hippopotamuses, chevrotain, giraffe, deer and bovids*: 214–219. Kingdon, J. & Hoffmann, M. (Eds). London: Bloomsbury Publishing.

Koster, S. H. & Hart, J. A. (1988): Methods for estimating ungulate populations in tropical forests. *African Journal of Ecology* **26**: 117–126.

Kovach W. L. (2011): Oriana—circular statistics for Windows, ver. 4. Kovach Computing Services Pentraeth, Wales, UK.

Lancia, R. A. & Bishir, J. W. (1996): Removal methods. In *Measuring and monitoring biological diversity: standard methods for mammals*: 210–217. Wilson, D. E., Cole, F. R., Nichols, J. D., Rudran, R. & Foster, M. S. (Eds.). Washington, DC: Smithsonian Institution Press.

Lunt, N., Bowkett, A. E. & Plowman, A. B. (2007): Implications of assumption violation in density estimates of antelope from dung-heap counts: a case study on grey duiker (*Sylvicapra grimmia*) in Zimbabwe. *African Journal of Ecology* **45**: 382–389.

Lyra-Jorge, M. C., Ciocheti, G., Pivello, V. R. & Meirelles S. T. (2008): Comparing methods for sampling large- and medium-sized mammals: camera traps and track plots. *European Journal of Wildlife Research* **54**(4): 739–744. doi:10.1007/s10344-008-0205-8.

MacKenzie, D. I., Nichols, J. D., Royle, J. A., Pollock, K. H., Bailey, L. L. & Hines J. E. (2006): *Occupancy estimation and modelling*. Amsterdam: Elsevier.

MacKenzie, D. I. & Royle, J. A. (2005): Designing occupancy studies: general advice and allocating survey effort, *Journal of Applied Ecology* **42**: 1105–1114.

Mills, L. S. (2013): *Conservation of wildlife populations: demography, genetics and management*. (2nd edn). Chichester: Wiley-Blackwell.

Nichols, J. D. & Dickman, C. R. (1996): Capture-recapture methods. In *Measuring and monitoring biological diversity: standard methods for mammals*: 217–226. Wilson, D. E., Cole, F. R., Nichols, J. D., Rudran, R. & Foster, M. S. (Eds.). Washington, DC: Smithsonian Institution Press.

Nielsen, M. R. (2006): Importance, cause and effect of bushmeat hunting in the Udzungwa Mountains, Tanzania: implications for community based wildlife management. *Biological Conservation* **128**: 509–516.

Noss, A. J. (1999): Censusing rainforest game species with communal net hunts. *African Journal of Ecology* **37**: 1–11.

Ntie, S., Johnston, A. R., Mickala, P., Bowkett, A. E., Jansen van Vuuren, B., Colyn, M., Telfer, P., Maisels, F., Hymas, O., Rouyer, R. L., Wallace, R. A., LeBlanc, K., van

Vliet, N., Sonet, G., Verheyen, E., Pires, D., Wickings, E. J., Lahm, S. A. & Anthony N. M. (2010a): A molecular diagnostic for identifying central African forest artiodactyls from faecal pellets. *Animal Conservation* **13**: 80–93.

Ntie, S., Soto-Calderon, I. D., Eaton, M. J. & Anthony, N. M. (2010b): Cross-species amplification of bovid microsatellites in central African duikers (genus *Cephalophus*) and other sympatric artiodactyls. *Molecular Ecology Resources* **10**: 1059–1065.

O'Brien, T. G. (2008): On the use of automated cameras to estimate species richness for large- and medium-sized rainforest mammals. *Animal Conservation* **11**: 170–181.

O'Connell, A. F., Nichols, J. D. & Karanth, K. U. (2011): Camera traps in animal ecology – methods and analyses. New York: Springer.

Plowman, A. B. (2002): Nutrient intake and apparent digestibility of diets consumed by captive duikers at the Dambari Field Station, Zimbabwe. *Zoo Biology* **21**: 135–147.

Plumptre, A. J. & Harris, S. (1995): Estimating the biomass of large mammalian herbivores in a tropical montane forest – a method of fecal counting that avoids assuming a steady-state system. *Journal of Applied Ecology* **32**: 111–120.

R Development Core Team (2015): R: a language and environment for statistical computing. R Foundation for Statistical Computing, Vienna. http://www.R-project.org (as at 16 February 2016).

Ratcliffe, N., Vaughan, D., Whyte, C. & Shepard, M. (1998): Development of playback census methods for storm petrels *Hydrobates pelagicus. Bird Study* **45**: 302–312.

Roberts, N. (2011): Investigation into survey techniques of large mammals: surveyor competence and camera-trapping vs. transect-sampling. *Bioscience Horizons*, **4**: 40–49.

Rohr, L. & Pfefferkorn, C. (2003): Captive management of duikers and how it can affect in situ conservation efforts. In *Ecology and conservation of small antelope: proceedings of an international symposium on duiker and dwarf antelope in Africa*: 247–253. Plowman, A. (Ed.). Fürth: Filander Verlag.

Rovero, F. & Marshall, A. R. (2004): Estimating the abundance of forest antelopes by line transect techniques: a case from the Udzungwa Mountains of Tanzania. *Tropical Zoology* **17**: 267–277.

Rovero, F., Martin, E., Rosa, M., Ahumada, J. A. & Spitale D. (2014): Estimating species richness and modelling habitat preferences of tropical forest mammals from camera trap data. *Plos One* **9**(7).

Rowcliffe, J. M., Field, J., Turvey, S. T. & Carbone, C. (2008): Estimating animal density using camera traps without the need for individual recognition. *Journal of Applied Ecology* **45**: 1228–1236.

Rowcliffe, J. M., Kays, R., Kranstauber, B., Carbone, C. & Jansen P. A. (2014): Quantifying levels of animal activity using camera trap data. *Methods in Ecology and Evolution* **5**: 1170–1179.

Royle, A. (2004): N-mixture models for estimating population size from spatially replicated counts. *Biometrics* **60**: 108–115.

Schmidt, J. L. (1983): A comparison of census techniques of common duiker and bushbuck in timber plantations. *South African Forestry Journal* **126**: 15–19.

Schwartz, M. K., Tallmon, D. A. & Luikart G. (1998): Review of DNA-based census and effective population size estimators. *Animal Conservation* **1**: 293–299.

Silveira, L., Jácomo, A. T. & Diniz-Filho, J. A. F. (2003): Camera trap, line transect census and track surveys: a comparative evaluation. *Biological Conservation*, **114**(3): 351–355.

Struhsaker, T. T. (1997): *Ecology of an African rain forest. logging in Kibale and the conflict between conservation and exploitation.* Gainesville: University Press of Florida.

Taberlet, P., Waits, L. P. & Luikart G. (1999): Non-invasive genetic sampling: look before you leap. *Trends Ecology and Evolution* **14**: 323–327.

Tobler, M. W., Carrillo-Percastegui, S. E., Leite Pitman, R., Mares, R. & Powell, G. (2008): An evaluation of camera traps for inventorying large- and medium-sized terrestrial rainforest mammals. *Animal Conservation* **11**: 169–178.

van Vliet, N., Kaniowska, E., Bourgarel, M., Fargeot, C. & Nasi, R. (2009): Answering the call! Adapting a traditional hunting practice to monitor duiker populations. *African Journal of Ecology* **47**: 393–399.

van Vliet, N., Nasi, R., Emmons, L., Feer, F., Mbazza, P. & Bourgarel, M. (2007): Evidence for the local depletion of bay duiker *Cephalophus dorsalis*, within the Ipassa Man and Biosphere Reserve, north-east Gabon. *African Journal of Ecology* **45**: 440–443.

van Vliet, N., Zundel, S., Miquel, C., Taberlet, P. & Nasi, R. (2008): Distinguishing dung from blue, red and yellow-backed duikers through non-invasive genetic techniques. *African Journal of Ecology* **46**: 411–417.

van Vuuren, B. (2003): Molecular phylogeny of duiker antelope. In *Ecology and conservation of small antelope: Proceedings of international symposium on duiker and dwarf antelope in Africa*: 262. Plowmanm, A. (Ed.). Fürth, Germany: Filander Verlag.

Viquerat, S. M. A., Bobo, S. S., Müller, M., Kiffner, C. & Waltert, M. (2012): Estimating forest duiker (Cephalophinae) density in Korup National Park: a case on the performance of three line transect methods. *South African Journal of Wildlife Research* **42**: 1–10.

Wacher, T. & Amin, R. (2015): *Kenyan north coastal forest mammal diversity surveys.* London: Zoological Society of London.

Waltert, M., Heber, S., Riedelbauch, S., Lien, J. L. & Muhlenberg, M. (2006): Estimates of blue duiker (*Cephalophus monticola*) densities from diurnal and nocturnal line transects in the Korup region, south-western Cameroon. *African Journal of Ecology* **44**: 290–292.

White, G. C., Anderson, D. R., Burnham, K. P. & Otis, D. L. (1982): Capture-recapture and removal methods for sampling closed populations. Los Alamos National Laboratory LA-8787-NERP, 235 pp.

White, G. C. & Burnham, K. P. (1999): Program MARK: Survival estimation from populations of marked animals. *Bird Study* **46** (Supplement): 120–138.

White, L. J. T. (1994): Biomass of rain forest mammals in the Lopé Reserve, Gabon. *Journal of Animal Ecology* **63**: 499–512.

Wilkie, D. S. & Carpenter, J. F. (1999): Bushmeat hunting in the Congo Basin: an assessment of impacts and options for mitigation, *Biodiversity and Conservation* **8**: 927–955.

Wilson, V. (2001): *Duikers of Africa: masters of the African forest floor: a study of duikers, people, hunting and bushmeat*. Chipangali Wildlife Trust. **10**, 798 pp.

Zhan, X. J., Li, M., Zhang, Z. J., Goossens, B., Chen, Y. P., Wang, H. J., Bruford, M. W. & Wei, F. W. (2006): Molecular censusing doubles giant panda population estimate in a key nature reserve. *Current Biology* **16**: R451–R452.

12

Reintroduction as an Antelope Conservation Solution

Mark R. Stanley Price

WildCRU, The Recanati-Kaplan Centre, Oxford University, United Kingdom

Introduction

The quantity and quality of reintroductions of plants and animals have increased greatly over the last 30 years (Soorae, 2008, 2010, 2011). Antelopes are of special interest here, for some of the first reintroductions of captive-bred stock involved the Arabian oryx (*Oryx leucoryx*), and many antelope species have large captive populations of imputed reintroduction potential. Since nearly two-thirds of all antelope species' populations are declining, and more than a quarter of the species are threatened with extinction (IUCN, 2015), the use of captive-bred animals to establish new wild populations is a natural and appealing prospect to counter local or total extinction in the wild. But while this is easy to propose, to what extent are antelope reintroductions an effective conservation solution? Here I review the extent to which reintroduction has been used in antelope conservation and the success of these initiatives. Antelopes provide a well-suited opportunity to understand the factors affecting the use and success of reintroduction as a conservation measure because the group exhibits great variation in form, behaviour, and ecology; and, in common with other large mammals, they face many challenges to their persistence in the wild. For the purpose of this review, I consider antelopes as comprising 35 genera and 92 species[1] (Table 1).

---

[1] The authoritative Wilson and Reeder (2008) lists 35 genera and 95 species. Subsequent work (Wronski *et al.*, 2010) shows four of their species (*Gazella arabica*, *G. gazella*, *G. erlangeri*, and *G. saudiya*) are merely northern and southern lineages of *G. gazella*. Hence the adjustment to 92 species.

Antelope Conservation: From Diagnosis to Action, First Edition. Edited by Jakob Bro-Jørgensen and David P. Mallon.

Table 1 **Antelope genera, extant species, weights (kg), Red List status and reintroduction status. Where a genus contains species with widely differing weights, the species have been grouped into separate categories, identified in Table 1 as 'Cephalophus 1', 'Cephalophus 2' etc. The apparent overlaps between some of these categories are due to extreme sexual dimorphism within their species. Reintroduction status: R = species has been reintroduced; r = species waiting to be reintroduced.**

Genus	No. spp.	Size range, kg	Species	Red List status[1]	Reintroduction status
Neotragus	3	2–3	*batesi*	LC	
			moschatus	LC	
			pygmaeus	LC	
Madoqua 1	3	2.5–5	*guentheri*	LC	
			piacentinii	DD	
			saltiana	LC	
Philantomba 1	1	4–7	*monticola*	LC	
Madoqua 2	1	5–6	*kirkii*	LC	
Philantomba 2	1	6–12	*maxwelli*	LC	
Raphicerus	3	6–16	*campestris*	LC	
			melanotis	LC	
			sharpei	LC	
Cephalophus 1	3	7–14	*adersi*	CR	
			natalensis	LC	
			rufilatus	LC	
Ourebia	1	8–17	*ourebi*	LC	
Sylvicapra	1	8–26	*grimmia*	LC	
Oreotragus	1	8–18	*oreotragus*	LC	
Dorcatragus	1	10–11	*megalotis*	VU	
Procapra	3	13–39	*gutturosa*	LC	
			picticaudata	NT	
			przewalskii	EN	
Cephalophus 2	9	14–27	*brookei*[2]		
			callipygus	LC	
			dorsalis	LC	
			leucogaster	LC	
			weynsi	LC	
			zebra	VU	
			niger	LC	
			nigrifrons	LC	
			ogilbyi	LC	
Eudorcas	2	14–37	*rufifrons*	VU	
			thomsonii	NT	

Table 1 (*continued*)

Genus	No. spp.	Size range, kg	Species	Red List status[1]	Reintroduction status
Tetracerus	1	15–25	*quadricornis*	VU	
Gazella	7	15–40	*gazella*	VU	R
			dorcas	VU	R
			subgutturosa	VU	R
			bennettii	LC	
			cuvieri	EN	
			leptoceros,	EN	
			spekei	EN	
Pelea	1	18–30	*capreolus*	LC	
Redunca 1	1	19–38	*fulvorufula*	LC	
Pantholops	1	20–40	*hodgsonii*	EN	
Tragelaphus 1	1	24–80	*scriptus*	LC	
Antilope	1	25–35	*cervicapra*	NT	
Antidorcas	1	26–48	*marsupialis*	LC	
Ammodorcas	1	27	*clarkei*	VU	
Saiga	2	31–41	*borealis*[3]		
			tatarica	CR	
Litocranius	1	29–52	*walleri*	NT	
Redunca 2	2	36–95	*arundinum*	LC	
			redunca	LC	
Nanger	3	38–81	*dama*	CR	R
			granti	LC	
			soemmerringi	VU	
Kobus 1	4	40–128	*kob*	LC	
			vardonii	NT	
			leche	LC	
			megaceros	EN	
Aepyceros	1	43–63	*melampus*	LC	
Cephalophus 3	3	45–80	*jentinki*	EN	
			spadix	EN	
			silvicultor	LC	
Tragelaphus 2	3	50–125	*angasii*	LC	
			spekei	LC	
			imberbis	NT	
Oryx 1	1	54–75	*leucoryx*	EN	R
Damaliscus 1	1	56–86	*pygargus*	LC	
Addax	1	60–125	*nasomaculatus*	CR	R

(*continued overleaf*)

Table 1 (*continued*)

Genus	No. spp.	Size range, kg	Species	Red List status[1]	Reintroduction status
Pseudoryx	1	70–100	*nghetinhensis*	CR	
Beatragus	1	73–118	*hunteri*	CR	
Damaliscus 2	3	90–168	*korrigum*[4]		
			lunatus,	LC	
			superstes		
Alcelaphus	3	105–218	*buselaphus*	LC	
			caama[5]		
			lichtensteinii		
Connochaetes 1	1	110–160	*gnou*	LC	
Oryx 2	3	116–240	*dammah*	EW	R
			beisa	NT	
			gazella	LC	
Boselaphus	1	120–288	*tragocamelus*	LC	
Connochaetes 2	1	141–295	*taurinus*	LC	
Tragelaphus 3	3	150–405	*buxtoni*	EN	
			eurycerus	NT	r
			strepsiceros	LC	
Kobus 2	1	160–275	*ellipsiprymnus*	LC	
Hippotragus 1	1	160–230	*niger*	LC	r
Hippotragus 2	1	215–300	*equinus*	LC	r
Taurotragus	2	390–950	*derbianus*	LC	r
			oryx	LC	

[1] Red List status from http://www.iucnredlist.org/initiatives/mammals (as at 16 January 2015).
[2] Treated by Red List as a sub-species of *C. ogilbyi.*
[3] Treated by Red List as a sub-species of *S. tatarica.*
[4] *D. korrigum* and *D. superstes* are combined with *D. lunatus* in Red List.
[5] *A. caama* and *A. lichtensteinii* are combined with *A. buselpahus* in Red List.

What is a reintroduction?

In 1998 the IUCN Species Survival Commission's Reintroduction Specialist Group defined a reintroduction as an attempt to establish a species in an area which was once part of its historical range, but from which it has been extirpated or become extinct (IUCN, 1998). This was amplified by the statement that the principal aim 'should be to establish a viable, free-ranging population in the wild, of a species, sub-species or race, which has become globally or locally extinct, or extirpated from the wild. It should be reintroduced within the species' former natural habitat and range and should require minimal long-term management'.

This sets a high standard of required conditions, and reflects thinking and aspiration from the early 1990s: reintroduction practice and experience have grown immensely since then. In 2012, IUCN adopted revised Guidelines on Reintroductions and Other Conservation Translocations to reflect this experience (IUCN/SSC, 2013). Notable changes between the two versions include:

1. Reintroductions are now placed within a typology of conservation translocations.
2. The primary objective of a conservation translocation should be a conservation benefit to either a population or a species or the ecosystem it occupies.
3. Reintroduction is defined as the intentional movement of an organism followed by its release within its indigenous range, from which it has disappeared, with the aim of re-establishing a viable population of the focal species. This is a conservation benefit, and consistent with the 1998 requirement.
4. 'Historical range' is open to arbitrary and inconsistent interpretation; accordingly, for a reintroduction, the 2013 Guidelines require release within 'indigenous range', which is defined as 'the known or inferred distribution generated from historical (written or verbal) records, or physical evidence of the species' occurrence; where direct evidence is inadequate to confirm previous occupancy, the existence of suitable habitat within ecologically appropriate proximity to proven range may be taken as adequate evidence of previous occupation' (IUCN/SSC, 2013).

The 2013 definition of a reintroduction is more liberal through removal of the phrase 'in the wild' and of the requirements of 'free-ranging' and 'minimal long-term management'. This reflects the current reality of many situations in which antelopes (and species of many other taxa) have been restored to indigenous range but are living within fenced areas, and receiving every degree of management support (in terms of feed, shade, water, veterinary care) from none to high dependence. Many of these situations are regarded as successful reintroductions, and the prospects of release onto unenclosed land may be impossible for the foreseeable future. The balance between enclosed and unenclosed populations is discussed later. But, there is an important issue: namely, at what point does an enclosed area become large enough, relative to any antelope species and its numbers within, and with what type and level of management support, to be counted as supporting a viable population? One significant reason for exploring this is the down-listing of the Arabian oryx from Extinct in the Wild to Vulnerable based on its 'wild' population exceeding 1000 individuals for 5 years, of which a high proportion are in enclosed areas, albeit that some are very large (IUCN Antelope Specialist Group, 2009).

There are two main ways in which new populations of antelopes are established through reintroduction:

1. Wild-to-wild translocation: the moving of ungulates from one site in their range to another (or indeed outside their known past range) has been a very common phenomenon for many years (Griffith *et al.*, 1989). It is a feasible option where there are adequate wild populations from which to draw individuals within reasonable proximity to the release site, and the source and destination sites are usually in the same country. Particularly in South Africa, there is great skill and professionalism in both protected area authorities and the private sector in the mass capture, transport, and release of antelopes: 50,000–60,000 animals are captured and moved every year by the 40 corporate members of the Wildlife Translocation Association (Wildlife Translocation Association, pers. comm., 2013). Much of this activity has not been systematically documented, as it is treated virtually as a farming operation, especially where the release area is fenced. Such wild-to-wild operations are not the subject of this chapter.
2. Animals have been born and reared under captive conditions, ranging from zoo-type pens to larger enclosures but in all of which they are protected from predation and illegal offtake, and are totally or very largely dependent on being provisioned; this may be either in their range state or in another country from which they will be repatriated. Released individuals of such species will usually have undergone a period of familiarisation or acclimatisation in the country of release, with a wide array of management procedures. Released individuals may have been bred out-of-country or they may be bred in-country as the descendants of captive-bred immigrants. This chapter considers as reintroduction only this latter situation of antelope founders from protective environments, and excludes the common situation of wild-to-wild translocations.

Releases of a few individuals into the wild are unlikely to succeed (Griffith *et al.*, 1989; Wolf *et al.*, 1998; Fischer & Lindenmayer, 2000). Thus, one-off releases of small numbers of individuals are not treated as reintroductions. For example, there were two releases of 38 captive-bred blackbuck, (*Antilope cervicapra*), into Nepal in 1980 and 1992 without recorded consequences (Shrestha, 2001). All such small, unsuccessful and/or undocumented efforts are excluded here.

For consideration here, an antelope population must be the subject of a clear statement that reintroduction is an intended objective; the population must be living in an ancestral or present range state, in indigenous habitat and with a number of individuals that comprise, or could increase to form, a viable population. Where enclosed, this will usually mean that the enclosure is large enough to meet a significant proportion of the animals' nutritional needs for most of the year, equating to light management only. The range of species and situations encountered prevent quantifying these thresholds, so subjectivity and exceptions must be accepted.

How common are antelope reintroduction efforts?

The species for which at least one reintroduction effort, in the sense above, has been made and documented are listed in Table 2. Any further transfers or releases to a second site in the same country, which may or may not have available documentation stating the objectives, are omitted here, although the present status of such populations may be reported later. Seven species fall into this category, with a further four for which release is intended but the animals are in the preliminary stages of holding in relatively small enclosures and/or are highly dependent on support, or have been released outside their indigenous range. These are labelled species-in-waiting.

The reintroduced antelopes comprise three *Gazella* species, the dama gazelle (*Nanger dama*), the addax (*Addax nasomaculatus*), and two oryx species, the Arabian and the scimitar-horned oryx (*Oryx dammah*). Thus, only seven out of 92 extant antelope species have been reintroduced, and these are from only four out of 35 genera. This suggests that, irrespective of success to date, reintroduction has not been a common conservation activity within the antelopes.

Do the antelope species of reintroduction efforts have anything in common?

To attempt to answer this question, a comparative analysis was performed on four key features of antelopes that could characterise each genus and its species. These were the species' body size, its main habitat, its diet, and its mobility pattern. The required information on each antelope genus and species was extracted from the compendium in Wilson and Mittermeier (2011). The information on each species was easily transposed into the species of the nomenclature used here (Wilson & Reeder, 2008), with the variation for *G. gazella* noted above.

Table 2 **Antelope species with a reintroduction effort**

Species	Country	Site	Reference
Dama gazelle, *Nanger dama*	Tunisia	Bou Hedma	Abaigar *et al.* (1997)
	Morocco	R'Mila	Wiesner & Muller (1998)
	Morocco	Souss Massa	Müller & Engel (2004)
	Senegal	Katane	Abaigar *et al.* (2009)
Dorcas gazelle, *Gazella dorcas*	Tunisia	Dghoumes	Anon. (unpublished)
	Tunisia	Oued Dekouk	Anon. (unpublished)
	Morocco	Sous Massa	Müller & Engel (2004)
	Senegal	Katane	Abaigar *et al.* (2009)
Sand gazelle, *Gazella subgutturosa*	Saudi Arabia	Mahazet as Sayid	Haque & Smith (1996)
	Saudi Arabia	Uruq Bani Ma'arid	Wacher & Robinson (2008)
Mountain gazelle, *Gazella gazella*	Saudi Arabia	Tuwaiq	Dunham *et al.* (1993)
	Saudi Arabia	Uruq Bani Ma'arid	Dunham (1997)
Addax, *Addax nasomaculatus*	Tunisia	Bou Hedma	Anon. (unpublished)
		Djebil	Anon. (unpublished)
		Senghar	Anon. (unpublished)
	Morocco	Souss-Massa/ Rokein	Engel & Brunsing (1999)
Scimitar-horned oryx, *Oryx dammah*	Tunisia	Dghoumes	Anon. (unpublished), Woodfine *et al.* (unpublished)
	Tunisia	Bou Hedma	Bertram (1988), Anon. (2001)
		Sidi Toui	Anon. (unpublished), Molcanova (2004)
		Oued Dekouk	Anon. (unpublished)
	Morocco	Souss Massa/ Arrouais	Müller & Engel (2004)
	Senegal	Katane	Abaigar *et al.* (2009)
Arabian oryx, *Oryx leucoryx*	Abu Dhabi	Umm al Zamoul	Kiwan, Al Kamy & Al Daheri (2008)
	Dubai	Dubai Desert Conservation Reserve	Simkins (2008)

Table 2 (*continued*)

Species	Country	Site	Reference
	Jordan	Wadi Rum	Al Zaidaneen & Al Hasaseen (2008)
	Oman	Al Wusta	Stanley Price (1989)
	Saudi Arabia	Mahazet as-Sayd	Islam *et al.* (2010a)
	Saudi Arabia	Uruq Bani Ma'arid	Islam *et al.* (2010a)
	Israel	Negev	Saltz (2008)
Species-in-waiting			
Bongo, *Tragelaphus eurycerus*	Kenya	Mt. Kenya	Veasey (2010)
Giant eland, *Taurotragus derbianus*	Senegal	Niokolo-Koba	Koláčková *et al.* (2011)
Roan, *Hippotragus equinus*	South Africa	Swaziland	Back to Africa: http://www.backtoafrica.co.za/operations/roan-antelope-project.html
Sable, *Hippotragus niger*	South Africa	Mokala	Back to Africa: http://www.backtoafrica.co.za/operations/sable-antelope-project.html

(a) Size

Size varies greatly within the antelopes. Table 1 shows the weight range found in each genus, displayed as the minimum recorded weight (usually female) and the maximum recorded weight (usually male). The species subject to a reintroduction effort are identified. The three *Gazella* spp. are the smallest to have been reintroduced, at 15–40 kg. There are 15 genera with 34 species that are smaller than this with no species reintroduced. The other four reintroduced species are medium-sized antelopes with weight ranges from 38–81 kg (*Nanger dama*) to 116–240 kg (*O. dammah*). There are five genera with 10 species that are larger than *O. dammah* but none has been reintroduced. It is notable that the four species-in-waiting are amongst the largest antelopes.

(b) Habitats occupied

The habitats used by antelopes were classified into:

(i) closed forest,
(ii) woodland and thicket
(iii) scrub, savanna, and thornbush
(iv) open grassland, desert, and steppe
(v) flooded grassland and swamp
(vi) rocky terrain/outcrops

Figure 1 shows use of these habitats by each genus. The reintroduced species are all occupants of open grassland-desert-steppe. While many of the 10 *Gazella* species also use woodland-thicket, the reintroduced *Gazella* species are all found in arid and semi-arid grasslands, and not in the more mesic grasslands and savannas. It is notable that the four species-in-waiting all occupy either dense forest (bongo, *Tragelaphus eurycerus*) or woodland-thicket (giant eland *Tragelaphus derbianus*, roan *Hippotragus equinus*, sable *H. niger*). Roan use open grassland vleis and woodland ecotones, while sable spend most time in woodlands (Wilson & Mittermeier, 2011). The giant eland is more catholic, occupying woodland, savanna, and open grassland (Wilson & Mittermeier, 2011). No species found exclusively in closed forest or in scrub, savanna or thornbush have been reintroduced; there has been no reintroduction effort for the relatively few species occupying the less common habitats such as flooded grassland and swamp or rocky outcrops. It appears, therefore, that reintroduced species use the wide range of open grassland types exclusively or to some extent.

(c) Diet

Diets of all antelope species were classified into one or more of four major diet types, with a yes/no class of two further types of (i) seeds, nuts, fruits in the diet; and (ii) roots and tubers. These are only broad categories, and individual species within genera with many species, such as *Gazella* or *Cephalophus*, may appear anomalous. Figure 2 shows the diet classifications of each genus. *Gazella*, of which three species have been reintroduced, comprise 10 species, which explains their broad classification as mixed feeders, in which browsing or grazing may predominate in certain species or at certain seasons; further, they eat fruits, nuts, and seeds, indicating a broad diet. The reintroduced *N. dama* probably browses more than it grazes (Wilson & Mittermeier, 2011).

R = reintroduction, r = species-in-waiting

Genus	No. species	Species and reintroduction status	Closed forest	Woodland, thicket	Scrub, savanna thornbush	Open grassland, desert, steppe	Flooded grassland, swamp	Rocky terrain/ outcrops
Neotragus	3	*batesi, moschatus, pygmaeus*	■					
Pantholops	1	*hodgsonii*	■					
Philantomba	2	*maxwellii, monticola*	■					
Pseudoryx	1	*nghetinhensis*	■					
Cephalophus	6	*brookei, dorsalis, leucogaster, spadix, zebra, ogilbyi,*	■					
	7	*adersi, callipygus, jentinki, weynsi, natalensis, niger, rufilatus,*	■	■				
	1	*sylvicultor*	■		■			
	1	*nigrifrons*	■	■			■	
Tragelaphus	1	*eurycerus [r]*	■					
	1	*buxtoni*	■	■	■			
	1	*angasii*		■	■			
	1	*strepsiceros*		■	■			■
	1	*imberbis*			■			
	1	*scriptus*	■	■		■	■	
	1	*spekii*	■				■	
Ammodorcas	1	*clarkei*		■				
Aepyceros	1	*melampus*		■	■			
Raphicerus	3	*campestris, melanotis, sharpei*		■	■			
Madoqua	4	*guentheri, kirkii, piacentinii, saltiana*		■	■			
Taurotragus	1	*derbianus [r]*		■	■	■		
	1	*oryx*		■	■	■		
Beatragus	1	*hunteri*		■	■	■		
Kobus	2	*megaceros, leche*				■	■	
	3	*ellipsiprymnus, kob, vardonii*		■	■	■		
Antilope	1	*cervicapra*		■		■		
Hippotragus	2	*equinus [r], niger [r]*		■		■		
Tetracerus	1	*quadricornis*		■		■		
Eudorcas	2	*rufifrons, thomsonii*		■	■	■		
Gazella	1	*gazella [R]*				■		■
	3	*dorcas [R], subgutturosa [R], leptoceros*				■		
	1	*bennettii*			■	■		■
	1	*cuvieri*		■	■			■
	1	*spekei*			■			
Alcelaphus	3	*buselaphus, caama, lichtensteinii*		■		■		
Connochaetes	2	*gnou, taurinus*		■		■		
Damaliscus	4	*korrigum, lunatus, pygargus, superstes*		■		■		
Redunca	3	*arundinum, fulvorufula, redunca*		■		■		
Antidorcas	1	*marsupialis*		■		■		
Ourebia	1	*ourebi*		■		■		
Dorcatragus	1	*megalotis*			■			
Litocranius	1	*walleri*			■			
Sylvicapra	1	*grimmia*			■			
Boselaphus	1	*tragocamelus*			■	■		
Saiga	2	*borealis, tatarica*			■	■		
Addax	1	*addax [R]*				■		
Nanger	1	*dama [R]*				■		
	1	*granti*			■	■		
	1	*soemmerringi*			■			
Oryx	1	*leucoryx [R]*				■		
	1	*dammah [R]*				■		
	1	*beisa*			■	■		
	1	*gazella*		■	■	■		
Pelea	1	*capreolus*				■		
Procapra	3	*gutturosa, picticaudata, przewalskii*				■		
Oreotragus	1	*oreotragus*						■

Figure 1 **Antelope habitat selection.**

R = reintroduction effort, r = species-in-waiting

Genus	No. spp	Reintroduction efforts	Near/ exclusive browse	Mixed: browse> graze	Mixed: graze> browse	Near/ exclusive graze	Seeds, nuts, fruit	Roots, tubers
Litocranius	1		■					
Madoqua	4		■					
Neotragus	3		■					
Pseudoryx	1		■					
Sylvicapra	1		■					
Cephalophus	15		■				■	
Oreotragus	1		■				■	
Philantomba	2		■				■	
Tragelaphus 1 *T. imberbis, T. scriptus, T. strepiceros*	3		■				■	
Aepyceros	1			■				
Ammodorcas	1			■				
Dorcatragus	1			■				
Nanger	3	R		■				
Ourebia	1			■				
Pelea	1			■				
Raphicerus	3			■				
Taurotragus	2	r		■				
Tragelaphus 2 *T. angasii, T. buxtonii, T. eurycerus*	3	r		■				
Pantholops	1			■	■			
Procapra	3			■	■			
Tragelaphus 3 *T. spekii*	1			■	■			
Saiga	2			■	■			
Gazella	7	RRR		■	■		■	
Antilope	1			■			■	
Boselaphus	1			■			■	
Tetracerus	1			■			■	
Alcelaphus	3				■			
Beatragus	1				■			
Eudorcas	2				■			
Hippotragus	2	rr			■			
Kobus	5				■			
Redunca	3				■			
Antidorcas	1				■	■		■
Addax	1	R				■		■
Oryx	4	RR			■			■
Connochaetes	2					■		
Damaliscus	4					■		

Figure 2 **Antelope diets.**

Species that are almost exclusive browsers, irrespective of whether their diet includes seeds-nuts-fruit, and those that browse more than graze have not been reintroduced. These also tend to be the species that occupy closed habitats (above), which exert selective pressures for smaller body size (Bro-Jørgensen, 2008). The oryx species are mixed feeders but with grass exceeding browse, and roots and tubers are also eaten. The addax is preferentially a grazer (T. Wacher, pers. comm., 2013), but also eats roots and tubers.

(d) *Movement patterns*

Each antelope species was classified as to whether it is nomadic (exhibiting large-scale and irregular movements in response to changing environmental conditions, not necessarily to a regular pattern), migratory (performing more or less regular and predictable seasonal changes in areas occupied, which may be widely separated), shows seasonal movements (on a smaller scale with predictable seasonal occupation of vegetation types), or is sedentary (occurring more or less equitably across a home range throughout the year), or some combination (Figure 3). Territoriality or lekking can be found in combination with some categories, but is not considered here.

Six of the seven reintroduced species are nomadic, but the two oryx species and dama gazelle also show migratory patterns (or did in the original wild populations), though clear distinction between nomadism and migration is difficult, and similarly between migratory and seasonal movements. The mountain gazelles at both release sites in Saudi Arabia occupy year-round territories where there is perennial vegetation (Wronski, 2010; Islam *et al.*, 2011c). The giant eland is also nomadic, but the roan, sable, and bongo perform seasonal movements and may tend to sedentarism. There are no reintroduced species within the 15 genera that are exclusively sedentary (counting a total of 37 species) or among the four *Tragelaphus* species that are sedentary.

(e) *Conservation status*

Table 1 shows the Red List status of 86 species. Within these, one species is Data Deficient and 51 are of Least Concern, with the remaining 34 species having some degree of endangerment. Of the seven species that have been reintroduced, three are Vulnerable, one is Endangered, two are Critically Endangered, and one is Extinct in the Wild. In contrast, of the four species-in-waiting, one is Near Threatened, and three are of Least Concern.

Do these observations adequately explain which species can be successfully reintroduced?

Combining 92 species into relatively few classes for each attribute of habitat use, diet, and mobility is necessarily crude, and may mask species-specific attributes that are relevant to reintroduction effort or success. However, some generalisations are possible about the species for which a reintroduction effort has been made:

R = reintroduction effort, r = species-in-waiting

Genus	No. species	Reintroduction effort	Nomadic	Migratory	Seasonal movements	Sedentary	
Addax	1	R	■				
Boselaphus	1		■				
Oryx	4	RR	■	■			
Nanger	3	R	■	■			
Procapra	3		■	■	■		
Taurotragus	2	r	■		■		
Beatragus	1		■		■		
Gazella	7	RRR	■	■		■	
Pantholops	1		■			■	
Saiga	2			■			
Antidorcas	1			■			
Damaliscus	4			■	■		
Connochaetes	2			■	■	■	
Eudorcas	2			■		■	
Pseudoryx	1				■		
Kobus	5				■		
Tragelaphus 1: *T. strepiceros, T. angasii, T. eurycerus*	3	r			■		
Hippotragus	2	rr			■	■	
Alcelaphus	3				■	■	
Redunca	3				■	■	
Ammodorcas	1					■	
Antilope	1					■	
Dorcatragus	1					■	
Pelea	1					■	
Tetracerus	1					■	
Aepyceros	1					■	
Litocranius	1					■	
Sylvicapra	1					■	
Cepahlophus	15					■	
Ourebia	1					■	
Raphicerus	3					■	
Tragelaphus 2: *T.imberbis, T.buxtonii, T. spekii, T. scriptus*	4					■	
Neotragus	3					■	
Oreotragus	1					■	
Madoqua	4					■	
Philantomba	2					■	

Figure 3 **Antelope movement patterns.**

- Antelope reintroductions have involved few genera (four out of 35) and few species (seven out of 92); thus most genera have no reintroduced species.
- The smallest species have **not** been reintroduced; otherwise, species with a wide range of sizes have been reintroduced.
- Species occupying open grassland-steppe-desert habitats to some extent have been reintroduced; those in closed habitats have **not**.
- As diet largely derives from habitat selection, reintroduced species tend to be grazers, and usually show wide diet composition.

- Reintroduced species are mobile, either migratory, nomadic, or with seasonal movements; sedentary species are **not** reintroduced, with the sole exception of the mountain gazelle.
- The seven reintroduced antelopes occupy semi-arid, arid, and even hyper-arid areas.
- With regard to extinction risk, while 40% of all antelope species are Near Threatened or worse, all the reintroduced species are Vulnerable or worse. This is not the case with the species-in-waiting. This would suggest that reintroduction efforts for antelopes are being directed at those species of relatively high degrees of endangerment. Further, the motivation or resources to reintroduce the species-in-waiting may be less because of their generally more secure conservation status.

These conclusions above suggest some aspects of antelope biology that may predispose species to reintroduction. But, do they explain why other species have not been reintroduced, and do they provide insights into the prospects for the species-in-waiting? The reasons as to why a species has not been reintroduced might include the following:

1. The conservation status of a species may mean there is no need for reintroduction. This is unlikely to be the prevalent situation given the status of antelopes (IUCN Antelope Specialist Group, 2009) and the generally observed reduction in effective range size of many species.
2. Reintroduction is not the most suitable or feasible conservation solution for a species. As the key reproductive parameters of birth rate, intrinsic rate of increase, and net reproductive rate scale closely with size in African mammals (Western, 1979), small antelopes have high fecundity and high potential rates of increase; small species also live at higher densities (Peters, 1983). Consequently, population recovery in small antelopes, many of which live in closed habitats and are subject to very high offtakes through hunting (Fa *et al.*, 2003; Fa, this volume), is probably most effectively achieved by reduction of hunting and allowing natural increase or recolonisation.
3. There are no adequate sources of stock for release; the reverse is probably true for all the reintroduced species, for which zoos and specialised facilities hold significant or even very large numbers. Successful reintroductions release larger numbers of founder individuals (Wolf *et al.*, 1998).
4. The threats which might have reduced antelope populations have not been identified and mitigated or removed, in which case a reintroduction is not justified (IUCN/SSC, 2013) or would be premature; this situation might apply to some of the species-in-waiting.

5. The four elements of antelope biology considered above are only a very crude profile of a genus, and even cruder for its component species: some antelope species probably have habitat requirements that are currently beyond our knowledge or appreciation; this is probably the case for some of the smaller browsing species living in forests. For such species, a reintroduction that is not based on in-depth knowledge of the species is unlikely to succeed.
6. Any reintroduced population must be monitored: this is infinitely easier for species that are large and occupy open habitats compared to small occupants of closed habitats. The seven reintroduced species all fall into the former category; thus the difficulties of post-release monitoring, an essential activity (IUCN/SSC, 2013), may deter efforts to reintroduce species of closed habitats.

Similarly, there is no reason to think that these reintroduced species necessarily represent the most 'reintroducible' species. They may have been chosen for reintroduction for diverse reasons:

- Their status in the wild made them a priority for re-establishment, or
- They were selected for political or, at least, non-biological reasons (including a pool of expertise in the zoo world), or
- A large captive population encouraged a reintroduction effort, or
- They may mistakenly have been considered most reintroducible.

But, whatever the motivation, antelopes for which a reintroduction effort has been made share certain characteristics.

How successful have antelope reintroductions been?

Table 3 shows the best estimate of current numbers of the seven species, and the four species-in-waiting. The numbers are likely to be minimal, and in some cases numbers have been higher in the past than reported most recently. Figure 4 shows the locations by country of antelope reintroductions, as defined here. Relatively few countries in Africa and west Asia have reintroduction efforts. But four countries (Morocco, Tunisia, Senegal, Saudi Arabia) are attempting to restore their antelope faunas through multiple species reintroductions.

Dama gazelle: Of the three recognised sub-species, *Nanger dama ruficollis*, *N. d. dama,* and *N. d. mhorr*, only the latter has been the subject of reintroduction efforts. The two males at Bou Hedma (M. Petretto, pers. comm., 2015) are the

Table 3 **Reintroduced antelopes: best estimates of locations and numbers. The area available, if fenced, may be part of a larger protected area.**

Species	Country	Site	Area available, ha[2]	Number	Year	Reference
Dama gazelle	Tunisia	Bou Hedma	2400	2	2014	M. Petretto (pers. comm., 2015)
	Morocco	R'Mila	470	600	2007	HCEF[1] (pers. comm., 2013)
		Safia	600	41	2013	HCEF (pers. comm., 2013)
		Souss Massa/Rokein	2000	0	2013	HCEF (pers. comm., 2013)
	Senegal	Katane	660	15–18	2013	Abaigar (pers. comm., 2013)
		Guembeul	720	13–15	2013	Abaigar (pers. comm., 2013)
Dorcas gazelle	Tunisia	Dghoumes	8000	80	2015	M. Petretto (pers. comm., 2015)
		Oued Dekouk	500	12	2015	M. Petretto (pers. comm., 2015)
		Bou Hedma	5000	214	2014	M. Petretto (pers. comm., 2015)
		Haddej	320	43	2015	M. Petretto (pers. comm., 2015)
		Jebil	7700	20–25	2014	M. Petretto (pers. comm., 2015)
		Sidi Toui	6315	150	2015	M. Petretto (pers. comm., 2015)
	Morocco	Souss Massa/Rokein	2000	450	2011	W. Oubrou (pers. comm., 2013)
		Souss Massa/Arrouais	1200	400	2011	W. Oubrou (pers. comm., 2013)
		Tizguine	250	100	2012	HCEF (pers. comm., 2013)
		Jbilet	280	115	2012	HCEF (pers. comm., 2013)
		Bou Assila	200	88	2012	HCEF (pers. comm., 2013)
		R'mila	2000	300	2007	HCEF (pers. comm., 2013)
		Bouznika	400	150	2007	HCEF (pers. comm., 2013)

(*continued overleaf*)

Table 3 ***(continued)***

Species	Country	Site	Area available, ha^2	Number	Year	Reference
	Senegal	Katane	660	28–34	2013	T. Abaigar (pers. comm., 2013)
		Guembeul	720	29	2013	T. Abaigar (pers. comm., 2013)
Sand gazelle	Saudi Arabia	Mahazet as Sayid	220,000	352	2010	Islam *et al.* (2011b)
		Uruq Bani Ma'arid	1,260,000	380	2008	Cunningham & Wacher (2009)
Mountain gazelle	Saudi Arabia	Jebel Tuwaiq	187,000	30–40	2011	Wronski *et al.* (2011)
		Uruq Bani Ma'arid	1,260,000	282	2011	Islam *et al.* (2011)
Addax	Tunisia	(Bou Hedma)/Haddej	5000	49	2014	M. Petretto (pers. comm., 2015)
		Djebil	7700	76	2014	M. Petretto (pers. comm., 2015)
		Senghar	2660	32	2014	M. Petretto (pers. comm., 2015)
	Morocco	Souss Massa/Rokein	2000	240	2011	W. Oubrou (pers. comm., 2013), HCEF (pers. comm., 2013)
		Safia	600	41	2013	HCEF (pers. comm., 2013)
Scimitar–horned oryx	Tunisia	Dghoumes	8000	63	2014	M. Petretto (pers. comm., 2015)
		Bou Hedma	5000	37	2014	M. Petretto (pers. comm., 2015)
		Bou Hedma/Haddej	320	18	2014	M. Petretto (pers. comm., 2015)
		Sidi Toui	6315	41	2014	M. Petretto (pers. comm., 2015)
		Oued Dekouk	5750	38	2014	M. Petretto (pers. comm., 2015)
	Morocco	Souss Massa/Arrouais	1200	270	2013	W. Oubrou (pers. comm., 2013), HCEF 2013

	Senegal	Katane	660	60–70	2013	T. Abaigar (pers. comm., 2013)
		Guembeul	720	110–120	2013	T. Abaigar (pers. comm., 2013)
Arabian oryx	Abu Dhabi	Um al Zomoul	790,000	347	2011	H. Al Qamy (pers. comm., 2011)
	Dubai	Dubai Desert Conservation Reserve	22,500	500	2013	G. Simkins (pers. comm., 2013)
	Jordan	Wadi Rum	4,300	75	2013	E. Eid (pers. comm., 2015)
	Oman	Al Wusta	282,400	60	2011	Al Jahdhami *et al.* (2011)
	Saudi Arabia	Mahazet as Sayid	220,000	324	2008	Islam *et al.* (2010a)
		Uruq Bani Ma'arid	1,260,000	200	2008	Islam *et al.* (2010a)
	Israel	Negev	200,000	130	2011	D. Saltz (pers. comm., 2011)
Species-in-waiting						
Roan	Swaziland	Mlilwane	35	35	2013	M. Reilly (pers. comm., 2013)
Sable	South Africa	Mokala	28,151	28	2011	Back to Africa (nd)
Bongo	Kenya	Nanyuki	<1000	61	2013	D. Bunge (pers. comm., 2015)
Giant eland	Senegal	Bandia, Fathala	1,790[2]	83	2013	Brandlová *et al.*, 2014)

[1] HCEF: Haut Commissariat aux Eaux et Forêts et à la Lutte Contre La Désertification, Morocco.

[2] Between five enclosures.

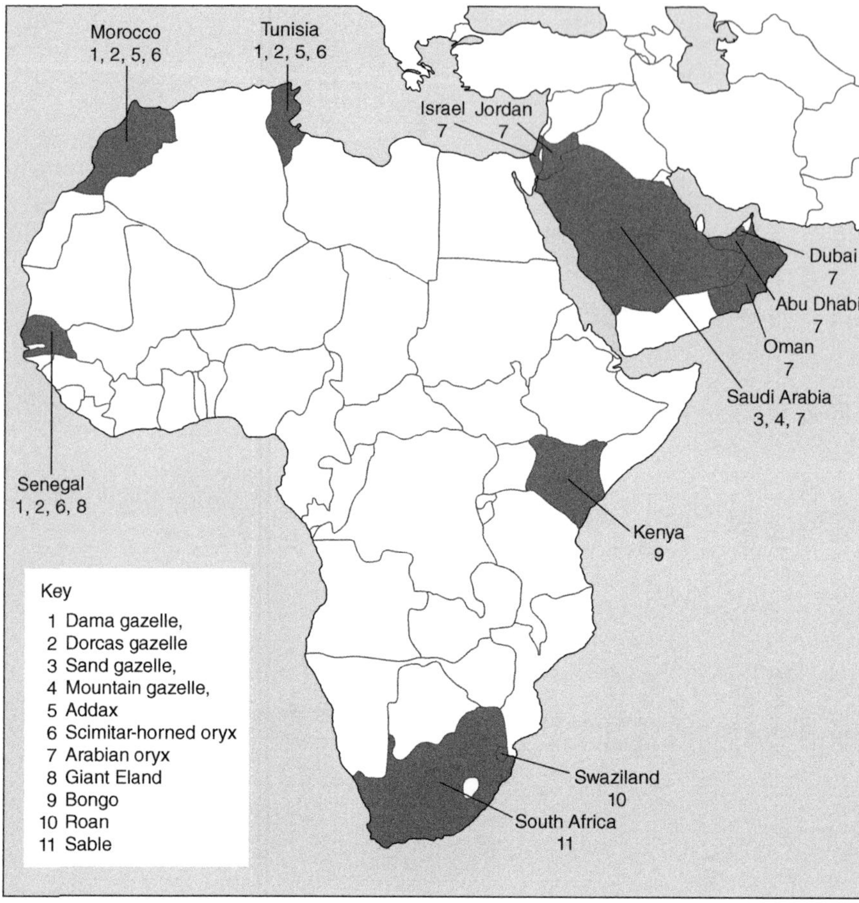

Figure 4 **The countries where antelopes have been reintroduced or which have species-in-waiting.**

remaining product of the release of a total of 22 animals in 1992–1994 and 1995 (Abaigar *et al.*, 1997).

Morocco hosted a deliberate effort to reintroduce the dama through repatriating animals from Europe to 'half-way' fenced reserves. Accordingly, 27 dama went to R'Mila Royal Reserve from 1992, increasing there to 41 in 1997 (Wiesner & Müller, 1998), and approximately 600 in 2007 (HCEF, pers. comm., 2013) under conditions of good security and provisioning (T. Abaigar, pers. comm., 2013) although in suburban surroundings. R'Mila provided 21 dama in 1996–1996 to Souss Massa; they numbered about 11 in 2002 (Müller & Engel, 2004), but none are reported in 2013 (HCEF, pers. comm., 2013).

In 2008, 16 dama from R'Mila were sent to Safia, where habitat and vegetation conditions are most suitable, allowing the animals to be independent of the food supplied to the addax there (T. Abaigar, pers. comm., 2013); they have increased to 41 in 2013 (HCEF, pers. comm., 2013). In 1984, eight founders from the captive centre of Almeria (EEZA, CSIC), Spain, went to Guembeul, Senegal (Cano *et al.*, 1993); although these were known to total 35–40 in 2005 (Jebali, 2008), their numbers have declined to 15–18, due probably to management standards and predation by dogs (T. Abaigar, pers. comm., 2013). In 2003, Guembeul provided five founders to Katane (Jebali, 2008), where numbers had reached 15–18 in 2013 (T. Abaigar, pers. comm., 2013).

The performance record of dama returned to Africa is patchy: while they have clearly increased greatly in R'Milia under exceptional conditions, they have failed to establish in Tunisia and in Souss Massa, Morocco; moreover, all three sites are outside its accepted indigenous range. The Safia population may have the best prospects for increasing to large numbers under more natural conditions.

This suggests that the dama is demanding in its ecological requirements and/or its social environment, with consequent implications for successful reintroduction. It is also notable that the population of dama in the Ouadi Rimé-Ouadi Achim Game Reserve, Chad, was assessed at 6,000–8,000 before the extreme hunting pressure of the late 1970s, following which the reserve was abandoned in 1984 (Grettenberger & Newby, 1986), and recent surveys there suggest only a very small population persists (Wacher & Newby, 2012).

Dorcas gazelle: Dorcas gazelles (*Gazella dorcas*) occur at 15 sites, which include three of at least 300 individuals each. It is notable that this species is the only reintroduced antelope whose wild numbers still exceed those in captivity: in Chad, more than 8,000 occur in the Manga area, and an estimated 10,000 based on a 5% sample survey in the Ouadi Achim-Ouadi Rimé Game Reserve (Wacher & Newby, 2012). The origins of some of the Moroccan populations of dorcas gazelles may be unclear (whether from confined wild individuals or from captive sources (Wacher, pers. comm., 2013)). Their abundance and number of populations in Morocco indicate successful increase followed by establishment of further herds; however, none have been released into unfenced areas.

Mountain gazelle: The mountain gazelle (*Gazella gazella*) has two reintroduced populations, and both are in unfenced areas. At Jebel Tuwaiq, 84 were released 1990–1995, with a further 21 released in 2007. Numbers reached 185 in 1995 but by 2011 only 30–40 in 2011 existed (Wronski *et al.*, 2011). In addition to immediate factors such as poor relations with local communities and poaching, the area was on the edge of the species' historic range and provided marginal habitat and low food availability (Wronski, 2010). At 'Uruq Bani Ma'arid, 73 gazelles were released in 1996–1997; censuses in 2009, 2010, and 2011 indicated populations of 224, 307, and 282 respectively, although there were suggestions that the true figure might be much higher (Islam *et al.*, 2011c).

Sand gazelle: Between 1990 and 1993, 135 sand gazelles (*Gazella subgutturosa*) were released into Mahazat as-Sayd Reserve, a fenced area of 220,000 ha in south-west Saudi Arabia (Islam *et al.*, 2011a). Over subsequent years, they exhibited great increases (reaching a maximum of 1537 in 2005) but also great losses due to extreme drought and lack of food (in 1999, 939 died) (Islam *et al.*, 2011a). In the unfenced 'Uruq Bani Ma'arid, 204 sand gazelles were released in 1995–1996 (Dunham *et al.*, 2001); the observed wide dispersal of some individuals made subsequent census difficult, but a survey in 2008 indicated a population of 380, with the likelihood that numbers may have been greater in previous years, and are likely to fluctuate according to climatic conditions (Cunningham & Wacher, 2009).

Addax: The addax number approximately 400 across four sites. While this might appear a significant total, the largest population, at Souss Massa, is subject to great fluctuations in numbers (T. Wacher, pers. comm., 2013). This total exceeds the estimated 200 known to remain in the wild in one area of Niger (Wacher *et al.*, unpublished).

Scimitar-horned oryx: There are no scimitar-horned oryx at free-range in the wild, but there are over 600 individuals between eight enclosed sites in three countries, principally in Tunisia. The large number in Morocco's Souss Massa National Park is evidence of the successful early stages of eventually reintroducing acclimatised animals to Saharan national parks, with subsequent dispersal into the Sahara zone (Müller & Engel, 2004), but these later steps have not yet been taken.

The more immediate prospect for free-living scimitar-horned oryx lies with their planned release into the Ouadi Rimé-Ouadi Achim Game Reserve of central Chad. Through a project of the governments of the Republic of Chad and Abu Dhabi, oryx will be provided by the latter for release in Chad. The project, which is managed by the Sahara Conservation Fund, will bring the first oryx to Chad in late 2015 or early 2016 for release soon thereafter (J. E. Newby, pers. comm., 2015).

Arabian oryx: Arabian oryx numbers exceed 1600 at seven sites across six countries. The largest numbers are in the fully-fenced area of the Dubai Desert Conservation Reserve (Simkins, 2008) and in the very large Umm Al Zamoul of Abu Dhabi (Kiwan *et al.*, 2008). The population in 'Uruq Bani Ma'arid is the only one not contained by fencing. Between 1995 and 2001, 15 groups totalling 174 oryx were released. The population numbered 200 in 2010 (Islam *et al.*, 2010, but numbers were much reduced by 2013 following a period of drought and illegal hunting (T. Wacher, pers. comm., 2013). The 60 oryx in Oman are the first to be released into the largely-fenced area of 282,400 ha; no oryx from the original reintroduction of the 1980's (Spalton *et al.*, 1999) now persist in the wild (M. Al Jahdhami, pers. comm., 2013). This total of more than 1000 adult Arabian oryx, maintained over five years at multiple sites within six countries, has allowed it to be downlisted from Extinct in the Wild to Vulnerable.

Table 3 shows that for dama and dorcas gazelles, scimitar-horned oryx and addax, 100% of the animals of these reintroduction efforts remain enclosed, albeit under varying conditions of dependence or independence of provisioning and other management support. Approximately half of the reintroduced sand gazelles are living unenclosed in the 'Uruq Bani Ma'arid of Saudi Arabia, while all the reintroduced mountain gazelles at two sites in Saudi Arabia are unenclosed. Currently, the only significant numbers of reintroduced and unenclosed Arabian oryx are in 'Uruq Bani Ma'arid. The present demographic status of most of these populations does not indicate successful reintroduction, even in the absence of clear criteria as what might constitute success or viable numbers, or progress towards these. The dorcas gazelle thrives in fenced conditions, evidenced by large populations at multiple sites, and thus remains the most widespread *Gazella* species in the wild (Wilson & Mittermeier, 2011). This is in marked contrast to the dama gazelle, which has more exacting requirements in semi-captivity, and less is known of its past or present ecology in the wild. While arid-land antelopes can exhibit great powers of increase under favourable conditions, they conversely can suffer high mortality under extreme, adverse conditions, and the overall performance of the Arabian oryx in the unfenced 'Uruq Bani Ma'arid does not yet indicate long-term viability convincingly. The mountain and sand gazelles have experienced mixed fortunes: both species are performing well in the 'Uruq Bani Ma'arid; however, the long-term persistence of mountain gazelles at Tuwaiq appears less certain.

Species-in-waiting

Giant eland: The semi-captive, western giant eland (*Tragelaphus derbianus derbianus*) in Senegal, a critically endangered subspecies, is divided between two sites into four breeding herds and one bachelor herd in enclosures between 70 and 1000 ha, numbering 83 animals in 2014 (Brandlová *et al.*, 2014). They derive from one male and five female wild founders, although genetic diversity remains high (Koláčková *et al.*, 2011). The only remnant wild population numbers 100–150 (Renaud *et al.*, 2006) in Niokolo-Koba National Park, where the release of any semi-captive eland as a reintroduction or reinforcement (IUCN/SSC, 2013) is not yet feasible because of rampant poaching, illegal tree-cutting, and massive incursions by domestic stock (Howard *et al.*, 2007).

Bongo: Captive-managed mountain bongo (*Tragelaphus eurycerus isaaci*) in Kenya, also a critically endangered subspecies, currently number 61 (D. Bunge, pers. comm. 2015). They are the products of a small number of animals taken from the wild in the 1970s and a repatriation in 2004 of 18 bongo from the USA, of Kenyan ancestry (P. Reillo, pers. comm., 2015). Eight of the latter succumbed to theileriosis in Kenya, but the remainder developed immunity through controlled exposure, and the managed population appears now to be epidemiologically resistant (P. Reillo, pers. comm., 2015). These bongo currently occupy a series of enclosures in forest habitat, where they are supplemented with natural browse. In preparation for release into the wild, weaned young are kept in a large paddock where human contact is minimised and discouraged (C. Musyoki, pers. comm., 2013).

The ultimate objective is the release of bongo into one or more montane forest areas, all of which contain very small remaining wild populations. Any release is pending the outcome of genetic analyses, understanding current management systems, and assessment of the risk of reinforcing a remnant native population. In anticipation, institutions in the USA are building up bongo numbers for further repatriations and subsequent in-Kenya breeding to provide an adequate supply of animals for release (P. Reillo, pers. comm., 2015).

Roan and sable antelope: The indigenous populations of roan and sable in Kruger National Park showed major declines, starting in the mid-1980s, as did those of other co-existing selective grazers (Owen-Smith & Mills, 2006). Various proximate factors have been suggested for these species, which are at the extreme of their current continent-wide ranges in Kruger (Ogutu & Owen-Smith, 2003). These include outbreaks of anthrax (De Vos *et al.*, 1973), competition with other grazers that immigrated in response to provision of water sources (Owen-Smith *et al.*, 2011), increased predation by lions as more grazers immigrated in response to the new water sources (Harrington *et al.*, 1999), and perimeter fencing that may have curtailed the potential for movements in response to changing conditions (Grant *et al.*, 2002). However, the decline in roan and sable, and of other selective grazing species, seems ultimately due to the impact of long-term climate changes on vegetation quantity and quality, to the detriment of selective grazers and to the benefit of bulk grazers (Seydack *et al.*, 2012 a, b).

The founder stock of sable were bred in European zoos and transported to South Africa. Because theileriosis was a threat for these animals (Sable Antelope Project, 2015), they have been released in Mokala National Park, which is a further 900 km south-west of Kruger and hence outside the accepted indigenous range (Skinner & Smithers, 1990). This is, therefore, technically a conservation introduction (IUCN/SSC, 2013). As its climate is cold enough to reduce the tick challenge, there is the prospect of this sable population thriving, with four new calves seen in early 2012 (Sable Antelope Project, 2015). But, as it is outside its indigenous range, it is treated as a species-in-waiting for reintroduction to indigenous range.

The roan reintroduction site in Mlilwane Game Reserve, Swaziland (Back to Africa, 2011), approximately 350 km due south of Kruger, is at the very limit of historic range (Skinner & Smithers, 1990). The captive-bred roan brought from Europe to southern Africa, for release in Swaziland, were surmised to be of geographically mixed provenance, and suffered greatly in South Africa from tick-borne theileriosis (Back to Africa, 2011). The reintroduction strategy is to build up adequate numbers over several generations under captive, but extensive conditions, with the intention of increasing adaptation to local conditions. This would include (i) reducing the incidence of births outside the traditional southern Africa season, which has resulted in calf mortality due to cold; and (ii) crossing with local males to reduce susceptibility to theileriosis (M. Reilly, pers. comm., 2013).

Given such a complex interaction of factors and challenges to the persistence of roan and sable in Kruger, the successful re-establishment of captive-bred roan and sable as viable, self-supporting populations at the extreme margins, or beyond, of their historic ranges may be very challenging.

Constraints to releasing antelopes on unenclosed areas

Because the antelopes released into unfenced areas all occupy arid areas of extreme climate and unpredictable rainfall, population establishment is intrinsically subject to many stochastic factors, both environmental (such as great variation in food and water availability between years) and demographic, such as non-breeding individuals or runs of male calves (Stanley Price, 1989). This becomes evident in dispersal patterns, survival and breeding rates, and individual performance.

The mountain gazelle at 'Uruq Bani Ma'arid in Saudi Arabia is an illustrative example of how intensive monitoring recorded yearly variations in breeding rates (Islam *et al.*, 2011c). Both Tuwaiq and 'Uruq Bani Ma'arid were at the extreme limits of the species' known past range (Wronski *et al.*, 2011; Islam *et al.*, 2011c). The performance at Tuwaiq was partially ascribed to lack of food (Wronski, 2010). In contrast, the released animals in 'Uruq Bani Ma'arid dispersed widely along an escarpment and evidently found adequate patches of habitat amidst large, relatively barren expanses (Islam *et al.*, 2011c).

But, in addition, reintroduction efforts show that the main threat to free-living antelopes is undoubtedly illegal hunting ('poaching'). The Arabian oryx in Oman is illustrative: a total of 33 (Spalton, 1993) were released in 1980–1984 and 1987–1989; these had increased to 359 in 1996 when the populations started to crash due to poaching and illegal capture efforts, leading to the return of approximately 120 individuals, predominantly female, to the release enclosure of 1 km^2 in 1998 (Spalton *et al.*, 1999). Numbers have subsequently built up in enlarged enclosures; 60 oryx have been released into a newly constructed fenced reserve of 2824 km^2, in anticipation of further large scale releases (Al Jahdhami, 2011).

Poaching of the Arabian oryx has been recorded even in the relatively inaccessible and unenclosed sand dune area of 'Uruq Bani Ma'Arid in Saudi Arabia where a population founded by 174 individuals released up to 1995 numbered only about 200 in 2010 (Islam *et al.*, 2011c). Between 1998 and 2001 poaching accounted for 30% of known deaths (Ostrowski & Bedin, 2001). The numbers of mountain gazelles at the unenclosed Jebel Tuwaiq in Saudi Arabia have also fluctuated greatly, with poaching a major problem, alongside issues of marginal habitat and attitude towards the reserve by the local people (Wronski *et al.*, 2011). Antelopes released from any of the fenced Tunisian populations risk illegal killing largely because of the density of surrounding human populations and the associated extent of land conversion (T. Woodfine, pers. comm., 2013). As these antelopes occupy open, and for the most part, easily accessible habitats, hunting is not surprisingly the major threat to population establishment and persistence.

Is fencing the solution? The performance and situation of both the reintroduced species and the species-in-waiting suggest that the risks facing released antelopes on unenclosed land may be unacceptable. Consequently, fenced reserves are perhaps necessary for the foreseeable future. Fencing is costly, and very large areas need to be enclosed if they are to support viable numbers of large antelopes in arid or semi-arid conditions. Fencing in arid lands is subject to the risks faced by any infrastructure in such an extreme environment. Floods, stochastic in nature in arid-lands, can destroy many kilometres of fencing (M. Al Jahdhami, pers. comm., 2011), and drifting sand may obliterate fences allowing, for example, addax in Tunisia to walk out (T. Woodfine, pers. comm., 2013). This indicates that great skill should be used to site fences and to ensure they are designed against multiple criteria and are cost-effective (Bode and Wintle, 2009).

Fenced enclosures have two further implications:

1. Fencing off areas in Guembeul, Senegal (Cano *et al.*, 1993) and Dubai (Simkins, 2010) demonstrated the remarkable recovery of vegetation: the release of grazing pressure through the removal of domestic livestock is greater than the subsequent impacts of stocked antelopes. The development of vegetation in the 2400 km^2 Mahazet as-Sayd, Saudi Arabia, is well documented; for example, the number of plant species recorded in the reserve increased from 56 at the time of fence completion in 1989, to 112 in 1990 and 173 in 1998 (Zafar-ul Islam, pers. comm., 2013). Monitoring at Dghoumes National Park in Tunisia indicates the same effect depending on time of year, previous rainfall and the relative grazing pressures from domestic stock outside and wild species inside (T. Woodfine, pers. comm., 2013). There is political opportunity in this phenomenon, through the argument for increasing the number of areas strictly protected from livestock in the interests of broad-based biodiversity conservation.

   It is also possible that vegetation succession within fenced areas may progress until the habitat inside has developed to being no longer suitable for its antelope inhabitants. This is a further consequence of preventing free movement by mobile species.
2. The potential rate of increase of antelopes may cause fenced areas to become over-stocked, leading to stress on animals and mortality. This has been well documented for Mahazet as-Sayd: a run of exceptionally dry years culminated in 939 out of 1034 sand gazelles dying in 1999, 644 from 1104 in 2000, and 679 from 1200 in 2006.

Also, 159 out of 378 Arabian oryx died in 2008 (Islam *et al.*, 2010b). This led to a systematic assessment of the situation and three primary strategies were developed: (i) monitoring, (ii) early warning, and (iii) emergency management programmes (Islam *et al.*, 2010c).

As antelope numbers increase to ceiling levels in fenced areas, the case for establishing further enclosed populations through extending or building new enclosures should be assessed against various criteria, which should certainly include the overall conservation status of the species. While enclosures at carrying capacity may be a stimulus to consider release into unfenced areas, one solution is to design enclosures to be partly porous. Dghoumes National Park, covering 800 km^2 in Tunisia, is fenced on only three sides (Woodfine *et al.*, 2009); the fourth boundary comprises a minor physical barrier of low hills, mountains, and less suitable habitat for scimitar-horned oryx. The oryx population will be monitored to see if it adjusts its density through voluntary emigration. Emigrant animals may be at greater risk, but as the value of each individual animal decreases, more costly conservation interventions become less needed or justifiable.

Overall, fencing avoids obvious problems for establishing antelope populations, but it comes with attendant issues for future years.

Discussion

Antelope reintroduction efforts have not been common considering the number of genera and species. Most serious efforts to reintroduce have involved species that are medium-size inhabitants of arid or semi-arid open habitats. Such species tend to be predominantly grazers but with broad diets, are relatively mobile, and occupy habitats with very low human densities. Very large and very small species, those of closed habitats and those that are sedentary (with the exception of the mountain gazelle in Saudi Arabia) and are predominantly browsers have not been reintroduced.

These generalisations suggest that successful establishment of the species-in-waiting may be more challenging. While this situation undoubtedly reflects many aspects of the biology of these species, antelopes of less arid environments may face further factors of competing land-uses, through which the opportunity cost of reintroduction and biodiversity conservation in general may be less favourable.

The large desert antelopes were reduced in number or driven to extinction by hunting, and this remains the greatest risk for new populations. It has been noted that species that are, in general, easy to remove through direct persecution are also the easiest to restore through active management (Stanley Price, 1989). While the renewed hunting of Arabian oryx in Oman in the late 1990s had different motivations and perpetrators from the hunting that caused the species' extinction from the wild in 1972, the threat of hunting remained. This suggests that while the reintroduction of this type of antelope is technically possible, its feasibility remains in doubt through persistent human threats. Hence, success requires political commitment and effective action rather than purely technical approaches.

The antelope reintroduction situation raises issues not restricted to antelopes, such as what constitutes successful population recovery. Depending on the context, recovery requires conservation interventions at one of the following five levels (Redford *et al.*, 2011):

1. Captive management
2. Intensive management
3. Light management
4. Conservation-dependence
5. Self-sustaining

The situations of 'reintroduced' antelopes cover a wide variety of both previous and current conditions. Most species started in Redford *et al.*'s (2011) category of 'Intensively managed' and are characterised by their captive populations outnumbering their small and remnant wild populations. Where antelopes occupy a large fenced area, in which breeding is not managed, and there is provisioning and protection from terrestrial predators, they have probably reached the 'conservation-dependent' stage. Some released antelope populations have moved further along the continuum: the mountain gazelle in 'Uruq Bani Ma'arid, Saudi Arabia, is probably truly self-sustaining, provided pressures from hunting and infrastructure development become no greater than at present (T. Wacher, pers. comm., 2013). The Arabian oryx in Uruq Bani Ma'arid are probably independent of supplied water but, despite active protection from poaching, their numbers have declined over recent years. For antelope reintroductions in general, truly self-sustaining status may be rarely attainable: 84% of all species under the US Endangered Species Act remain 'conservation-reliant' despite specific conservation interventions to improve their status (Scott *et al.*, 2010).

We must recognise that current reintroductions of antelopes are unlikely to restore their populations to an earlier, utopian condition. Therefore, at a practical level, antelope conservationists need to be both visionary and pragmatic: do we see large fenced areas as an interim stage, both in the conservation of the species and also in the evolution of society? Will the communities that share lands with antelopes again tolerate, or welcome back these species? Will they be able to use them sustainably? Is there potential for using the current model of antelope conservation through enclosures as a means to demonstrate the recovery power of natural vegetation with its attendant benefits for man and biodiversity? Can new technologies help us create cheap or virtual enclosures that can be moved around, thereby increasing the size of the recovering area?

Can we not exploit the fact that for the arid-land antelopes that are being reintroduced, absolute lack of habitat is not the main problem, unlike the situation for so many other species? Should antelope conservation join forces more effectively with carnivore conservation, for many problems of conflict between large carnivores and human interests are probably the result of depleted prey populations, many of which would be antelopes?

Many antelope populations need recovery. To promote further successful antelope reintroductions, should there be an evidence-based review of the factors that have promoted reintroduction success, exploring in greater depth antelope biology and 'reintroducibility'? Generalising, can the antelope reintroduction experience assist with further exploration of the Red List criteria, which are silent on the extent of containment and of management inputs for species considered 'in the wild'? If populations that are either enclosed or supported or both can be treated as 'wild', there must be thresholds for the permitted extent or intensity of management support. Antelope reintroduction to date has been an uncommon tool in conservation, even if a high profile one. However, the pointers for greater adoption and success are clear, and experience with antelopes can contribute much to the advancement of reintroduction science.

References

Abaigar, T., Cano, M., Espeso, G. & Ortiz, J. (1997): Introduction of Mhorr gazelle, *Gazella dama mhorr*, in Bou-Hedma National Park, Tunisia. *International Zoo Yearbook* **35**: 311–316.

Abaigar, T., Youm, B., Niaga, M., Ensenyat, C., & Cano, M. (2009): The role of Senegal in the recovery of Sahelo-Saharan antelope species: the case of the reintroduction of Dorcas gazelle. *Gnusletter* **28**(1): 6–8.

Al Jahdhami, M., Al-Mahdhoury, S. & Al Amri, H. (2011): The re-introduction of the Arabian oryx to the Al Wusta Wildlife Reserve in Oman: 30 years on. In *Global Re-introduction Perspectives: More case studies from around the globe:* 194–198. Soorae, P. S. (Ed.). IUCN/SSC Re-introduction Specialist Group, Gland, Switzerland and Environment Agency, Abu Dhabi, UAE.

Al Zaidaneen, J. & Al Hasaseen, A. A. (2008): Re-introduction of Arabian oryx into Wadi Rum Protected Area, Jordan. *Global Re-Introduction Perspectives: Re-introduction case-studies from around the globe:* 181–184. Soorae, P. S. (Ed.). IUCN/SSC Re-introduction Specialist Group, Gland, Switzerland and Environment Agency, Abu Dhabi, UAE.

Anon. (unpublished): Stratégie Nationale Tunisienne pour la Conservation et la Restauration des Antilopes Sahélo-Sahariennes et de leurs habitats. Ministère de l'Agriculture, Tunisia, 2001.

Back to Africa. (2011): Back to Africa. http://www.backtoafrica.co.za/operations.htm (as at 2 March 2015).

Bertram, B. C. R. (1988): Re-introducing the Scimitar-horned oryx into Tunisia. *Conservation and Biology of Desert Antelopes:* 136–145. Dixon, A. & Jones, D. (Eds). London: Zoological Society of London; Bromley, Kent: Christopher Helm (Publishers) Ltd.

Bode, M. & Wintle, B. (2009): How to build an efficient conservation fence. *Conservation Biology* **24** (1): 182–188. DOI: 10.1111/j.1523-1739.2009.01291.x

Brandlová, K., Mallon, D., Hejcmanová, P., Regnaut, S., Jůnková Vymyslická, P., Fedorova, T., Žáčková M. & Ndiaye, S. (2013): Western Derby eland (*Taurotragus derbianus derbianus*) Conservation Strategy. Czech University of Life Sciences, Prague.

Brandlová, K., Vymyslicka, P. J., Žačkova, M., Fedorova T. & Hejcmanová, P. (Eds). (2014): African Studbook. Western Derby eland, *Taurotragus derbianus derbianus* (Gray, 1847). **7**. Czech University of Life Sciences, Prague, 2014, 83 pp.

Bro-Jørgensen, J. (2008): Dense habitats selecting for small body size: a comparative study on bovids. *Oikos* **117**: 729–737.

Cano, M., Abaigar, T. & Vericad, J.-R. (1993): Establishment of a group of dama gazelles *Gazella (= Nanger) dama* for reintroduction in Senegal. *International Zoo Yearbook* **32**: 98–107.

Cunningham, P. L. & Wacher, T. (2009): Changes in the distribution, abundance and status of Arabian Sand Gazelle (*Gazella subgutturosa marica*) in Saudi Arabia: a review. *Mammalia* **73**: 203–210. DOI 10.1515/MAMM.2009.041

De Vos, V., van Rooyen, G. L. & Kloppes, J. J. (1973): Anthrax immunization of free-ranging roan antelope, *Hippotragus equinus*, in the Kruger National Park. *Koedoe* **16**: 11–25.

Dunham, K. M. (1997): Gazelles and oryx in Saudi Arabia. *Re-introduction News* **13**: 5–6.

Dunham, K. M. (2001): Status of a reintroduced population of mountain gazelles *Gazella gazella* in central Arabia: management lessons from an aridland reintroduction. *Oryx* **35**(2): 111–118.

Dunham, K. M., Kichenside, T. B., Lindsay, N., Rietkirk, F. & Williamson, D. T. (1993): The reintroduction of mountain gazelle *Gazella gazella* in Saudi Arabia. *International Zoo Yearbook* **32**: 107–116.

Dunham, K. M., Williamson, D. T. & Joubert, E. (2001): Saudi Arabia. In *Antelopes. Part 4: North Africa, the Middle East, and Asia. Global survey and action* plans: 55–62. Mallon, D. P. & Kingswood, S. C. (Compilers). Gland, Switzerland/ Cambridge: SSC Antelope Specialist Group, IUCN.

Engel, H. & Brunsing, K. (1999): European studbook: *Addax nasomaculatus*. Zoo Hannover, 158 pp.

Fa, J. E., Currie, D. & Meeuwig, J. (2003): Bushmeat and food security in the Congo Basin: linkages between wildlife and people's future. *Environmental Conservation* **30** (1): 71–78.

Fischer, J. & Lindenmayer, D. B. (2000): An assessment of the published results of animal relocations. *Biological Conservation* **96**: 1–11.

Grant, C. C., Davidson, T., Funston, P. J. & Pienaar, D. J. (2002): Challenges faced in the conservation of rare antelope: a case study on the northern basalt plains of the Kruger National Park. *Koedoe* **45**(2): 45–62.

Grettenberger, J. F. & Newby, J. E. (1986): The status and ecology of the Dama Gazelle in the Aïr and Ténéré National Reserve, Niger. *Biological Conservation* **38**: 207–216.

Griffith, B., Scott, J., Carpenter, J., & Reed, C. (1989): Translocation as a species conservation tool: Status and strategy. *Science* **245**: 477–480.

Harrington, R., Owen-Smith, N., Viljoen, P. C., Mason, D. R. & Funston, P. J. (1999): Establishing the causes of the roan antelope decline in the Kruger National Park, South Africa. *Biological Conservation* **90**: 69–78. 10.1016/S0006-3207(98)00120-7

Haque, M. N. & Smith, T. R. (1996) Reintroduction of Arabian sand gazelle *Gazella subgutturosa marica* in Saudi Arabia. *Biological Conservation* **76**: 203–207.

Howard, P., Wangari, E. & Rakotoarisoa, N. (2007): Mission Report: UNESCO/IUCN joint monitoring mission to Niokola-Koba National Park, Senegal. UNESCO. 22 pp.

IUCN (2015): The IUCN Red List of Threatened Species. Version 2015.1. http://www.iucnredlist.org (as at 23 February 2016).

Islam, M. Z., Boug, A., Ismail, K., Robinson, E. R. & Al Hafedh, Y. S. (2010a): Over 20 years of re-introducing the Arabian oryx in fenced and free-range protected areas of Saudi Arabia. *Global Re-Introduction Perspectives: Additional case-studies from around the globe*: 269–273. Soorae, P. S. (Ed). IUCN/SSC Re-introduction Specialist Group, Gland, Switzerland and Environment Agency, Abu Dhabi, UAE.

Islam, M. Z., Ismail, K. & Boug, A. (2010b): Catastrophic die-off of globally threatened Arabian oryx and sand gazelle in the fenced protected area of the arid central Saudi Arabia. *Journal of Threatened Taxa* **2**(2): 677–684.

Islam, M. Z., Knutson, C. & Boug A. (2010c): Strategy and action plan to reduce the risk of mass mortalities of reintroduced ungulates in the Mahazat as-Sayd protected area in Saudi Arabia. *Gnusletter* **28**(2), 9–15.

Islam, Z. M., Boug, A., Ismail, K. & Shah, M. S. (2011a): Status of a reintroduced population of threatened sand gazelles in SW Saudi Arabia: management lessons from an aridland reintroduction. *Global Re-introduction Perspectives: 2011. More case studies from around the globe*: 185–189. Soorae, P. S. (Ed.). IUCN/SSC Re-introduction Specialist Group, Gland, Switzerland and Environment Agency, Abu Dhabi, UAE.

Islam, M. Z., Ismail. K. & Boug, A. (2011b): Restoration of the endangered Arabian oryx, *Oryx leucoryx*, Pallas 1766 in Saudi Arabia: lessons learnt from the twenty years of reintroduction in arid fenced and unfenced protected areas. Biodiversity Conservation in the Arabian Peninsula. *Zoology in the Middle East* (Supplement 3): 125–140.

Islam, Z. M., Wacher, T. J., Boug, A. & Wronski, T. (2011c): Population development of mountain gazelle in the western Empty Quarter (Uruq Bani Ma'arid Protected Area), Saudi Arabia. *Global re-introduction perspectives: more case studies from around the globe*:180–184. Soorae, P. S. (Ed.). IUCN/SSC Re-introduction Specialist Group, Gland, Switzerland and Environment Agency, Abu Dhabi, UAE.

IUCN (1998): Guidelines for reintroductions. Prepared by the IUCN/SSC Reintroduction Specialist Group. IUCN, Gland, Switzerland and Cambridge, UK. 10 pp.

IUCN Antelope Specialist Group (2009): Quarter of antelope species in danger of extinction. *Gnusletter* **28**(1): 2–5.

IUCN/SSC (2013): *Guidelines for reintroductions and other conservation translocations*. Version 1.0. Gland, Switzerland: IUCN Species Survival Commission, viiii + 57 pp. ISBN: 978-2-8317-1609-1.

Jebali, A. (2008): Déclin de la faune Sahelo-Saharienne et tentative de réintroduction d'antilopes dans des habitats restaurés: cas de l'oryx algazelle (*Oryx dammah*) et de la gazelle dama (*Gazella dama mhorr*) dans la Reserve de Faune du Ferlo Nord (Senegal). PhD Thesis, Museum National d'Histoire Naturelle, Paris, France.

Kiwan, K., El Al Kamy, H. & Al Daheri, A. (2008): Re-introduction of Arabian oryx into Um Al Zomoul, Abu Dhabi Emirate, United Arab Emirates (2008). *Global Re-introduction Perspectives: re-introduction case-studies from around the globe*: 177–180. Soorae, P. S. (Ed.). IUCN/SSC Re-introduction Specialist Group, Gland, Switzerland and Environment Agency, Abu Dhabi, UAE.

Koláčková, K., Hejcmanová, P., Antonínová, M. & Brandl, P. (2011a): Population management as a tool in the recovery of the critically endangered Western Derby eland *Taurotragus derbianus* in Senegal, Africa. *Wildlife Biology* **17**: 299–310.

Koláčková, K., Hejcmanová, P., Brandl, P., Žáčková, M. & Vymyslická, P. (2011b): African studbook. Western Derby eland, *Taurotragus derbianus derbianus* (Gray, 19847). Koláčková, K., Hejcmanová, P., Brandl, P., Žáčková, M. & Vymyslická, P. (Eds). Czech University of Life Sciences, Prague, pp. 80.

Molcanova, R. (2004): The reintroduction of scimitar-horned oryx to Sidi Toui National Park, Tunisia. In *The biology, husbandry and conservation of scimitar-horned oryx, Oryx dammah*: 72–76. Gilbert, T. & Woodfine, T. (Eds). Marwell Preservation Trust.

Müller, H. P. & Engel, H. (2004): The reintroduction of herbivores to Souss Massa National Park, Morocco. In *The biology, husbandry and conservation of scimitar-horned oryx, Oryx dammah*: 77–81. Gilbert, T. & Woodfine, T. (Eds). Marwell Preservation Trust.

Ogutu, J. O. & Owen-Smith, N. (2003): ENSO, rainfall and temperature influences on extreme population declines among African savannah ungulates. *Ecology Letters* **6**: 412–419.

Ostrowski, S. & Bredin, E. (2001): Arabian oryx re-introduction in 'Uruq Bani Ma'arid, Saudi Arabia: a summary and update: January 2001. *Reintroduction News* **20**: 16–17.

Owen-Smith, N. & Mills, M. G. L. (2006): Manifold interactive influences on the population dynamics of a multispecies ungulate assemblage. *Ecological Monographs* **76** (1): 73–92.

Owen-Smith, N., Chirima, G. J., Macandza, V. & Le Roux, E. (2011):Shrinking sable antelope numbers in Kruger National Park: what is suppressing population recovery? *Animal Conservation*. doi:10.1111/j.1469-1795.2011.00504.x p.1–10

Peters, R. H. (1983): The ecological implications of body size. Cambridge Studies in Ecology. Cambridge: Cambridge University Press, xii + 329 pp.

Redford, K. H., Amato. G, Baillie, J., Beldomenico, P., Bennett, E. L., Clum, N., Cook, R., Fonseca, G., Hedges, S., Launay, F., Lieberman, S., Mace, G. M., Murayama, A., Putnam, A., Robinson, J. G., Rosenbaum, H., Sanderson, E. W., Stuart, S. N., Thomas, P. & Thorbjarnarson, J. (2011): What does it mean to successfully conserve a (vertebrate) species? *BioScience* **61**(1): 39–48.

Renaud, P. C., Gueye, M. B., Hejcmanová, P., Antonínová, M. & Samb, M. (2006): Inventaire aerien et terrestre de la faune et relevé des pressions au Parc National du Niokolo Koba: Plan of Urgence, Report Annex A, August 2006. 74 pp. *Wildlife Biology* **17**(3): 309.

Sable Antelope Project. (2015): Back to Africa. http://www.backtoafrica.co.za/operations/sable-antelope-project.html (as at 2 March 2015).

Saltz, D. (2008): Re-introduction of Arabian oryx into the Negev Desert, Israel. In *Global re-introduction perspectives: re-introduction case-studies from around the globe*: 185–187. Soorae, P. S. (Ed.). IUCN/SSC Re-introduction Specialist Group, Gland, Switzerland and Environment Agency, Abu Dhabi, UAE.

Scott, J. M., Goble, D. D., Haines, A. M., Wiens, J. A. & Neel, M. C. (2010): Conservation-reliant species and the future of conservation. *Conservation Letters* **3**: 91–97.

Seydack, A. H., Grant, C. C., Smit, I. P., Vermeulen, W. J., Baard, J. & Zambatis, N. (2012a): Climate and vegetation in a semi-arid savanna: development of a climate–vegetation response model linking plant metabolic performance to climate and the effects on forage availability for large herbivores. *Koedoe* **54**(1): Art. #1046, 12 pages. http:// dx.doi.org/10.4102/koedoe. v54i1.1046

Seydack, A. H., Grant, C. C., Smit, I. P., Vermeulen, W. J., Baard, J. & Zambatis, N. (2012b): Large herbivore population performance and climate in a South African semi-arid savannah. *Koedoe* **54**(1), Art. #1047, 20 pp. http://dx.doi.org/10.4102/koedoe. v54i1.1047

Shrestha, T. K. (2001): Nepal. *Antelopes. Part 4: North Africa, the Middle East, and Asia. Global survey and regional action plans*: 188–191. Mallon, D. P. & Kingswood, S. C. (Compilers). SSC Antelope Specialist Group. IUCN, Gland, Switzerland and Cambridge, UK.

Simkins, G. (2008): Re-introduction of Arabian Oryx into the Dubai Desert Conservation Reserve, Dubai, UAE. In *Global re-introduction perspectives: re-introduction case-studies from around the globe*: 188–191. Soorae, P. S. (Ed.). IUCN/SSC Re-introduction Specialist Group, Gland, Switzerland and Environment Agency, Abu Dhabi, UAE.

Skinner, J. D. & Smithers, R. H. N. (1990): The mammals of the Southern African subregion (2nd edn). University of Pretoria, Pretoria.

Soorae, P. S. (Ed.) (2008): *Global reintroduction perspectives: re-introduction case-studies from around the globe.* IUCN/SSC Re-introduction Specialist Group, Abu Dhabi, UAE. viii + 284 pp.

Soorae, P. S. (Ed.). (2010): *Global re-introduction perspectives: additional case-studies from around the globe.* IUCN/SSC Re-introduction Specialist Group, Abu Dhabi, UAE, xii + 352 pp.

Soorae, P. S. (Ed.) (2011): *Global re-introduction perspectives: 2011. more case studies from around the globe.* Gland, Switzerland: IUCN/SSC Re-introduction

Specialist Group and Abu Dhabi, UAE: Environment Agency-Abu Dhabi. xiv + 250 pp.

Spalton, J. A. (1993): A brief history of the reintroduction of the Arabian oryx (*Oryx leucoryx*) into Oman 1980–1992. *International Zoo Yearbook* **32**: 81–90.

Spalton, J. A., Lawrence, M. W. & Brend, S. A. (1999): Arabian oryx reintroduction in Oman: successes and setbacks. *Oryx* **33**(2): 168–175.

Stanley Price, M. R. (1989): *Animal reintroductions: the Arabian oryx in Oman.* Cambridge: Cambridge University Press, 291 pp.

Veasey, J. (2010): Re-introduction and translocation of eastern bongo in Kenya; is it necessary and if so what are likely to be the most effective methodologies of achieving meaningful conservation goals? In *Global re-introduction perspectives: additional case-studies from around the globe*: 274–277. Soorae, P. S. (Ed.). IUCN/SSC Re-introduction Specialist Group, Gland, Switzerland and Environment Agency, Abu Dhabi, UAE.

Wacher, T., Rabeil, T. & Newby, J. E. (unpublished): *Monitoring survey of Termit and Tin Toumma (Niger)* - 2008. Sahara Conservation Fund, 2008.

Wacher, T. & Robinson, E. R. (2008): Re-introduction of sand gazelle into the Uruq Bani Ma'arid Protected Area, Saudi Arabia. In *Global re-introduction perspectives: re-introduction case-studies from around the globe*: 172–176. Soorae, P. S. (Ed.): IUCN/SSC Re-introduction Specialist Group, Gland, Switzerland and Environment Agency, Abu Dhabi, UAE.

Wacher, T. & Newby, J. E. (2012): Summary of results and achievements of the Pilot Phase of the Pan Sahara Wildlife Survey 2009–2012. SCF PSWS Technical Report 12. Sahara Conservation Fund. ii + 26 pp. + Annexes.

Western, D. (1979): Size, life history and ecology in mammals. *African Journal of Ecology* **17**: 185–204.

Wiesner, H. & Müller, P. (1998): On the reintroduction of the Mhorr Gazelle in Tunisia and Morocco. *Naturwissenschaften* **85**: 553–555.

Wilson, D. E. & Reeder, D. M. (Eds). (2008): Mammal species of the world. Smithsonian Institution. (3rd edn). http://vertebrates.si.edu/msw/mswCFApp/msw/index.cfm (as at 23 February 2016).

Wilson, D. E. & Mittermeier, R. A. (Eds). (2011): *Handbook of the mammals of the world. 2. Hoofed mammals.* Barcelona: Lynx Edicions.

Wolf, C. M., Garland, T. & Griffith, B. (1998): Predictors of avian and mammalian translocation success: reanalysis with phylogenetically independent contrasts. *Biological Conservation* **86**: 243–255.

Woodfine, T., Zahzah, K., Chetoui, A., Gilbert, T. & D'Alterio, G. L. (unpublished): Reintroduction of scimitar-horned oryx to Dghoumes National Park, Tunisia. Report prepared for Direction Générale des Forêts. Tunis. 2009.

Wronski, T. 2010. Population density and home range size of re-introduced mountain gazelles (*Gazella gazella*) in relation to resource availability in the Ibex Reserve, central Saudi Arabia. *Journal of Arid Environments* **74**: 1427–1434.

Wronski, T., Wacher, T., Hammond, R. L., Winney, B., Hundertmark, K. J., Blacket, M. J., Mohammed, O. B., Flores, B., Omer, S. A., Macasero, W., Plath, M., Tiedemann, R. & Bleidorn, C. (2010): Two reciprocally monophyletic mtDNA lineages elucidate the taxonomic status of mountain gazelles (*Gazella gazella*). *Systematics and Biodiversity* **8**(1): 119–129. DOI: 10.1080/14772001003613192.

Wronski, T., Sandouka, M. A. & Butynski, T. M. (2011): Twenty years of conservation and monitoring of re-introduced mountain gazelle in the Ibex Reserve, Saudi Arabia. In *Global re-introduction perspectives: 2011: more case studies from around the globe*: 175–179. Soorae, P. S. (Ed.). IUCN/SSC Re-introduction Specialist Group, Gland, Switzerland and Environment Agency, Abu Dhabi, UAE.

13

Desert Antelopes on the Brink: How Resilient is the Sahelo-Saharan Ecosystem?

John Newby[1], Tim Wacher[2], Sarah M. Durant[3], Nathalie Pettorelli[3] and Tania Gilbert[4]

[1]*Sahara Conservation Fund, L'Isle, Switzerland*
[2]*Conservation Programmes, Zoological Society of London, United Kingdom*
[3]*Institute of Zoology, Zoological Society of London, United Kingdom*
[4]*Marwell Wildlife, Colden Common, Winchester, United Kingdom*

Introduction

The large mammal and bird fauna of the Sahara and bordering sub-desert and Sahelian grasslands is one of the most threatened on earth. Several species and subspecies from this vast Sahelo-Saharan region of Northern Africa are listed as threatened by the IUCN Red List of Threatened Species (IUCN, 2013; Table 1). Of particular concern are the large ungulates, represented by five species of gazelle (dorcas, slender-horned, Cuvier's, red-fronted, dama) and two species of large antelope (addax, scimitar-horned oryx) (UNEP/CMS, 2006). A further subspecies of large antelope, the bubal hartebeest, was also present in the Mediterranean, sub-desert regions of North Africa but became extinct in the early part of the 20th century (Harper, 1945). Table 1 lists these species, their scientific names, and those of others mentioned in the text, together with their respective IUCN Red List status.

Although hunting has without doubt been the principal cause for the decline and disappearance of desert ungulates throughout their range, new threats are rapidly growing in significance, notably habitat loss through desertification, climate change, resource extraction, agricultural encroachment, and competition from livestock development (Durant *et al.*, 2013; Newby, 1988, 2014).

Concern for the Sahara's threatened mega-fauna has been expressed at both national and international levels. In 1999, the Convention on the Conservation

Antelope Conservation: From Diagnosis to Action, First Edition. Edited by Jakob Bro-Jørgensen and David P. Mallon.

Table 1 **IUCN Red List status of ungulates and selected other taxa of the Sahelo-Saharan zone.**

Species	**Scientific name**	**IUCN Red List status**
Dorcas Gazelle	*Gazella dorcas*	Vulnerable (VU)
Cuvier's Gazelle	*Gazella cuvieri*	Endangered (EN)
Slender-horned Gazelle	*Gazella leptoceros*	Endangered (EN)
Red-fronted Gazelle	*Eudorcas rufifrons*	Vulnerable (VU)
Dama Gazelle	*Nanger dama*	Critically Endangered (CR)
Bubal Hartebeest	*Alcelaphus buselaphus buselaphus*	Extinct (E)
Addax	*Addax nasomaculatus*	Critically Endangered (CR)
Scimitar-horned Oryx	*Oryx dammah*	Extinct in the Wild (EW)
Korrigum	*Damaliscus lunatus korrigum*	Vulnerable (VU)
Cheetah (Saharan)	*Acinonyx jubatus hecki*	Critically Endangered (CR)
Caracal	*Felis caracal*	Not yet assessed

of Migratory Species of Wild Animals (CMS) put in place an action plan, 'Concerted Action', for the six Sahelo-Saharan ungulates (CMS, 1999a, 1999b). More recently, IUCN adopted a resolution calling for "*Support for national and regional initiatives for the conservation of large mammals in the Sahara*" (IUCN, 2012). This was echoed by Durant *et al.* (2012, 2013), expressing the same concerns and focussing on the massive range reductions the larger mammals of the Sahara have suffered.

The objective of this chapter is to highlight challenges facing conservation of the most threatened antelopes by focusing on the situation in the Sahara desert and the bordering Sahelian grasslands. We here provide a broad overview of the current conservation status and distribution of the gazelles and larger antelopes in the region, present the threats they face, and discuss options for their conservation and future valorization.

The Sahelo-Saharan zone

The Sahara desert and bordering sub-desert xeric shrub and grasslands, the Sahelo-Saharan zone, cover over 10.8 mio km^2 or approximately one quarter of the surface area of the African continent. The area is classified as arid to hyper-arid, with annual rainfall varying between 350 mm in the south to less than 10 mm in parts of the central Sahara (UNEP, 2006; UNCCD, 2011; IUCN,

2012) (Figure 1). North of the Sahara, the sub-desert band is relatively narrow, merging into the higher country of the Atlas and associated mountain ranges of Morocco, Algeria, and Tunisia, or the coastal Mediterranean habitats of Libya and Egypt. South of the Sahara, the sub-desert band, or Sahel, is considerably broader in many parts and interfaces gradually with better-watered savanna habitats (Mali, Niger, Burkina Faso, Nigeria, Chad, and Sudan). In terms of its biogeography and ecoregion characteristics, the area under review consists of two main habitat types: (i) Deserts and xeric shrublands (Plate 5i); and (ii) Tropical and sub-tropical grasslands, savannas, and shrublands (Plate 5ii) (Olson & Dinerstein, 2001, 2002; WWF, 2013).

This enormous yet relatively homogenous area comprises part of the territories of 13 nations[1] and extends from the Atlantic seaboard in the west to the Red Sea in the east. Politically, parts of the countries of Ethiopia, Eritrea, and Djibouti may also be included in the greater Sahel, but these have been excluded from this review because of their far greater affinities with the ecoregions and biodiversity of the Horn of Africa.

The geology, geomorphology, and landscapes of the Sahara and Sahel are varied and include extensive fixed dune systems, sand seas and sandsheets, stony plains and plateaux, lava flows and highly eroded sandstones, deeply incised valleys and wadis, and significant high mountain systems, culminating in the Atlas, Hoggar, Aïr, and Tibesti mountains, with altitudes up to 3445 m in Emi Koussi in Northern Chad (Buckle, 1978). Vegetation types are also diverse, varying from wooded grasslands and high altitude pseudo-steppes on the perimeter, through more central sub-desert and desert habitats with hardy perennials and ephemeral annual plant communities (Ozenda, 1983). Extensive drainage patterns, made up of seasonal pans and wadis, especially in and around the highlands, exhibit increased species richness, including plants from more tropical and/or more temperate floras (Quézel, 1965; Benchaleh *et al.*, 2011; Sayre *et al.*, 2013).

Sahelo-Saharan ungulates are uniquely adapted, physiologically, morphologically, and behaviourally, to deal with the ecology and climate of the region (Newby, 1984). They ably manage the extremes of heat and drought, as well as coping with the high degree of spatial, temporal, and quantitative variability intrinsic to Sahelo-Saharan rainfall and the resulting vegetation through mobility. This is characterized by seasonal movements to and from areas of suitable grazing and shade; migrations that shadow the development of the wet season, rainfall, and resulting pastures, and in general a tendency towards nomadism in response to the variable quantity, quality, and distribution of the available vegetation. These movements may be quite specific in nature and associated with the occurrence of a particular plant or plant community,

[1] Morocco, Algeria, Tunisia, Libya, Egypt, Sudan, Chad, Nigeria, Niger, Mali, Burkina Faso, Senegal, and Mauritania.

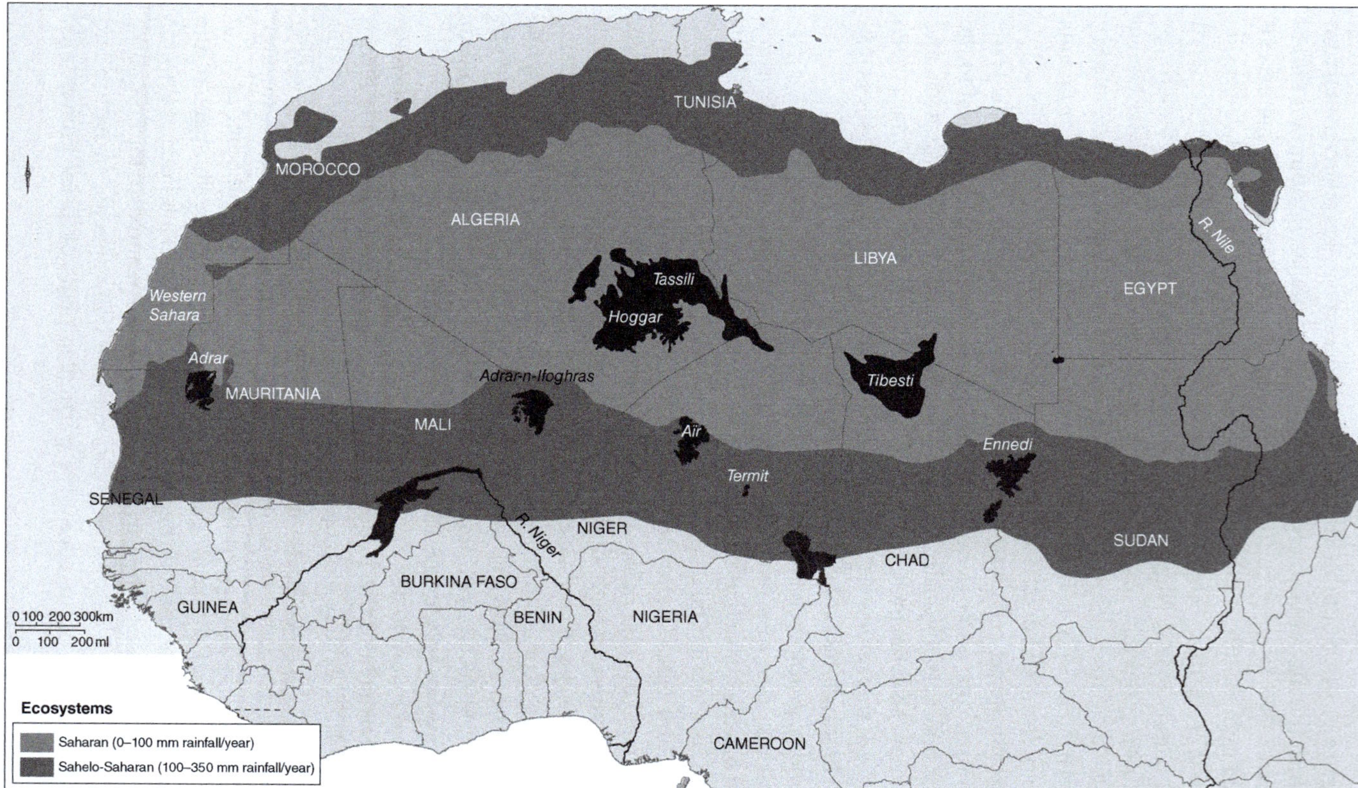

Sources: FAO (2008).

Figure 1 **Map of the Sahelo-Saharan zone of Africa. The Sahara desert is shown in light grey, the Sahel is the bordering dark grey area to the south, and the dark grey area to the north is normally referred to as "sub-desert steppe" (see WWF ecoregions in WWF 2013).**

especially water-rich species during the hotter months of the year, or known areas of shade in the form of rocky outcrops or groups of regularly frequented trees (Dragesco-Joffé, 1993; SCF 2007, 2009).

Status and distribution of Sahelo-Saharan wildlife

The gazelles

Five species of gazelle occur within the region covered by this chapter, dorcas (*Gazella dorcas*) (Plate 5ii), Cuvier's (*Gazella cuvieri*) (Plate 3vi), slender-horned (*Gazella leptoceros*) (Plate 3vii), dama (*Nanger dama*) (Plate 3viii), and red-fronted (*Eudorcas rufifrons*) (Figure 2). Of these, Cuvier's and red-fronted gazelles can be considered as having ranges only partially covered by true Sahelo-Saharan habitats, in this case defined as lying roughly between the 0–350 mm isohyets. In the case of Cuvier's gazelle, its range lies mostly to the north of the Sahara in the stony piedmonts, plateaux, and Mediterranean dry forests of the Atlas, Anti-Atlas, and associated ranges stretching from western Morocco, through Algeria to northern Tunisia (Cuzin, 1996; Aulagnier *et al.*, 2008; Beudels *et al.*, 2013). The red-fronted gazelle, on the other hand, inhabits the grasslands and wooded savannas of the southern Sahel and northern Sudan vegetation belts. It can only be found today in the Sahelo-Saharan zone where suitable riparian habitat occurs or as a rare and temporary wet season visitor to the more densely vegetated wooded grasslands of the northern Sahel (Scholte & Hashim, 2013).

Figure 2 **Female red-fronted gazelle in Zakouma National Park, Chad (© John Newby).**

Of the three remaining species, only the dorcas gazelle can today be considered widespread, existing in highly variable quantities in most of the Sahelo-Saharan countries (UNEP/CMS, 2006; Wilson & Mittermeier, 2011; Scholte & Hashim, 2013). The dorcas gazelle is also the most versatile of the group, inhabiting not only the typical grasslands of the Sahel and sub-desert but also open desert and mountain systems where suitable habitat occurs. In Chad and Algeria, they have been recorded at altitudes above 2500 m. In many areas, though, the dorcas gazelle has been wiped out or relegated to marginal habitats on the edge of desert country, where opportunities for flight amongst the dunes in the case of harassment exist.

Data gathered recently by the Sahara Conservation Fund (SCF) in the framework of its Pan Sahara Wildlife Survey (Wacher & Newby, 2012b) in the gazelle's prime Sahelian grassland habitats in Niger and Chad show it present over very large areas but varying widely in number. Populations faced with high levels of livestock development, such as Gadabeji in Niger, persist in remnants of less than one gazelle to every 10 km^2. In other areas, quite extensive populations of at least 1–3 dorcas/km^2 can still be found, such as Termit in Niger and the Manga in Chad. The largest known remaining populations are found in the Sahelian grasslands of the Ouadi Rimé-Ouadi Achim Game Reserve in central Chad (Plate 5ii). Here gazelles are present at local densities of at least 6–8 gazelles/km^2 over substantial areas of several hundred square kilometres, especially where good shade and grazing are available in the dry season (April–July) (Wacher & Newby, 2011).

Along the edges of the larger sand seas, such as the Great Ergs of Algeria and Tunisia, dorcas gazelles can be found in association with the larger and more desert-adapted slender-horned gazelle. An inhabitant of the major Saharan sand seas, the slender-horned gazelle is primarily North African in distribution with few or no confirmed sightings of the species south of the Tropic of Cancer (Beudels & Devillers, 2013). The species is currently considered to exist in two geographically distinct sub-species, *Gazella l. loderi* in the west (Morocco, Algeria, Tunisia, W. Libya) (also known as Loder's gazelle) and *G. l. leptoceros* in the east (E. Libya, Egypt). Recent surveys have confirmed the gazelle to be present across most of the Erg Occidental in Algeria, at least on its northern and eastern margins, and on both western and eastern sides of the Erg Oriental in Algeria and Tunisia (de Smet *et al.*, 2009; Fellous *et al.*, 2009; Wacher *et al.*, 2011). Both populations are listed as Endangered and have suffered the sustained impact of uncontrolled hunting (El Alqamy & Baha el Din, 2006; SCF, 2012a).

The dama gazelle is the largest of the gazelles inhabiting the Sahelo-Saharan region. Typically divided into three geographic races, *N. d. mhorr* (mohor gazelle), *N. d. dama* (dama gazelle), and *N. d. ruficollis* (red-necked or addra gazelle) (Cano-Pérez, 1984; Groves & Grubb, 2011), recent findings based on DNA analysis of samples from both wild and captive animals indicate a rather less clear-cut situation with more mixing of races than previously expected (Senn *et al.*, 2014). Phenotypically, however, and natural variation notwithstanding, a cline from west to east (*mhorr* to *ruficollis*) can be broadly distinguished based on an increasing amount and extent of red in the pelage.

With the exception of the extension of its range northwards into the Drâa Valley of central Morocco via the Atlantic coastal deserts (Cuzin, pers. comm., 2013), there are few or no recent records of the dama gazelle occurring north of the Sahara. The dama gazelle is overwhelmingly a Sahelian species, whose historical range coincided largely with that of the dorcas gazelle and the now extinct-in-the-wild scimitar-horned oryx. This range comprises wooded steppes, sparsely wooded grasslands, and wadi habitats (Newby, 1978). As a result of long-term persecution, the current range of the species is widely spread in small isolated populations across the central Sahelian nations of Mali, Niger, and Chad, the whole probably not amounting to more than 300 individuals (RZSS & IUCN Antelope Specialist Group, 2014). Furthermore, in response to hunting pressure these remaining groups are largely located in atypical mountain habitats (Termit and Aïr mountains in Niger) (Rabeil, 2014) or in those tiny parts of its former grassland range the least disturbed by cattle and the development of permanent wells, roads, and so on (Manga and Ouadi Haddad in Chad, Tamesna in Mali) (Newby *et al.*, 2014).

The three best known remaining wild populations are extraordinarily isolated and restricted in distribution relative to their vast former ranges. In the Termit Massif, almost completely surrounded by an inhospitable sand sea, a minimum convex polygon around all sightings made since 1997 measures close to 900 km^2 (Newby, Wacher, Rabeil, pers. obs.). In the Chadian Manga, all dama seen since 2001 were found in an area encompassing about 6000 km^2 (Wacher & Newby, 2010; Wacher *et al.*, 2014). And in Chad's Ouadi Rimé-Ouadi Achim Game Reserve, a small number of remnant herds are thinly scattered through 1,100 km^2 (Wacher & Newby, 2012b; Newby *et al.*, 2014). In both the latter sites, the dama are mainly located in vegetated fixed dune systems or sparse woodlands, where vehicles are less nimble than on the surrounding flatter areas. In Termit, the gazelles occupy rocky plateaux and secluded valleys where vehicles have trouble going and where flight is an easy option. To give an idea of the scarcity of dama gazelles today relative to the 1970s, comparative data on their presence and encounter rates in the Ouadi Rimé-Ouadi Achim Game Reserve (Newby, 1978; Wacher & Newby, 2012b; Newby *et al.*, 2014), show a 98% reduction in distribution (c.1,000 km^2 today vs. c. 50,000 km^2 in the 1970s), and within that remaining range, an encounter rate 10 times lower than before (0.02 damas per kilometre vs. 0.2 damas per kilometre).

The larger antelopes

In addition to the gazelles already mentioned, the Sahara and bordering Sahel are also home to two larger antelopes, the addax (*Addax nasomaculatus*) (Plate 2iv) and the scimitar-horned oryx (*Oryx dammah*) (Plate 2v). The former is an inhabitant of true desert country, while the latter occupies the grasslands of the sub-desert and northern Sahel. The addax occupies the gravelly plains, firmer sands, and inter-dune valleys of the Sahara, with a range that formerly stretched from Morocco and Mauritania in the west, across the central Sahara and into Egypt and Sudan as far as the Nile valley (Newby, 2013). It was to be found both north and south of the great sand seas (UNEP/CMS, 2006). Addax are a mobile, nomadic species exploiting the sparse pastures of desert-adapted plants, such as *Stipagrostis vulnerans* and *Cornulaca monacantha*, as well as ephemeral grazing brought on by the occasional rainfall. Where available, addax will also make use of areas of shade (rocky outcrops, trees, tussocks of grass) during the hotter months (Newby, 2013).

Over the past two centuries, hunting pressure has reduced the addax's once vast range to a couple of isolated populations, the main one of which lies in the Tin Toumma desert of eastern Niger (Plate 5i). Addax here now number less than 100 head (Wacher *et al.*, 2010a; T. Rabeil, pers. comm.). A further small population, possibly at one time contiguous, still exists in western Chad's Eguey region to the far north of Lake Chad (Monfort, 2003; Fay, 2005). Since 2000, the only other credible, isolated observations of addax have come from central Mauritania (Vernet, pers. comm., 2005) and the far north of Chad in the Erdis that border Libya (Kroepelin, pers. comm., 2010). The species' survival relies very much on efforts currently underway to protect the remaining animals in the newly created Termit & Tin Toumma National Nature Reserve and ongoing efforts to raise support for its conservation in Chad.

Once abundant across the whole Sahelian belt of Africa, the scimitar-horned oryx is currently classified as Extinct in the Wild by IUCN (IUCN, 2008) having disappeared from its last haunts in Chad and Niger sometime between the late 1980s and early 1990s (Gilbert *et al.*, 2012; Morrow *et al.*, 2013). The scimitar-horned oryx provides a classic example of a species having suffered the multiple effects of natural and anthropogenic impacts on its Sahelian grassland habitat. This large-bodied antelope has been a prey of choice for hunters for millennia. Scimitar-horned oryx being hunted are a feature of Saharan rock art dating back to the Neolithic and proto-historic periods. During the Middles Ages, it was also much sought after for its durable hides, used for battle shields and strong leather, with records of this in trans-Saharan trade; see Leo Africanus (aka al-Hasan ibn Muhammad al-Wazzan al-Fasi), an Andalusian diplomat and explorer who visited large parts of North Africa in the 16th century (Leo Africanus, 1896). With the arrival of firearms and increased numbers of horses and latterly, motorized vehicles, the scimitar-horned oryx suffered heavy losses across its entire range, disappearing from North Africa during the 19th century (Gilbert, 2011). South of the Sahara, they fared a little better, but accounts written in the 1930s already point to the species' increasing rarity (In Tanoust, 1930; Brocklehurst, 1931; Henriques, 1938). By the 1970s, the only sizeable populations left were in Niger and Chad, with tiny numbers remaining elsewhere in Mali, and possibly southern Algeria and Sudan (Newby, 1978; Newby, 1982). With numbers reduced drastically by over-hunting, the additional pressure from the severe droughts of the 1970s and 1980s, coupled with civil war, desertification, pastoral development, and agricultural encroachment finally extirpated the species from its last remaining haunts in Chad and Niger (Freemantle *et al.*, 2013). During the past decade, the most important former strongholds in both countries have been regularly surveyed and only heavily weathered horn fragments have been found.

A third large antelope, the korrigum (*Damaliscus lunatus korrigum*) used to be a regular wet season visitor to the Sahel, migrating into the region from more southerly savanna latitudes in response to rainfall and grazing (In Tanoust,

Figure 3 **Maps showing the current distribution of the addax (*Addax nasomaculatus*), the scimitar-horned oryx (*Oryx dammah*), the dama gazelle (*Nanger dama*), the dorcas gazelle (*Gazella dorcas*), the slender-horned gazelle (*Gazella leptoceros*), the Cuvier's gazelle (*Gazella cuvieri*), and the red-fronted gazelle (*Eudorcas rufifrons*). Drawn by Mike Hoffmann based on data from the IUCN Red List, v 2015.3, www.iucnredlist.org.**

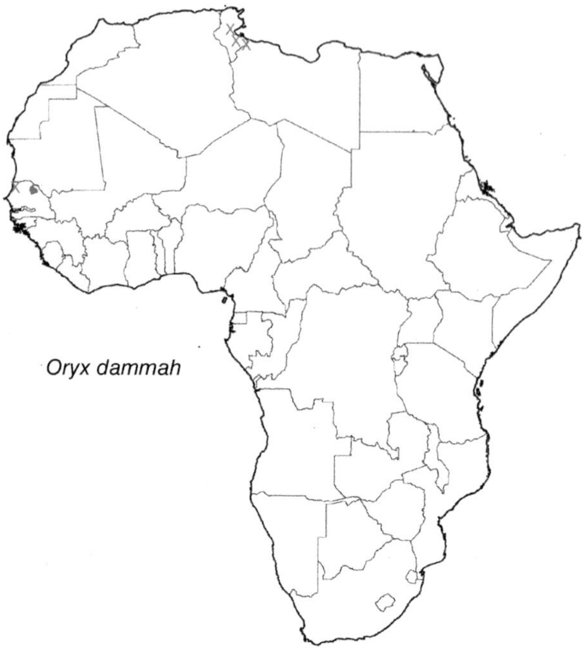

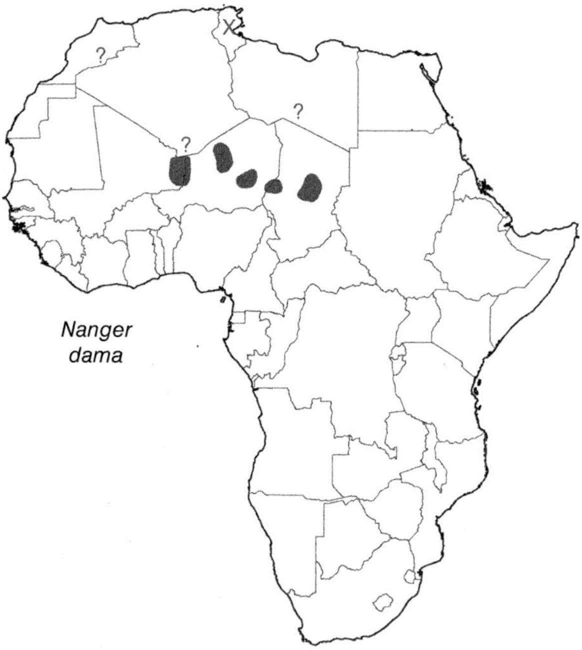

Figure 3 (***continued***)

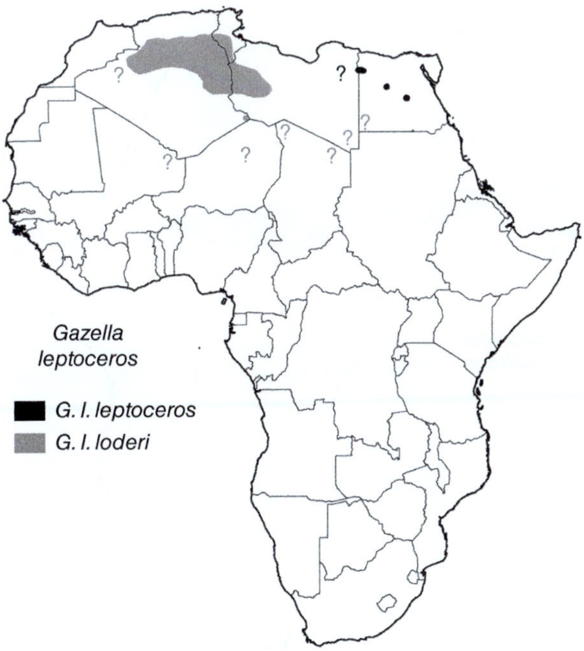

Figure 3 (***continued***)

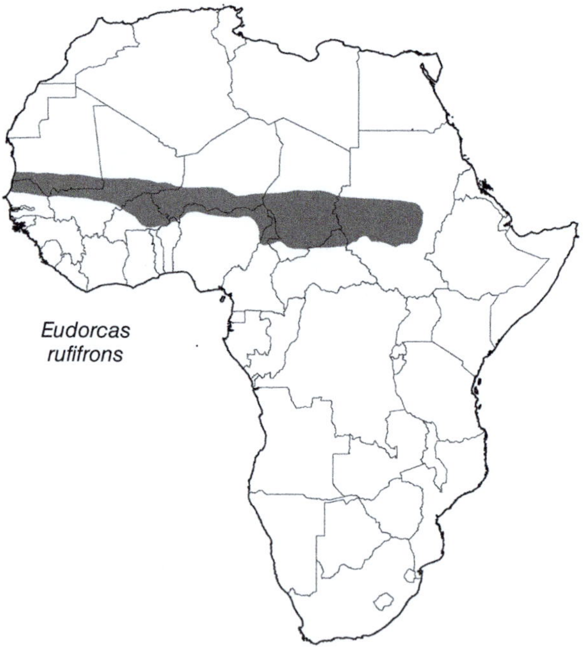

Figure 3 (***continued***)

1930; Malbrant, 1952; Duncan, 2013). Today, however, this is rarely if ever the case. Korrigum numbers have plummeted, and large tracts of former habitat are now occupied by livestock, villages, and fields.

Distribution maps for the addax, scimitar-horned oryx, dama gazelle, dorcas gazelle, slender-horned gazelle, Cuvier's gazelle, and the red-fronted gazelle can be found in Figure 3.

Threats to Sahelo-Saharan wildlife

Unlike many endangered species, where habitat loss has been a significant cause of decline and range reduction, the antelopes and gazelles of the Sahara and Sahel have suffered primarily from overhunting, and this quite probably for centuries (In Tanoust, 1932; Malbrant, 1952). Most of the species occupy open, accessible spaces, which make them for the most part both visible and vulnerable. And while hunting on foot or horseback with dogs or primitive weapons or traps is effective but largely sustainable, the arrival of firearms and latterly, motorized vehicles, revolutionized hunting, making it unsustainable. Easily accessible populations were the first to go and by the end of the 19th century the larger ungulates of North Africa had already largely disappeared (UNEP/CMS, 2006). Colonization of North and West Africa facilitated this process, with wildlife making up a sizeable proportion of the meat rations of far flung garrisons and expeditionary forces (Brosset, 1904; Gautier, 1923). South of the Sahara, the situation was little different, but with larger numbers and greater spaces the process was more protracted. Although the late colonial and early post-colonial periods saw the establishment of a few game reserves, these were by and large poorly funded or ill-respected by the administration and the military and mining personnel that continued to plunder what little wildlife remained.

Today, the threat of hunting on the Sahara's antelopes and gazelles continues, with conservation efforts thwarted by lack of national and international resources coupled with chronic regional insecurity. Especially vulnerable are the remaining gazelle populations (there are virtually no more larger antelopes to hunt), and these are being tracked down even in difficult-to-access areas, such as the great sand seas, thanks to modern off-road motorbikes and quads (SCF, 2012). Evidence of this proliferates on the Internet and social media, as hunters post images of themselves and their bags and gloat over their successes. With the search for new oil fields, petroleum exploration and exploitation in countries such as Mauritania, Niger, and Chad is also a serious threat, not so much from the activity itself but from the disturbance caused and the hunting carried out by the military detachments sent to protect the oil workers from terrorism and the like (SCF, 2010).

With the major development of the Sahelian livestock industry over the past 30–40 years and the sinking of numerous permanent wells in previously unoccupied rangelands, Sahelo-Saharan wildlife is facing an increasing and unprecedented threat from competition with livestock and from habitat loss through occupation and degradation from overgrazing (Newby, 2014). This latter aspect is particularly worrying because current trends towards overstocking and associated land degradation prejudice not only wildlife conservation and recovery but may also threaten the long-term sustainability of pastoralism itself (IIED, 2013). These grasslands were previously largely abandoned by herders during the hot season for lack of water for their livestock and as such constituted a major source of dry-season grazing and shade for antelopes and gazelles (Newby, 1978).

Similarly, critical habitat is also being lost to agricultural encroachment, especially in vital hot-season ranges, where access to shade and good grazing is critical for survival, or along temporarily flooded wadi habitats that formerly provided both green vegetation and shade (Newby *et al.*, 2014). These factors were less significant when wildlife was more abundant; however, they are critically so when one considers restoration and reintroduction initiatives. As more and more of the wildlife's former habitat is occupied, so other threats appear, including increased risk of bushfires and disease.

At the scale of the Sahara and the Sahel, desertification is probably only having a very local impact. Desert ungulates are well adapted to cope with both drought and widely dispersed pastures. During prolonged drought, however, and lack of adequate pasture, especially water-rich plants on which to feed during the hotter months of the year, die-offs do occur even amongst the most resilient and adaptable species (Newby, pers. obs.). As with disease and bush fires, drought and desertification have a disproportionate impact on small residual populations.

As for the impact of climate change, again, desert-adapted ungulates should be capable of significant resilience. Long-term scenarios for the Sahara and Sahel are still being developed, but some recent work focussing on Chad (Freemantle *et al.*, 2013) indicates a drying trend for the Sahara over the past 20 years or so but by contrast, increased rainfall and vegetation for the Sahel (see also Hermann *et al.*, cited in Freemantle *et al.*, 2013). If this holds true, the likely increase in pastoral and agricultural development in the Sahel could have a negative impact on conservation and wildlife recovery.

Conservation of Sahelo-Saharan ungulates

Conservation of the remaining wild populations

The conservation of Sahelo-Saharan gazelles and antelopes has been consistently beggared by a lack of resources and interest on the part of the international conservation community. Drylands in general are not seen to be a priority, and this stems in part from high biodiversity "hotspot" prioritization, which selects against drylands, together with a fundamental lack of awareness of deserts and their remarkable wildlife (Durant *et al.*, 2012, 2013).

In the region itself, many countries are extremely poor, and the conservation that does occur is severely limited. Furthermore, the region has suffered long bouts of civil unrest and insecurity that have simultaneously impacted wildlife and prevented conservation action taking place. With growing democratization, however, the number of wildlife-focussed non-governmental organizations is increasing. Coupled with the rapid growth in access to the Internet and social media, networking enables them to communicate their messages to both internal and international audiences.

Up until 2004 and the establishment of the Sahara Conservation Fund, there was no international NGO with a mission dedicated specifically to saving imperilled Saharan and Sahelian wildlife, nor any substantial wildlife-related effort by any of the existing international NGOs. SCF's approach has been to focus as far as possible on the remaining centres of high biodiversity, favouring the creation and rehabilitation of protected areas and the generation of up-to-date data on wildlife abundance, threats and distribution necessary for governments to prioritize action and seek sources of support. Key areas identified include the Tassili and Hoggar massifs in Algeria, the Aïr and Termit Tin Toumma areas of Niger, the Anti-Atlas and Drâa Valley in Morocco, and the Manga, Ennedi, Tibesti and Ouadi Rimé regions of Chad. Of international significance has been the recent establishment of a vast protected area in Niger, the Termit & Tin Toumma National Nature Reserve (97,000 km^2). The reserve harbours addax, dama gazelles, Barbary sheep, dorcas gazelles, several carnivores, including the Saharan cheetah, the caracal, and several species of small foxes and cats, as well as important populations of vultures and bustards. Survey work has been instrumental in highlighting the existence of addax populations in Niger and Chad, as well as the presence of residual populations of dama gazelles in those two countries. SCF's Pan Sahara Wildlife Survey is also the first of its kind in the area to collect data on wildlife simultaneously with livestock and human land-use, providing a unique insight into human-wildlife dynamics (Wacher & Newby, 2012b). All survey operations are carried out within the framework of cooperation with the local wildlife agency, and training of wildlife staff is a key feature.

With protection, Sahelo-Saharan wildlife can recover. This has been seen in Chad, where a presidential ban on hunting has had a positive impact on dorcas gazelle numbers, and in Morocco, where land management in favour of houbara bustards has led to an increase in the numbers of Cuvier's gazelles in the same area (Hingrat, pers. comm., 2013). For both the dorcas gazelle and the even more threatened slender-horned gazelle, action to address the growth in poaching with off-road vehicles, quads, and motorbikes is a high priority. Far greater synergy of purpose is required to ensure that modern threats, such as desert-based tourism and livestock development, are compatible with wildlife conservation. Similarly, significant efforts are still required to rein in organized hunting parties, often originating in the Middle East and operating under high-level protection, into line with both the law and sustainable practices.

Captive breeding and reintroduction

With so many species threatened with extinction or absent from vast parts of their former ranges, the conservation of the remaining centres of diversity may on its own be insufficient and call for the adoption of more integrated "One Plan" approaches, bringing together captive breeding, habitat restoration, and reintroduction (Byers *et al.*, 2013). Coordinated captive breeding programmes, especially in the United States, Europe, and the Middle East, maintain populations of threatened Sahelo-Saharan ungulates as insurance against species extinction and are increasingly involved in reintroduction projects in North Africa. However, captive populations also face challenges as many have few founders, suffer from poor historical studbook records, carry the risk of adaptation to captivity, and are not of sufficient size for the long-term retention of genetic diversity without supplementation. Maintenance of genetic variation is necessary for future adaptation to changing environmental conditions and avoidance of inbreeding depression (Frankham *et al.*, 2010; Gilbert, 2011; Lees & Wilcken, 2009, 2011). While intensive population management can counter some of the negative effects of small closed populations, eventually the erosion of genetic diversity and demographic robustness will reduce long-term viability (Frankham *et al.*, 2010; Junhold & Oberwemmer, 2011). Furthermore, economic fragmentation caused by legislation, disease restrictions, and financial considerations mean that coordinated captive breeding programmes cannot always be managed optimally (Gilbert, 2011). The slender-horned gazelle held in formal breeding programmes in the US Association of Zoos and Aquariums (AZA) and the European Association of Zoos and Aquaria (EAZA) collections is an extreme example of the challenges faced by captive populations. Both zoo populations are very small and isolated, resulting in an increased extinction risk from demographic stochasticity (Table 2) (Ballou *et al.*, 2010), and descend from just one male and two females acquired from Tunisia in the 1960s (Correll, 2010), resulting in low levels of genetic diversity and high levels of inbreeding.

Despite these challenges, intensively managed captive populations of species such as addax, scimitar-horned oryx, and dama gazelle are still of sufficient size, genetic diversity, and demographic robustness to be a source of animals for reintroduction and augmentation of wild populations. Efforts to reintroduce Sahelo-Saharan species began as early as the mid-1980s, when a

Table 2 **Sahelo-Saharan ungulates in formal zoo breeding programmes in the United States (AZA) and Europe (EAZA). Source: Abáigar (pers. comm. 2013), Espeso Pajares (pers. comm. 2013), Fischer (pers. comm. 2013), Goodwin (pers. comm. 2013), Moreno (pers. comm. 2013). SSP = Species Survival Plan, EEP = European Endangered Species Programme. The numbers in brackets use the convention of number of males followed by the number of females e.g. 2.7 translates as two males and seven females.**

Species/Sub-species	No. in AZA	AZA zoos	AZA Strategy	No. in EAZA	EAZA zoos	EAZA Strategy
Gazella dorcas	1 (1.0)	1	No programme. Phase out	242 (109.133)	12*	EEP, target 290
Gazella cuvieri	34 (14.20)	3	Target 125 but declining	187 (78.109)	6	EEP, target 250
Gazella leptoceros	57 (21.36)	7	SSP, Target >75	17 (5.12)	1	Monitored only
Eudorcas rufifrons	26 (10.16)	1	New programme	0	0	No programme
Nanger dama mhorr	9 (2.7)	4	No programme. Phase out	208 (91.117)	13	EEP, target 300
Nanger dama ruficollis	134 (54.80)	22	SSP, Target >150	10 (7.3)	2	No programme. Phase out
Addax nasomaculatus	199 (78.121)	19	SSP, Target 250	214 (74.140)	33	EEP
Oryx dammah	186 (71.115)	20	SSP, Target 250	358 (101.257)	55	EEP

* including dorcas gazelle at Guembeul Fauna Reserve (Senegal)

group of scimitar-horned oryx was released into a fenced enclosure in Bou Hedma National Park in Tunisia (Bertram, 1988). This was followed by repatriation of more scimitar-horned oryx, as well as addax, slender-horned, Cuvier's, dorcas, and dama gazelles to fenced protected areas in Morocco, Tunisia, Libya, and Senegal, although the success of these endeavours has varied (Abáigar *et al.*, 1997, 2005, 2009; Gilbert & Woodfine, 2004, 2008; Moreno *et al.*, 2012; Woodfine *et al.*, 2009). While the repatriation of animals to countries within their historical range is highly laudable, so far these populations have made a limited contribution to wider species and habitat restoration across the vast Sahelo-Saharan landscape. Genetic and demographic challenges apart, the re-establishment of self-sustaining, free-ranging ungulate populations may also be hampered by lack of suitable habitat in the country concerned and a general reluctance to release animals into an environment where controls on potential illegal offtake or migration into neighbouring countries are lacking (Gilbert 2011; Molcanova & Wacher, 2011; Godinho *et al.*, 2012). This said, the process has nonetheless provided practical insights and experience and generated much support from both former range states and the global zoo community. A new initiative by the Chadian government, the Environment Agency of Abu Dhabi and the Sahara Conservation Fund seeks to reintroduce large numbers of scimitar-horned oryx to one of their former strongholds, the Ouadi Rimé-Ouadi Achim Game Reserve in central Chad (Newby, 1978; SCF, 2012b). The Chad initiative is the first reintroduction explicitly designed to re-establish a sustainable free-ranging population of scimitar-horned oryx in the wild following a short period of quarantine and adaptation to these new surroundings (Bemadjim *et al.*, 2012). The first release of 25 oryx into temporary holding pens in Chad took place in March, 2016, with an expected release of animals into the wild planned for later in the year.

For a species like the scimitar-horned oryx, which is now extinct in the wild, restoration is entirely dependent on animals held in captivity. For others, like the addax and dama gazelle, whose population sizes in the wild are extremely small, captive-bred animals could be used for both genetic and demographic augmentation if certain conditions can be met (IUCN, 2012). Given the relatively small population sizes of formal breeding programmes and the vast task of restoring lost or critically endangered species across such a huge range, looking beyond traditional scenarios becomes important, including the use of animals from substantial private collections that fall outside formal breeding programmes. Scimitar-horned oryx, addax, and dama gazelles are relatively numerous in private hands (Johnson, pers. comm., 2010).

Given the multiple challenges facing Sahelo-Saharan ungulates, it is now time to adopt an integrated approach to their conservation linking global captive, semi-captive, and wild populations to enhance population sustainability and maximise the chance of success in conserving these species (Byers *et al.*,

2013). This "One Plan Approach" comes with its own challenges, but if the threatened antelopes and gazelles of the Sahelo-Saharan region are to be saved from extinction, then these challenges must be met.

The future of Sahelo-Saharan antelopes

The physiological, behavioural, and ecological adaptations of Sahelo-Saharan ungulates make them a unique and admirable resource for exploiting low density, ephemeral desert pastures that are high in quality but invariably unpredictably sparse and discontinuous in distribution. Considering the vast spaces still available to many of the species, as well as the uncertainty in climate change scenarios for this part of the world, these attributes are extremely valuable and could contribute in many ways to the alleviation of poverty and to ensuring environmental sustainability as resumed in the Millennium Development Goals (UN, 2014a, 2014b).

Exploitation of this resource as part of sustainable land management and natural resource use strategies has often been mooted but rarely if ever put into practice (CMS, 1999a, 1999b). While the potential for sustainable use is certainly present, there are some major practicalities to be addressed. To begin with, most of the ungulates concerned are now extremely rare, rendering meaningful offtake totally unrealistic until numbers increase. This group includes the addax, dama gazelle, slender-horned gazelle, and scimitar-horned oryx when released back to the wild. The smaller and more resilient dorcas and Cuvier's gazelles probably offer the best opportunities for some form of near-term sustainable consumptive use. This could be for lower-value subsistence purposes through low cost and locally managed schemes or higher-value returns through trophy hunting. There is a strong market for this type of desert-based hunting safari, often combined with falconry for bustards, but it currently occurs in a largely uncontrolled, unsustainable, and often illegal manner.

Harnessing the wildlife potential and coupling this with well thought out schemes that are managed by and benefit local communities could provide powerful economic incentives to conserve and manage the wildlife resource in ways currently not seen. A good example of just what can be done in broadly similar aridlands is the community-based natural resource management (CBNRM) scheme implemented in Namibia (NACSO, 2014). Tens of thousands of Namibians have received economic and social benefits from jobs, training, game meat, and cash dividends, including employment and related opportunities from a healthy, wildlife-based tourism industry. Thanks to increased economic benefits from wildlife conservancies, wildlife poaching has decreased, and land management in favour of wild animals has seen a reduction in habitat destruction and overgrazing by cattle (Brown & Bird, 2011; NACSO, 2014; UN, 2014).

Were it not for the current security issues in much of the Sahara and Sahel, desert-based tourism would no doubt present a growing opportunity in which wildlife and nature could play a significant role. Again, increased efforts are required to ensure tourism really does benefit local communities in a way significant enough for it to be an incentive to conserve. Unfortunately, the rarest and most charismatic species are currently so endangered that any kind of disturbance, however benign in intention, would not be in the interests of their recovery until numbers increase and tourists and tourism operators adopt responsible codes of practice.

In Niger, work with the local community living in and around the newly gazetted Termit & Tin Toumma National Nature Reserve is opening up some useful avenues (Damagram, 2010, 2013; Rabeil *et al.*, 2014). Instead of focussing on use of the still largely threatened wildlife resources, support for conservation action, protection of wildlife and zonation, is being built through alternative socio-economic development, including the construction of environmentally-friendly wells, increased access to health care, support for education, and resolution of livestock predation problems. Employment opportunities are also created through the recruitment of community-nominated game guards. Thanks largely to them, together with the support of local leaders and elected councils, soldiers were successfully brought to trial for the first time ever in Niger for hunting addax (Le Damagaram, 2014a, 2014b).

With growing interest in the phenomenon of resilience or the ability of societies, both natural and human, to recover from change, shock or misfortune, the ungulate fauna of the Sahara and Sahel offers immense potential for research in some of the most vulnerable areas of Africa. Promising fields of research that could contribute to raising the profile and value of desert ungulates include their role in ecosystem services, plant germination, regeneration, and dispersal, as well as habitat maintenance, stability, and productivity, including comparison with livestock and the risk of land degradation and desertification. Assessing the potential of wildlife to cope with changing vegetation cover and composition as a result of climate change or biodiversity loss is also pertinent.

Given the dire circumstances faced by many Sahelo-Saharan species, halting and reversing the declines will call for a mix of traditional and more modern conservation methods in order to address and take advantage of rapidly changing economic, political, and social environments, as well as climate change. Extraordinary efforts to conserve what remains in the field are required, along with restoration ecology, captive breeding, and community engagement. With foresight, the new paradigms imposed by population increase, climate change, and desertification need not necessarily be seen as supplementary threats to an already beleaguered group of animals but might just offer the opportunities needed for the full value of desert ungulates to be recognized and harnessed for the good of both wildlife and mankind alike.

References

Abaigar, T., Cano, M., Espeso, G. & Ortiz, J. (1997): Introduction of Mhorr gazelle *Gazella dama mhorr* in Bou-Hedma National Park, Tunisia. *International Zoo Yearbook* **35**: 311–316.

Abaigar, T., Cano, M. & Sakkouhi, M. (2005): Evaluation of habitat use of a semi-captive population of Cuvier's gazelles *Gazella cuvieri* following release in Boukornine National Park, Tunisia. *Acta Theriologica* **50**(3): 405–415.

Abáigar, T., Youm, B., Niaga, M., Ensenyat, C. & Cano, M. (2009): The role of Senegal in the recovery of the Sahelo-Saharan antelope species: the case of the reintroduction of dorcas gazelle. *Gnusletter* **28**: 6–8.

Ballou, J. D., Lees, C., Faust, L. J., Long, S., Lynch, C., Bingaman Lackey, L. & Foose, T. J. (2010): Demographic and genetic management of captive populations. In *Wild mammals in captivity: principles and techniques for zoo management:* 219–252. Kleiman, D. G., Thompson, K. V. & Kirk Baer. C. (Eds). Chicago: University of Chicago Press.

Bemadjim, N. E., Newby, J., Desbiez, A., Lees, C. & Miller, P. (Eds): (2012): Atelier technique sur la réintroduction de l'oryx algazelle dans la Réserve de Faune de l'Ouadi Rimé-Ouadi Achim, Tchad. IUCN/SSC Conservation Breeding Specialist Group, Apple Valley, MN, USA and Sahara Conservation Fund, L'Isle, Switzerland.

Benchaleh, A.-C., Bouziane, H., Maka, M. & Ouahès, C. (2011): *Fleurs du Sahara*. Paris: Editions Ibis Press.

Bertram, B. C. R. (1988): Re-introduction of scimitar-horned oryx to Tunisia. In *Conservation biology of desert antelopes:* 136–145. Dixon, A & Jones, D. (Eds). Including the proceedings of the 25th anniversary of 'Operation Oryx' symposium. Zoological Society of London. London: Christlopher Helm.

Beudels, R. C. & Devillers, P. (2013): *Gazella leptoceros* Slender-horned Gazelle. In *Mammals of Africa.* 1. *Pigs, hippotamuses, chevrotain, giraffes, deer and bovids*: 352–355. Kingdon, J. & Hoffmann, M. (Eds). London: Bloomsbury Publishing.

Beudels, R. C. & Devillers, P. & Cuzin, F. (2013): *Gazella cuvieri* Cuvier's Gazelle. In *Mammals of Africa: Volume I: Pigs, Hippotamuses, Chevrotain, Giraffes, Deer and Bovids:* 349–352. Kingdon, J. & Hoffmann, M. (Eds). London: Bloomsbury Publishing.

Brocklehurst, H. C. (1931): *Game animals of the Sudan*. Gurney & Jackson.

Brosset, L. H. (1904): *Reconnaissance au Sahara d'In Salah à Amguid et à Tukhammar*. Publication Comité de l'Afrique française.

Brown, J. & Bird, N. (2011): *Sustainable natural resource management in Namibia: successful community-based wildlife conservation. Development Progress*. London: Overseas Development Institute.

Buckle, C. (1978): *Landforms in Africa: introduction to geomorphology*. London: Longman.

Byers, O., Lees, C., Wilcken, J. & Schwitzer, C. (2013): The One Plan Approach: the philosophy and implementation of CBSG's approach to integrated species conservation planning. *WAZA Magazine* **14**: 2–5.

Cano-Pérez, M. (1984): Revision der Systematik von *Gazella (Nanger) dama. Zeit. Kölner Zoo* **27**: 103–107.

CMS. (1999a): *Proceedings of the seminar on the conservation and restoration of Sahelo-Saharan antelopes*. CMS Technical Series Publication No. 3. Bonn, Germany: UNEP/CMS Secretariat.

CMS. (1999b): *Conservation measures for Sahelo-Saharan Antelopes*. Action Plan and status reports. CMS Technical Series Publication No. 4. Bonn, Germany: UNEP/CMS Secretariat.

CMS. (2006): *Sahelo-Saharan antelopes. Status and perspectives. Report on the conservation status of the six Sahelo-Saharan antelopes*. Beudels-Jamar, R.C., Devillers, P., Lafontaine, R.-M., Devillers-Terschuren, J., & Beudels, M.-O. (Eds). CMS Technical Series Publication N° 11, 2005. Bonn, Germany: UNEP/CMS Secretariat.

Correll, T. (2010): *Slender-horned gazelle Gazella leptoceros international studbook data set in SPARKS*. Current to 31/12/2010. Tulsa Zoo.

Damagram. (2010): No. 9, December 2010 (Niamey, Niger).

Damagram. (2013): No. 15, December 2013 (Niamey, Niger).

De Smet, K., Maziz, S. B., Fellous, A., Belbachir, F., Belbachir-Bazi, A., Commizoli, P. & Wacher, T. (2009): *Inventaire de la faune sauvage des zones désertiques en Algérie (Grand Erg Occidental)*. Rapport de mission 3–15 mars, 2007. L'Isle, Switzerland: Sahara Conservation Fund.

Dragesco-Joffé, A. (1993): *La vie sauvage au Sahara*. Lausanne: Delachaux et Niestlé.

Duncan, P. (2013): *Damaliscus lunatus Topi; In Mammals of Africa*. **1**. *Pigs, hippotamuses, chevrotain, giraffes, deer and bovids:* 502–510. Kingdon, J. & Hoffmann, M. (Eds). Bloomsbury London: Publishing.

Durant, S. M., Pettorelli, N., Bashir, S., Woodroffe, R., Wacher, T., De Ornellas, P., Ransom, C., Abáigar, T., Abdelgadir, M., El Alqamy, H., Beddiaf, M., Belbachir, F., Belbachir-Bazi, A., Berbash, A. A., Beudels-Jamar, R., Boitani, L., Breitenmoser, C., Cano, M., Chardonnet, P., Collen, B., Cornforth, W. A., Cuzin, F., Gerngross, P., Haddane, B., Hadjeloum, M., Jacobson, A., Jebali, A., Lamarque, F., Mallon, D., Minkowski, K., Monfort, S., Ndoassal, B., Newby, J., Ngakoutou, B. E., Niagate, B., Purchase, G., Samaïla, S., Samna, A. K., Sillero-Zubiri, C., Soultan, A. E., Stanley Price, M. R. & Baillie, J. E. M. (2012): Forgotten Biodiversity in Desert Ecosystems. *Science* **336**: 1379.

Durant, S. M., Wacher, T., Bashir, S., Woodroffe, R., De Ornellas, P., Ransom, C., Newby, J., Abaigar, T., Abdelgadir, M., El Alqamy, H., Baillie, J., Beddiaf, M., Belbachir, F., Belbachir-Bazi, A., Berbash, A. A., Bemadjim, N. E., Beudels-Jamar, R., Boitani, L., Breitenmoser, C., Cano, M., Chardonnet, P., Collen, B., Cornforth, W. A., Cuzin, F., Gerngross, P., Haddane, B., Hadjeloum, M., Jacobson, A., Jebali, A., Lamarque, F., Mallon, D., Minkowski, K., Monfort, S., Ndoassal, B., Niagate, B., Purchase, G., Samaila, S., Samna, A. K., Sillero-Zubiri, C. A., Soultan, E., Stanley Price, M. R. & Pettorelli, N. (2013): Fiddling in biodiversity hotspots while deserts burn? Collapse of the Sahara's megafauna. *Diversity and Distributions* **20**: 114–122.

El Alqamy, H. & Baha el Din, S. (2006): Contemporary status and distribution of gazelle species (*Gazella dorcas* and *Gazella leptoceros*) in Egypt. *Zoology in the Middle East* **39**.

Fay, M. (2005): *Aerial Reconnaissance Survey, BET*, Tchad, 1–17 November 2005. African Parks Network.

Fellous, A., Wacher, T., De Smet, K. & Commizoli, P. (2009): *Inventaire de la faune sauvage des zones désertiques en Algérie. Grand Erg Oriental.* L'Isle, Switzerland: Sahara Conservation Fund.

Frankham, R., Ballou, J. D. & Ballou, D. A. (2010): *Introduction to conservation genetics.* Cambridge: Cambridge University Press.

Freemantle, T. P., Wacher, T., Newby, J. & Pettorelli, N. (2013): Earth observation: overlooked potential to support species reintroduction programmes. *Afr. J. Ecol.* **51**(3): 1–11.

Gautier, E. F. (1923): *Le Sahara*. Paris : Collection Payot.

Gilbert, T. (2011): *The sustainability of endangered species under intensive management: the case of the scimitar-horned oryx (Oryx dammah).* PhD Thesis. University of Southampton, Southampton, UK.

Gilbert, T. & Woodfine, T. (Eds). (2004): *The Biology, Husbandry and Conservation of Scimitar-horned Oryx (Oryx dammah).* Winchester, UK: Marwell Preservation Trust.

Gilbert, T. & Woodfine T. (2008): *The re-introduction of scimitar-horned oryx Oryx dammah to Dghoumes National Park, Tunisia.* Report to members of the European Endangered Species Programme for scimitar-horned oryx. Winchester, UK: Marwell Preservation Trust, December 2008.

Gilbert, T., Woodfine, T., Houston, E. W., Newby, J. & Zahzah, K. (2012): The fall and rise of the scimitar-horned oryx. *WAZA Magazine* **13**: 25–28.

Godinho, R., Abàigar, T., Lopes, S., Essalhi, A., Ouragh L., Cano, M. & Ferrand, N. (2012): Conservation genetics of the endangered Dorcas gazelle (Gazella dorcas spp.) in Northwestern Africa. *Conservation Genetics* **13**: 1003–1015.

Groves, C. & Grubb, P. (2011): *Ungulate taxonomy.* Baltimore: Johns Hopkins University Press.

Hammond, R. L., Macasero, W., Flores, B., Mohammed, O. B., Wacher, T. & Bruford, M. (2001): Phylogenetic reanalysis of the Saudi Gazelle and its implications for conservation. *Conservation Biology* **15**: 1–11.

Harper, F. (1945): *Extinct and vanishing mammals of the Old World*, New York, American Committee for International Wild Life Protection.

Henriques, R. (1938): *A Darfur journey: death by moonlight.* London: Collins.

IIED. (2013): Global public policy narratives on the drylands and pastoralism. http://pubs.iied.org/pdfs/10040IIED.pdf (as at 19 July 2013).

In Tanoust. (1930): *La chasse dans le pays saharien et sahélien.* Paris: Editions du Comité Algérie-Tunisie-Maroc.

IUCN. (2008): *The nature of drylands: diverse ecosystems, diverse solutions.* Nairobi: IUCN Eastern and Southern Africa Regional Office.

IUCN. (2013): SSC Antelope Specialist Group (2008): *Oryx dammah.* In IUCN 2013. IUCN Red List of Threatened Species. Version 2013.1. www.iucnredlist.org. Downloaded on 19 July 2013.

IUCN. (2012): *Resolutions and recommendations.* World Conservation Congress, Jeju, Republic of Korea, 6–15 September 2012. Gland, Switzerland: IUCN.

IUCN/SSC. (2012): *Guidelines for Reintroductions and Other Conservation Translocations. Version 1.0*. Gland, Switzerland: IUCN Species Survival Commission.

IUCN. (2013): *The IUCN Red List of Threatened Species. Version 2013.1*. http://www.iucnredlist.org (as at 4 July 2013).

Junhold, J. & Oberwemmer, F. (2011): How are animal keeping and conservation philosophy of zoos affected by climate change? *International Zoo Yearbook* **45**: 99–107.

Le Damagaram. (2014a): No. 69–70, 1–15 February 2014, p. 2.

Le Damagaram. (2014b): No. 71–72, 1–31 March 2014, p. 4.

Lees, C. M. & Wilcken, J. (2009): Sustaining the ark: challenges faced by zoos in maintaining viable populations. *International Zoo Yearbook* **43**: 6–18.

Lees, C. M. & Wilcken, J. (2011): Global programmes for sustainability. *WAZA Magazine* **12**: 2–5.

Leo Africanus. (1896): *The history and description of Africa* (3 vols). Brown, R. (Ed.). London: Hakluyt Society.

Malbrant, R. (1952): *Faune du Centre africain français*. Paris: Lechevalier.

Molcanova, R. & Wacher, T. (2011): *Scimitar-horned oryx monitoring, Sidi Toui National Park, Tunisia*. November 2011. SCF/Pan Sahara Wildlife Survey. Technical Report No. 10. November 2011. L'Isle, Switzerland: Sahara Conservation Fund.

Monfort, S., Newby, J., Wacher, T., Tubiana, J. & Moksia, D. (2003): Sahelo-Saharan interest group wildlife surveys. part 1: western and central Chad (September–October 2001). *Sahelo-Saharan Interest Group*. L'Isle, Switzerland.

Moreno, E., Sane, A., Benzal, J., Ibáñez, B., Sanz-Zuasti, J. & Espeso, G. (2012): Changes in habitat structure may explain Decrease in reintroduced mohor gazelle population in the Guembeul Fauna Reserve, Senegal. *Animals* **2**: 347–360.

Morrow, C., Molcanova, R. & Wacher, T. (2013): *Oryx dammah Scimitar-horned Oryx. In Mammals of Africa. 1. Pigs, Hippotamuses, chevrotain, giraffes, deer and bovids: 586–592*. Kingdon, J. & Hoffmann, M. (Eds). London: Bloomsbury Publishing.

NACSO. (2014): Namibian Association of CBNRM Support Organizations. http://www.nacso.org.na/what_is_cbnrm.php (as at 14 August 2014).

Newby, J. E. (1978): *The ecological resources of the Ouadi Rimé-Ouadi Achim Game Reserve, Chad*. Unpublished technical report.

Newby, J. E. (1982): *Action plan for the Sahelo-Saharan fauna of Africa*. Gland: Switzerland: WWF/IUCN.

Newby, J. E. (1984): Larger mammals of the Sahara. In *Sahara Desert*. Cloudsley-Thompson & J. L. (Ed.). Key Environments Series. Oxford: Pergamon Press.

Newby, J. E. (1988): Aridland wildlife in decline. In *Conservation and biology of desert antelopes*. Dixon, A. & Jones, D. M. (Eds). London: Christopher Helm.

Newby, J. E. (2013): *Addax nasomaculatus* Addax. In *Mammals of Africa. 1. Pigs, hippotamuses, chevrotain, giraffes, deer and bovids: 566–571*. Kingdon, J. & Hoffmann, M. (Eds). London: Bloomsbury Publishing.

Newby, J. E. (2014): Home, home on the range… *Oryx* **48**(2): 157–158, DOI: 10.1017/S0030605314000143.

Newby, J. E., Wacher, T., Hatcha, M. H., Ligneureux, L., Pusey, R. & Rabeil, T. (2014): *Reintroduction of the scimitar-horned oryx Oryx dammah to Chad*. Project development and preparatory fieldwork in the Ouadi Rimé-Ouadi Achim Game Reserve, March & September 2013. L'Isle, Switzerland: Sahara Conservation Fund.

Newby, J. E., Wacher, T. & Hatcha, M. H. (2014): *Conservation of the dama gazelle.* Survey of the Manga & Western Chad. January-February 2014. L'Isle, Switzerland: Sahara Conservation Fund.

Olson, D. M. & Dinerstein, E. (2002): The Global 200. Priority ecoregions for global conservation. *Annals of the Missouri Botanical Garden* **89**(2): 199–224.

Olson, D. M., Dinerstein, E., Wikramanayake, E. D., Burgess, N. D., Powell, G. V. N., Underwood, E. C., D'Amico, J. A., Itoua, I., Strand, H. E., Morrison, J. C., Loucks, C. J., Allnutt, T. F., Ricketts, T. H., Kura, Y., Lamoreux, J. F., Wettengel, W. W., Hedao, P. & Kassem, K. R. (2001): Terrestrial ecoregions of the world: a new map of life on Earth. *Bioscience* **51**(11): 933–938.

Ozenda, P. (1983): *Flore du Sahara*. Paris : Editions du CNRS.

Quézel, P. (1965): *La végétation du Sahara du Tchad à la Mauritanie*. Stuttgart: Gustav Fischer Verlag.

Rabeil, T. (2014): *Evaluation de l'état de conservation de la faune, des habitats et des menaces dans la Réserve Naturelle Nationale de l'Aïr et du Ténéré (RNNAT), Niger*. Niger : Ministère de l'environnement, de la salubrité urbaine et du Développement durable.

RZSS & IUCN Antelope Specialist Group. (2014): *Dama gazelle, Nanger dama, Conservation Review*. Edinburgh: Royal Zoological Society of Scotland.

Sayre, R., Comer, P., Hak, J., Josse, C., Bow, J., Warner, H., Larwanou, M., Kelbessa, E., Bekele, T., Kehl, H., Amena, R., Andriamasimanana, R., Ba, T., Benson, L., Boucher, T., Brown, M., Cress, J., Dassering, O., Friesen, B., Gachathi, F., Houcine, S., Keita, M., Khamala, E., Marangu, D., Mokua, F., Morou, B., Mucina, L., Mugisha, S., Mwavu, E., Rutherford, M., Sanou, P., Syampungani, S., Tomor, B., Vall, A., Vande Weghe, J., Wangui, E. & Waruingi, L. (2013): *A new map of standardized terrestrial ecosystems of Africa*. Washington, DC: Association of American Geographers.

SCF. (2007): The miracle of life. *Sandscript* **2**: 1–2.

SCF. (2009): Mad dogs and Englishmen. *Sandscript* **6**: 3.

SCF. (2010): Niger's addax hit by drought and oil. *Sandscript* **8**: 5.

SCF. (2012a): Massive wildlife slaughter. *Sandscript* **12**: 4.

SCF. (2012b): Poised for re-entry. *Sandscript* **11**: 4.

Schattenberg, P. (2013): Researchers: second phase of dama gazelle study will focus on social interactions. Department of Wildlife and Fisheries Sciences, Texas A&M University. http://wfsc.tamu.edu/2015/07/30/researchers-second-phase-of-dama-gazelle-study-will-focus-on-social-interactions (as at 20 February 2016).

Scholte, P. (2013): *Nanger dama Dama Gazelle. In Mammals of Africa 1. Pigs, hippotamuses, chevrotain, giraffes, deer and bovids: 382–387*. Kingdon, J. & Hoffmann, M. (Eds). London: Bloomsbury Publishing.

Scholte, P. & Hashim, M. (2013): *Eudorcas rufifrons* Red-fronted Gazelle. In *Mammals of Africa 1. Pigs, hippotamuses, chevrotain, giraffes, deer and bovids: 357–359.* Kingdon, J. & Hoffmann, M. (Eds). London: Bloomsbury Publishing.

Scholte, P. & Hashim, M. (2013): *Gazella dorcas* Dorcas Gazelle. In *Mammals of Africa 1. Pigs, hippotamuses, chevrotain, giraffes, deer and bovids: 340–346*. Kingdon, J. & Hoffmann, M. (Eds). London: Bloomsbury Publishing.

Sclater, P. L. & Thomas, O. (1898): *The book of Antelopes. III.* London, R.H. Porter (pp. 147–148).

Senn, H., Banfield, L., Wacher, T., Newby, J., Rabeil, T., Kaden, J., Kitchener, A. C., Abaigar, T., Luisa da Silva, T., Maunder, M. & Ogden, R. (2014): Splitting or lumping? A conservation dilemma exemplified by the critically endangered dama gazelle (*Nanger dama*). *PLOS One* **9**(6): 1–13.

Serre, G. (2000): *Une nomadisation d'hivernage dans l'Ouadi Rimé (Tchad 1956).* Paris: Le Harmattan.

UNCCD. (2011): *Desertification: a visual synthesis.* United Nations Convention on Combatting Desertification.

UNEP. (2006): *Global deserts outlook.* United Nations Environment Programme.

UNEP/CMS. (2005): *Les antilopes sahélo-sahariennes. Statut et perspectives. Rapport sur l'état de conservation des six antilopes sahélo-sahariennes.* CMS Technical Series Publication N° 11, 2005. Beudels-Jamar, R. C., Devillers, P., Lafontaine, R.-M., Devillers-Terschuren, J. & Beudels, M.-O. (Eds). Bonn, Germany: UNEP/CMS Secretariat.

UN Millennium Development Goals. (2014): http://www.un.org/millenniumgoals/(as at 14 August 2104).

UN Sustainable Development Platform. (2014): http://sustainabledevelopment.un.org/index.php?page=view&type=1006&menu=1348&nr=2027 (as at 14 August 2014).

Wacher, T., Rabeil, T. & Newby, J. (2010a): *Monitoring survey of Termit and Tin Toumma (Niger) & review of monitoring results December 2008 – December 2009.* L'Isle, Switzerland: Sahara Conservation Fund.

Wacher, T. & Newby, J. (2010b): *Wildlife and land use survey of the Manga and Eguey regions, Chad.* Pan Saharan Wildlife Survey. Technical Report No. 4. August 2010 L'Isle, Switzerland: Sahara Conservation Fund.

Wacher, T., Newby, J., Bourtchiakbe, S. & Banlongar, F. (2010c): *Wildlife survey of the Ouadi Rimé- Ouadi Achim Game Reserve, Chad (Part I).* SCF/Pan Saharan Wildlife Survey. Technical Report No. 5. March 2011. L'Isle, Switzerland: Sahara Conservation Fund.

Wacher, T., Mahdi, S., Guidara, H. & Tahri, J. (2011): *A reconnaissance survey of Jabbes-Senghar National Park, Tunisia.* SCF Pan Sahara Wildlife Survey. Technical Report No. 9. March 2012.L'Isle, Switzerland: Sahara Conservation Fund.

Wacher, T. & Newby, J. (2011): *Summary of results and achievements of the Pilot Phase of the Pan Sahara Wildlife Survey 2010–2011.* PSWS Technical Report 6. L'Isle, Switzerland: Sahara Conservation Fund.

Wacher, T., Newby, J., Molcanova, R., Bourtchiakbe, S. & Hassan, M. (2012): *Wildlife and land use survey of the Ouadi Rimé- Ouadi Achim Game Reserve, Chad (Part II). September 2011* SCF/Pan Sahara Wildlife Survey. Technical Report No. 7. November 2011. L'Isle, Switzerland: Sahara Conservation Fund.

Wacher, T. & Newby, J. (2012a): *Reconnaissance survey of the Ouadi Rimé- Ouadi Achim Game Reserve, Chad. April 2012.* SCF/Pan Sahara Wildlife Survey. SCF/PSWS Technical Report No. 11. July 2012. L'Isle, Switzerland: Sahara Conservation Fund.

Wacher, T. & Newby, J. (2012b): *Summary of results and achievements of the Pilot Phase of the Pan Sahara Wildlife Survey 2009–2012.* SCF PSWS Technical Report 12, November 2012. L'Isle, Switzerland: Sahara Conservation Fund.

Woodfine, T., Zahzah, K., Chetoui, A., Gilbert, T. & d'Alterio, G. L. (2009): *Reintroduction of scimitar-horned oryx to Dghoumes National Park, Tunisia.* Report prepared for Direction Générale des Forêts, Tunis. Winchester, UK: Marwell Wildlife.

Wilson, D. E. & Mittermeier, R. A. (Eds). (2011): *Handbook of the mammals of the world. 2. Hoofed mammals.* Barcelona: Lynx Edicions.

WWF. (2013): Ecoregions. http://worldwildlife.org/biomes (as at 2 July 2013).

14

The Fall and Rise of the Scimitar-Horned Oryx: A Case Study of Ex-Situ Conservation and Reintroduction in Practice

Tim Woodfine and Tania Gilbert

Marwell Wildlife, Colden Common, Winchester, United Kingdom

Once occurring in large migratory herds, over vast tracts of the Sahara and bordering Sahelian grasslands, the scimitar-horned oryx (*Oryx dammah*) suffered massive and catastrophic decline, and was extirpated from nature by the late 20th century. Ironically, the species now has the paradoxical status of being Extinct in the Wild (IUCN SSC Antelope Specialist Group, 2008) and yet one of the most abundant species of antelope in captivity. With the availability of *ex situ* populations, the scimitar-horned oryx is one of only six species of large mammals to have become extinct in the wild to then be reintroduced into indigenous range solely from captive-bred stock. For these reasons, it has become an iconic conservation species and an important case study.

From abundance to extinction in the wild

The scimitar-horned oryx is a large, arid-adapted and gregarious hippotraginae antelope named for its long, elegant backward-curving horns (Devillers & Devillers-Terschuren, 2005; Gilbert & Woodfine, 2004). It is likely that the ecological range of the scimitar-horned oryx once matched that of the Sahara's peripheral and interior grasslands, with populations expanding and contracting according to changes in climatic conditions (Iyengar *et al.*, 2007; Le Houerou, 1997). Hence, the expansive Saharan savanna of the more humid early Holocene (9500–4500 B.P.) probably represented the pinnacle for scimitar-horned oryx, with contiguous populations of at least one million animals (Iyengar *et al.*, 2007).

Antelope Conservation: From Diagnosis to Action, First Edition. Edited by Jakob Bro-Jørgensen and David P. Mallon.

The historic distribution and decline of scimitar-horned oryx has been previously well documented, drawing on a wealth of original sources (Devillers & Devillers-Terschuren, 2005; Wakefield *et al.*, 2004; Newby, 1988). Even with the more recent expansion of the desert, the range of the species remained vast, covering much of North Africa from the Atlantic Ocean to the Nile and extending south to 15^0 of latitude. At its height, the past distribution of the scimitar-horned oryx is likely to have mirrored the occurrence of the North and South Saharan steppe, dry woodlands, and savanna, together with refuges within the true desert such as accessible oases and highlands. Indeed, documented historical sightings, including references to specific locations and broader zones of presence, overlap with this potential ecological range (Figure 1).

Evidence of the presence and suggested abundance of scimitar-horned oryx includes realistic rock art dating back several thousand years, the frescoes and reliefs of ancient Egypt, and the depictions and writings of ancient

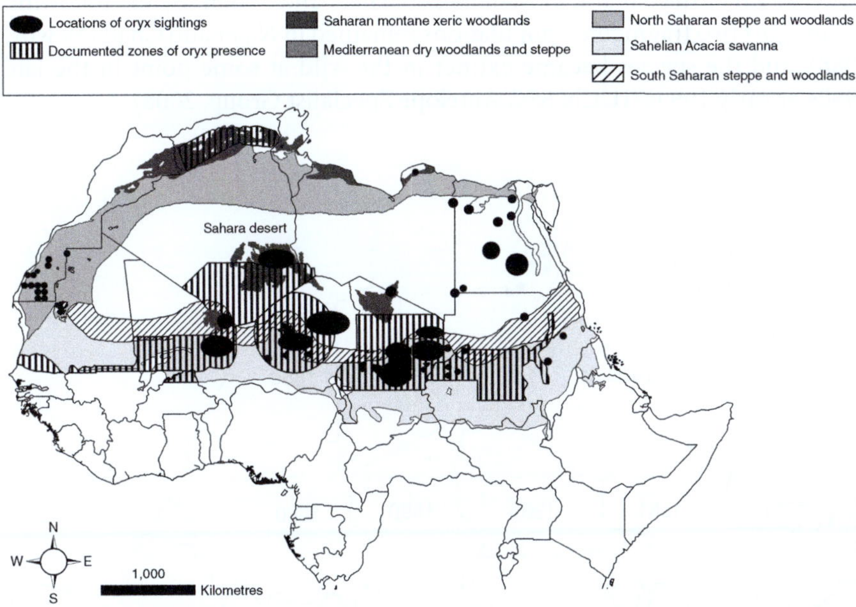

Figure 1 **Assumed indigenous distribution of scimitar-horned oryx across the Saharan region based on documented sightings and zones of presence, and likely ecological range (Barzdo, 1982; Bonnet, 1909; Brehm, 1920; H. Brahim, pers. comm.; Dixon *et al.*, 1991; Flower, 1932; Gillet, 1965; Gillet, 1966; Gillet, 1971; Gordon, 1991; Haltenorth, 1963; Heringa, 1989; IUCN. 1987; Maydon, 1923; Newby, 1975; Newby, 1984; Olsen *et al.*, 2001; Rode, P. 1943; Schomber, H.W. 1963; Sclater & Thomas, 1899/1900; Wilson, 1980).**

Greek and Roman civilisations (Engel, 2004; Newby, 1988). It is therefore likely that the species was in gradual decline throughout history under steadily increasing human pressure (Wakefield *et al.*, 2004). The scimitar-horned oryx disappeared from opposite ends of its longitudinal range in Senegal and Egypt by the mid-19th century (Figure 2). A congregation of at least 10,000 animals was reported in Chad as recently as the 1930s (Bassett, 1975), but it is during the following decades of the 20th century that numbers fell dramatically as over-exploitation and competition with domestic livestock took their toll, leaving the species vulnerable to natural catastrophes such as periodic drought (Newby, 1988). The advent and availability of four-wheel-drive vehicles and automatic weapons brought large-scale hunting to formerly remote and inaccessible places; the situation exacerbated by insecurity and civil war (Newby, 1988). As outlined by Newby *et al.* (Chapter 13), these are all factors that remain pertinent today as overexploitation continues to threaten extant Saharan antelopes and other wildlife.

By the 1970s, scimitar-horned oryx populations were reduced to several nuclei south of the Sahara, in Mali, Niger, Chad, and Sudan. By the following decade, just two fragmented populations remained in Niger and Chad (Newby, 1988), and the species became extinct in the wild at some point in the late 1980s or early 1990s (IUCN SSC Antelope Specialist Group, 2008).

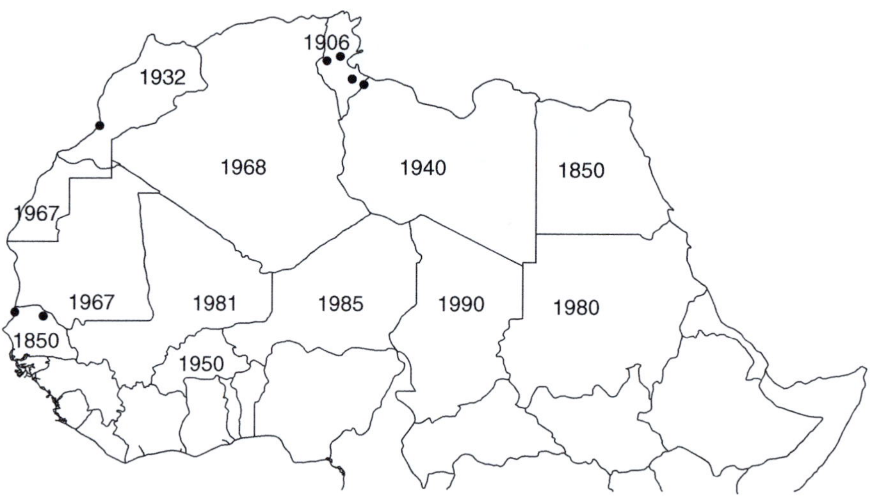

Figure 2 **Approximate dates marking the local extinction of scimitar-horned oryx by country (Devillers & Devillers-Terschuren, 2005; Wakefield *et al.*, 2004; Newby, 1988). Dots mark the locations of current scimitar-horned oryx populations in Tunisia, Morocco, and Senegal.**

There was, however, a parallel story as captive populations of scimitar-horned oryx were established and reintroduction efforts began. Around the time that the last wild individuals disappeared, releases of captive bred animals began in Tunisia, and then Morocco and Senegal during the subsequent quarter of a century.

The safety net: the development of modern captive populations

Scimitar-horned oryx have been held in zoological collections since the 1800s, although the origins of those early animals are unclear. Most of the recorded founders of the modern global population originated from Chad in the mid-1960s (Gilbert, 2015), making those capture operations the most significant events in the formation of captive herds documented in the International Studbook (Gilbert, 2015). Anecdotally, further animals were acquired from elsewhere in the wild as recently as the 1970s, with the herd at Cairo Zoo reported to have originated from animals caught at that time in Sudan (Abdoon, pers. comm., 2005). Although not formally documented, scimitar-horned oryx captured for private collections in the Middle Eastern Gulf States apparently originated from different parts of North Africa and may yield lineages unrelated to those represented in the International Studbook.

Detailed records of the important Chad capture operations are not available, but some information can be gleaned from correspondence at that time and the International Studbook, which indicates that two missions took place. The first, in 1963, resulted in the capture of four individuals, three of which were sent to Catskill, USA, and the fourth to Hannover Zoo, Germany. The second, larger-scale operation happened in 1967 when 44 scimitar-horned oryx were captured. Twenty-six of these animals went to the USA to form the basis of what became the North American Species Survival Plan (SSP) under the auspices of the Association of Zoos & Aquaria (AZA); 16 were quarantined at Aalborg Zoo, Denmark, before being distributed to other collections in the region and latterly forming the basis of the European Endangered Species Programme (EEP) under the auspices of the European Association of Zoos & Aquaria (EAZA); and two were sent to Tama Zoo in Tokyo, Japan (Van den Brink, pers. comm., 1979; Gilbert, 2015). The International Studbook lists an extra animal imported into the USA not detailed in the capture records, but presumed to have been part of one of the Chad operations. Since the 1960s, captive-bred scimitar-horned oryx have been distributed across Europe and North America, and then to South Africa, and throughout Central and South America, Asia, and Australasia. A few were also sent to Israel, Saudi Arabia, and the UAE.

The scimitar-horned oryx SSP has approximately 25 founders; 23 from the Chad operations, one from Europe and one from the Middle East of which the wild origins are unknown. While more scimitar-horned oryx were received from the wild, they either did not produce offspring, or if they did, still failed to leave any living descendants in the current population. The situation in Europe is more complicated due to a lack of historical records. In total there were about 21 founders to the EEP population as scimitar-horned oryx were well established in Europe prior to the 1960s (Gilbert, 2015).

There are thus a number of animals listed in the International Studbook with unknown origins, and some of these have contributed to the SSP and EEP populations. It is possible that at least a few of them may be additional founders, but this cannot be verified. Additionally, poor record keeping during transport, quarantine, and the initial period in captivity means that it is not possible to identify specifically which individuals are founders to the global captive population, and which animals were perhaps part of the first generation to be born after the capture operations. Consequently, the global population of scimitar-horned oryx documented in the International Studbook is thought to be founded on 45–50 individuals (Gilbert, 2015).

All the animals imported from Chad in 1967 were captured during a single expedition over a period of two months. Information on the capture location is imprecise, but described as the "plane desert" in the northern part of Chad (Van den Brink, pers. comm., 1980), which is likely to be between the Ouadi Achim and the Ouadi Hawach (Newby, pers. comm., 2006). Scimitar-horned oryx were relatively abundant in this area in 1967, and it was a key trophy-hunting region due to the large number of high-quality animals and accessible terrain (Newby, pers. comm., 2006). As a result, the global captive population probably represents only a fraction of the species' potential genetic diversity in the 1960s. Moreover, animals captured in close proximity may have been at least partially related to each other; an assumption supported by genetic analyses of recent tissue samples, which indicate some degree of relatedness within the early captive population (Iyengar *et al.*, 2007).

Genetically, the North American, European, and Middle Eastern scimitar-horned oryx populations all have some unique lineages, although the majority of this diversity is represented in the EEP and SSP (Iyengar *et al.*, 2007). More recent analyses revealed additional diversity in previously unrecorded groups in the Middle East (Ogden & Senn, pers. comm., 2011), while some herds that could make a further genetic contribution (e.g., at Cairo Zoo) have yet to be tested.

To date, scimitar-horned oryx have been distributed to 444 institutions in 48 countries across the world. The International Studbook that predominantly tracks animals in zoological institutions currently lists 1,749 individuals in 221 institutions (Gilbert, 2015), but the entire global population has grown to an estimated 15,000–19,000 animals when private collections and ranches are also taken into account. This includes approximately 11,000 scimitar-horned oryx on ranches in Texas (Johnson, pers. comm., 2010), and at least 4,000 in the UAE (Chuven, pers. comm., 2012), making them one of the most abundant species of antelope held in captivity. Despite these numbers, the fact remains that the modern global population is derived from a limited number of founders unevenly distributed across the world (Gilbert, 2015). There is also variable application of genetic, demographic, and infectious disease management between locations, which are important considerations for the selection of reintroduction stock (IUCN SSC, 2013).

Returning to nature: the initial steps

While the remaining vestiges of scimitar-horned oryx were coming to an end in the wild, attempts to reintroduce the species had already begun. A pioneering but trial release of 10 animals (5♂.5♀) from zoos in the UK took place in Bou Hedma National Park, Tunisia, in 1985 (Gordon & Gill, 1993; Kacem *et al.*, 1994), marking the return of a species that had been absent from that country for almost 80 years. This set the model for subsequent initiatives characterised by releases of limited founder representation into relatively small, fenced protected areas in otherwise human and livestock dominated landscapes.

In 1999, a further 14 scimitar-horned oryx from across Europe were released into two additional protected areas in Tunisia: 10 (1♂.9♀) animals were released into Sidi Toui National Park, three (1♂.2♀) into the Oued Dekouk National Reserve, and a single male was sent to augment the population at Bou Hedma (Gilbert, 2015). A planned meta-population of scimitar-horned oryx spread across four protected areas in Tunisia was completed with the release of animals in Dghoumes National Park in 2007. This initiative included the translocation of stock from Bou Hedma National Park (2♂.5♀), augmented with animals from the SSP and EEP (5♂.4♀) (Woodfine *et al.*, 2009).

Morocco began repatriation of scimitar-horned oryx between 1995 and 1997 (Muller & Engel, 2004) during which time a total of 25 animals (15♂.10♀) from European zoos became the founder population for a release into the Arrouais Reserve within Souss Massa National Park (Gilbert, 2015). These animals continue to be maintained at artificially high density outside of historical range. According to Morocco's strategy for antelope conservation (Cuzin *et al.*, 2008), they are regarded as secure semi-captive stock for eventual reintroduction at sites to be determined around Msissi and in the region between Dakhla National Park and Boujdour. To this end, 10 animals (4♂.6♀) were translocated to a reserve at Msissi in 2009.

In 1999, eight (3♂.5♀) scimitar-horned oryx from the Hai Bar Reserve in Israel were transferred to the Guembeul Faunal Reserve in Senegal, followed by a further two (0♂.2♀) from Paris Zoo in 2001. This represented the first stage of an initiative to reintroduce the species to the Ferlo-North Faunal Reserve which subsequently began with the transfer of eight animals from Guembeul in 2003.

Because of variation in the size of reserves, fluctuating local conditions, and differences in management intervention, it is difficult to examine the performance of reintroduced scimitar-horned oryx populations. Nevertheless, some broad conclusions can be drawn from annual census data available from the largest sites.

Following release, the number of consecutive years during which scimitar-horned oryx populations continually grew prior to stabilising or decreasing varied according to location (mean = 9, min = 5, max = 14). During this initial phase, the mean annual growth rate of newly established herds ranged between 21% and 35% per annum (weighted mean = 27%), with the higher growth rates associated with greater numbers of founders. Contrary to arithmetic expectation, high proportions of females in founder herds did not result in greater initial population growth rates compared to groups with balanced sex ratios. This is a likely artefact of the reproductive behaviour of the species whereby females leave the main herd to give birth and may become pregnant again during post-partum oestrus depending on the availability of males to accompany them. Meanwhile, population growth in the first 3 years after release was influenced by the mean age of founder females ($r_s = 0.90$, $n = 5$, $p < 0.05$). Hence, despite being sexually mature at point of release, the strategy of selecting and transporting younger animals for reintroduction projects appeared to result in lower initial growth rates compared to selection of older and reproductively experienced females.

In the four Tunisian protected areas subject to similar management regimes, initial post-release population growth phases displayed linear trends. Thereafter, no common pattern emerges, and comparisons become difficult because of the varying time frames since releases occurred. The fortunes of the scimitar-horned oryx are therefore dependent on site-specific interpretation.

Based on pedigree analysis of the International Studbook (Gilbert, 2015) between 8 and 35 of the 45–50 global founders are represented in each of the reintroduced herds. Tunisia's meta-population of four sites includes animals descended from 40 global founders, while Morocco's scimitar-horned oryx represent less than half that number (17). Tracing the origins of Senegal's animals is rather more complex. This is partly because records of the dozen scimitar-horned oryx sent to create the herd in the Hai Bar Reserve in Israel are incomplete, and partly because closed breeding in that facility for over 30 years prior to the transfer of animals to Senegal could have resulted in loss of founder representation through bottlenecks and genetic drift.

The reintroduced scimitar-horned oryx populations are small and connected only through human-mediated translocation. It is likely that inbreeding and genetic drift have resulted in a loss of founder genetic diversity to varying degrees at each site, although the extent to which this has occurred has not been established. Nevertheless, these factors remain potential threats to the persistence of these herds in the absence of active meta-population strategies. Indeed, given the small number of animals, limited founder representation, and current lack of space for viable population sizes, all of the current scimitar-horned oryx herds within indigenous range remain intrinsically vulnerable.

To date, combined efforts across Tunisia, Morocco, and Senegal have resulted in a total population of around 600 semi-free ranging scimitar-horned oryx, although this varies according to site-specific fluctuations. These animals are spread across several isolated protected areas totalling less than 30,000 ha in size. Assuming a census versus effective population size (N_e) ratio of 0.19 (Gilbert, 2011), current numbers represent N_e of little over 100 animals for the whole of this existing but fragmented protected area network. To achieve the generally recommended minimum N_e of 500 individuals theoretically needed to achieve a long-term viable population and evolutionary adaptation (Franklin, 1980), at least 2,630 animals are required. Recent challenges to conventional wisdom suggest that at least double the number of animals may be needed (Frankham *et al.*, 2014). Regardless, the process of reintroducing scimitar-horned oryx should be considered to be in its formative stages with a need to create far greater carrying capacity for the species and connectivity of populations across its former range.

Fences enclosing the current scimitar-horned oryx protected areas both alleviate threats to re-establishing populations and create challenges for their management. On the one hand, the model appears to have been largely successful in deterring poachers and preventing further land degradation and overgrazing by livestock, thus addressing underlying causes of the species' decline in nature. On the other hand, concentration of animals within fences may affect intra- and inter-specific competition, social behaviour, and disease transmission that exacerbate the intrinsic challenges of managing these small populations.

To date there is insufficient empirical evidence to truly elucidate these influences, and it is likely that the size and health of scimitar-horned oryx populations are shaped by a combination of factors. Nevertheless, there have been some notable observations. For example, single point supplementary feeding reportedly led to heightened conflict amongst scimitar-horned oryx males in Sidi Toui National Park (Molcanova & Wacher, 2007), while overt aggression between scimitar-horned oryx and addax (*Addax nasomaculatus*) occurred in Bou Hedma National Park (Gordon, 1989). Here, cases of hybridisation between scimitar-horned oryx males and addax females were also documented due to the restricted cohabitation of these hippotraginae antelope. The two species were separated following a later review of the national strategy for antelope conservation in Tunisia (DGF, 2001). Known hybrids were also removed and although not necessarily possible to fully rule out, there has been no subsequent genetic indication of their presence or influence in the Tunisian scimitar-horned oryx population (unpublished data).

Scimitar-horned oryx founders from EEP and SSP populations were subject to rigorous health screening prior to transport (e.g., Flach *et al.*, 2000; Woodfine *et al.*, 2009). However, once established in-country it seems likely that infectious disease transmission has occurred, given close proximity to livestock and likely movements of other species of wildlife in and out of protected areas. Despite no documented epidemics associated with released scimitar-horned oryx, this remains a critical issue for the security of those populations and local pastoral economies.

Whether releases of scimitar-horned oryx undertaken so far should be regarded as true 'reintroductions' is open to debate. According to the IUCN definition, reintroduction refers to releases inside indigenous range from which the species has disappeared, with the aim of re-establishing a viable population (IUCN SSC, 2013). In one or two cases, sites are unlikely to be within indigenous range, so technically these may be better described as 'conservation introductions'. And, although the aim and intent may have been implicit, the reality is that no single site has the carrying capacity to support a viable population of large ungulate like the scimitar-horned oryx. With its network of four protected areas for scimitar-horned oryx and the feasibility of managing these sites collectively, Tunisia currently has the most practicable strategy, but would still need to create additional space for the species to achieve a viable meta-population. Regardless, the survival of released animals, breeding of founders and their offspring, and persistence of populations provide some measures of success (Seddon, 1999; Woodfine *et al.*, 2011).

During this time, the scimitar-horned oryx became a flagship for Sahelo-Saharan conservation and an important case study. It is one of only six species of large mammals that have been successfully bred in captivity after their extinction in the wild and then released into their former range; the others being the Arabian oryx (*Oryx leucoryx*) (Stanley Price, 1989), Przewalski's horse (*Equus ferus przewalskii*) (van Dierendonck & Devries, 1996), European wisent (*Bison bonasus*) (Olech, 2014), Red wolf (*Canis rufus*) (Kelly *et al.*, 2014), and Père David's deer (*Elaphurus davidianus*) (Hu & Jiang, 2002).

The future of the scimitar-horned oryx

Less than 5% of the global scimitar-horned population is currently found within its indigenous range. Whether viable numbers can be re-established and sustained is dependent on addressing a series of complex challenges.

The primary causes of the species' extinction in the wild (Newby, 1988) are still prevalent across the Saharan region. In addition, there are increasing threats to wildlife from climate change, insecurity, and disturbance from oil exploration and extraction (Brito *et al.*, 2014; Duncan *et al.*, 2014; Loarie *et al.*, 2009) in a part of the world where conservation is particularly under-resourced and neglected (Durant *et al.*, 2014).

Because of the need to draw on captive founder populations for more expansive reintroduction initiatives and uncertainty about such outcomes, the species as a whole will remain highly dependent on *ex situ* management for the foreseeable future. However, the goal of maintaining demographically stable and genetically diverse populations is hindered by limited space and because of constraints on the ability to physically move animals between locations (Hibbard *et al.*, 2011; Junhold & Oberwemmer, 2011; Leus *et al.*, 2011). Indeed, the effective economic fragmentation of the world's captive scimitar-horned oryx population results from the prohibitive costs of animal transport, and legislative restrictions designed to mitigate spread of infectious diseases that may otherwise impact livestock production (Gilbert, 2011). Notwithstanding these issues, an integrated and global approach to the management of scimitar-horned oryx would be better than isolating regional subpopulations from each other including those within indigenous range (Gilbert, 2011).

In nature, it is likely that the model for the conservation of scimitar-horned oryx and other large herbivores across the Sahelo-Saharan region will mirror that found elsewhere in Africa. There may be opportunities for reintroduction of unrestricted populations (e.g., Chapter 13), although this assumes that large enough tracts of habitat can be identified where methods of sustainable wildlife management can be implemented. Other populations will no doubt be maintained in smaller, isolated protected areas subject to more intensive interventions, but ideally as part of wider meta-populations such as the model developed in Tunisia. Each approach brings advantages and disadvantages that statutory authorities of range states and reintroduction practitioners will need to consider, although options in some countries are likely to be dictated by available space and resources.

Among existing sites, the vulnerability of small populations to environmental, demographic, and genetic stochasticity, and the influence of population density (Bonenfant *et al.*, 2009; Mills & Johnson, 2013) require further attention, especially as effects will be particularly acute with limited founders in such erratic arid environments (Woodfine *et al.*, 2009). Management intervention such as the provision of supplementary food and water may offset losses during periods of drought, but equally maintain artificially high numbers of animals to the longer-term detriment of the habitat and the population. Instead, a better understanding of carrying capacity and efforts to maintain scimitar-horned oryx numbers below the transitional phase of the population curve (e.g., Islam *et al.*, 2010) could create a buffer against episodic drought while maintaining fecundity. This will inevitably require policy on population control and options for offtake such as culling or translocation (Gordon *et al.*, 2004).

There are good examples of vegetation improving within existing scimitar-horned oryx protected areas compared to the surrounding landscape, following the exclusion of livestock and active habitat restoration (Belgacem *et al.*, 2008; Dhaou *et al.*, 2010; Karem *et al.*, 1993.; Tarhouni *et al.*, 2014; Woodfine *et al.*, 2009). While this is likely to have mainly resulted from resting the land following previous over-grazing, scimitar-horned oryx reintroductions have added impetus to the protection of these areas. Moreover, this species like other large herbivores should be considered integral to site management, helping to shape the structure and diversity of vegetation communities (Danell *et al.*, 2006) and re-establishing ecosystem processes like seed dispersal and germination (e.g., Polak *et al.*, 2014), assuming the numbers of these animals are managed in balance with the environment.

Benefits of restoration such as increased vegetation cover, abundance of perennial grasses, habitat heterogeneity, and soil nutrients seen in protected areas may have wider applications for drought resilience among pastoral communities (e.g., Belgacem *et al.*, 2008; Glew *et al.*, 2010). This is important because it seems unlikely that populations of large herbivores numbering in their thousands over a vast landscape can be achieved without species reintroduction contributing directly to human development goals and creating an explicit link between biodiversity conservation and poverty alleviation (Millennium Ecosystem Assessment, 2005), and security (Rotshuizen & Smith, 2013).

Socio-economic justifications of current scimitar-horned oryx reintroductions have yet to be realised, while living close to these protected areas may also be seen as a disadvantage if there are restrictions on access and land use (e.g., Lewis, 2014). Although there will be challenges in translating successful initiatives from other parts of Africa to specific and remote locations in the Sahelo-Saharan region, such examples at least indicate what can be achieved through indigenous community-conserved areas (Barnes *et al.*, 2002; Davies *et al.*, 2012; Glew *et al.*, 2010), while public-private partnerships may also provide options for effective protected area management (Hatchwell, 2014).

With these wider considerations in mind, contemporary efforts to reintroduce viable populations of scimitar-horned oryx and to manage the species from a truly global perspective were initiated during a seminal discussion at the Sahelo-Saharan Interest Group meeting hosted by Al Ain Zoo in 2008. Representatives of the major source populations in North America, Europe, and the Middle East met for the first time to share information and explore the possibility for international cooperation to ensure the future of the species

(Woodfine & Wacher, 2008). The following year, the Sahara Conservation Fund and Al Ain Zoo hosted a stakeholder workshop to create a strategy for conservation and reintroduction of scimitar-horned oryx. A site selection matrix was then developed to identify suitable scimitar-horned oryx reintroduction sites across the Sahelo-Saharan region, including consideration of biological, strategic, and socio-economic factors (SCF, 2012).

This process re-focussed attention on the Ouadi Rimé-Ouadi Achim Game Reserve in central Chad (Bemadjim *et al.*, 2012). An ambitious initiative by the Chadian government, the Environment Agency–Abu Dhabi and the Sahara Conservation Fund now seeks to re-establish a sustainable, free-ranging population of scimitar-horned oryx in this area (Chapter 13). Established in 1969, the 78,000 km^2 Ouadi Rimé-Ouadi Achim Game Reserve was a former stronghold for scimitar-horned oryx and the source of the majority of the founders of modern global captive breeding programs (Gilbert, 2015). It would therefore seem a fitting place for a new chapter in the remarkable history of the scimitar-horned oryx.

References

Barnes, J. I., MacGregor, J. & Weaver, L. C. (2002): Economic Efficiency and Incentives for Change within Namibia's Community Wildlife Use Initiatives. *World Development* **30** (4): 667–681.

Barzdo, J. (1982): New hope for the scimitar. *Country Life* **27**: 1552–1553.

Bassett, T. H. (1975): Oryx and addax in Chad. *Oryx* **13**: 50–51.

Belgacem, A., Ben Salem, H., Bouaicha, A. & El-Mourid, M. (2008): Communal rangeland rest in arid area, a tool for facing animal feed costs and drought mitigation: the case of the Chenini community, Southern Tunisia. *Journal of Biological Sciences.* **8**(4): 822–825.

Bemadjin, N. E., Newby, J., Desbiez, A., Lees, C., & Miller, P. (2012): *Technical workshop on the reintroduction of scimitar-horned oryx to the Ouadi Rimé-Ouadi Achim Game Reserve, Chad.* IUCN/SSC Conservation Breeding Specialist Group.

Belgacem, A., Tarhouni, M. & Louhaichi, M. (2013): Effect of protection on plant community dynamics in the Mediterranean Arid Zone of southern Tunisia: a case study from Bou Hedma National Park. *Land Degradation & Development* **24**: 57–62.

Bonenfant, C. Gaillard, J. Coulson, T. Festa-Bianchet, M., Loison, A., Garel, M., Loe, L., Blanchard, P., Pettorelli, N., Owen-Smith, N, DuToit, J. & Duncan, P. (2009): Empirical evidence of density dependence in populations of large herbivores. *Advances in Ecological Research* **41**: 313–357.

Bonnet, A. (1909): l'oryx dans l'ancienne Egypte. *Archives Museum d'Histoire naturelle de Lyon* **10**: 159–173.

Brehm, A. (1920): *Brehms Tierleben* (4th edn). Leipzig & Vienna: Bibliographisches Institut.

Brito, J., Godinho, R., Martínez-Freiría, F., Pleguezuelos, J., Rebelo, H., Santos, X., Vale, C., Velo-Anton, G., Boratynski, Z., Carvalho, S., Ferreira, S., Goncalves, D., Silva, T., Tarroso, P., Campos, J., Leite, J., Nogueira, J., lvares, F., Sillero, N., Sow, A., Fahd, S., Crochet, P. & Carranza, S. (2014): Unravelling biodiversity, evolution and threats to conservation in the Sahara-Sahel. *Biol. Rev.* **89**: 215–231.

Cuzin, F. Sehhar, E. A. & Wacher T. (2008): *Strategic action plan for the conservation, restoration and management of ungulates in Morocco* (English supplement to Vol.1). Haut commissariat aux eaux et forêts et à la lutte contre le désertification (HCEFLCD), Projet de Gestions des Aires Protégées (PGAP) and World Bank Global Environment Facility (GEF).

Danell, K., Bergstrom, R., Duncan, P. & Pastor, J. (Eds). (2006): *Large herbivore ecology, ecosystem dynamics and conservation*. Cambridge: Cambridge University Press.

Davies, J., Poulsen, L., Schulte-Herbrüggen, B., Mackinnon, K., Crawhall, N., Henwood, W.D., Dudley, N., Smith, J. & Gudka, M. (2012): *Conserving dryland biodiversity*. IUCN/UNEP-WCMC/UNCCD.

Devillers, P. & Devillers-Terschuren, J. (2006): *Oryx dammah. In Sahelo-Saharan antelopes status and perspectives; report on the conservation status of the six Sahelo-Saharan antelopes:* 13–38. Beudels, R. C., Devillers, P., Lafontaine, R. M., Devillers-Terschuren, J. & Beudels, M. O. (Eds). CMS SSA Concerted Action. (2nd edn). CMS Technical Series Publication N°11, 2005. Bonn, Germany: UNEP/CMS Secretariat.

DGF. (2001): *Stratégie nationale tunisienne pour la conservation et la restauration des antilopes sahélo-sahariennes et de leurs habitats 2001–2120.* Ministère de l'Agriculture, Direction Générale des Forêts, Direction de la Conservation des Forêts.

Dhaou, S. A., Abdallah, F., Belgacem, A. O. & Chaieb, M. (2010): The protection effects on floristic diversity in a North African pseudo-savanna. *Pakistan Journal of Botany* **42**(3): 1501–1510.

Dixon, A. M., Mace, G. M., Newby, J. E. & Olney, P. J. S. (1991): Planning for the re-introduction of scimitar-horned oryx (*Oryx dammah*) and addax (*Addax nasomaculatus*) into Niger. *Symposia of the Zoological Society of London* **62**: 201–216.

Duncan C., Kretz D., Wegmann M., Rabeil T. & Pettorelli, N. (2014): Oil in the Sahara: mapping anthropogenic threats to Saharan biodiversity from space. *Phil. Trans. R. Soc. B.* **369**: 20130191. http://dx.doi.org/10.1098/rstb.2013.0191

Durant, S., Wacher, T., Bashir, S., Woodroffe, R., De Ornellas, P., Ransom, C., Newby, J., Abaigar,T., Abdelgadir, M., El Alqamy, H., Baillie, J., Beddiaf, M., Belbachir, F., Belbachir-Bazi, A., Berbash, A., Bemadjim, M., Beudels-Jamar, R., Boitani, L., Breitenmoser, C., Cano, M., Chardonnet, P., Collen, B., Cornforth, A., Cuzin, F., Gerngross, P., Haddane, B., Hadjeloum, M., Jacobson, A., Jebali, A., Lamarque, F., Mallon, D., Minkowski, K., Monfort, S., Ndoassal, B., Niagate, B., Purchase, G., Samaıla, E., Samna, A., Sillero-Zubiri, C., Soultan, A., Stanley Price, M. & Pettorelli, N. (2014): Fiddling in biodiversity hotspots while deserts burn? Collapse of the Sahara's megafauna. *Diversity and Distributions* **20**: 114–122.

Engel, J. (2004): Cultural history. In *The biology, husbandry and conservation of scimitar-horned oryx (Oryx dammah)*: 12–13. Gilbert, T. & Woodfine, T. (Eds). Winchester, UK: Marwell Preservation Trust.

Flach, E. J., D'Alterio, G.-L., Zahzah, K., Molconova, R., Wakefield, S., Barossi, D., Eulenberger, K., Kopcok, M., Ollivet, F., Petit, T., Vahala, J. & Frolich, K. (2000): Veterinary monitoring of captive-bred scimitar-horned oryx (*Oryx dammah*) prior to reintroduction in Tunisia. *Proceedings of the European Association of Zoo & Wildlife Veterinarians.* Third Scientific Meeting, May 31–June 4, 2000, Paris.

Flower, S. S. (1932): Notes on the recent mammals of Egypt, with a list of the species recorded from that kingdom. *Proceedings of the Zoological Society of London:* 369–450.

Franklin, I. R. (1980): Evolutionary change in small populations. In *Conservation biology: an evolutionary-ecological perspective*: 135–149. Soulé, M. E. & Wilcox, B. A. (Eds.). Sunderland, MA: Sinauer.

Frankham, R., Bradshaw, C. J. A. & Brook, B. W. (2014): Genetics in conservation management: revised recommendations for the 50/500 rules, Red List criteria and population viability analyses. *Biol. Conserv.* **170** (2014): 56–63.

Gilbert, T. (2011): *The sustainability of endangered species under intensive management: the case of the scimitar-horned oryx (Oryx dammah).* PhD Thesis, University of Southampton, Southampton, UK.

Gilbert, T. (2015): *International studbook for scimitar-horned oryx (Oryx dammah).* (10th edn). Winchester, UK: Marwell Wildlife.

Gilbert, T. & Woodfine, T. (Eds) (2004): *The biology, husbandry and conservation of scimitar-horned oryx (Oryx dammah).* Winchester, UK: Marwell Preservation Trust.

Gillet, H. (1965): L'oryx algazelle et l'addax au Tchad. *Revue d'écologie (la terre et la vie)* **3**: 257–272.

Gillet, H. (1966): The scimitar oryx and the addax in the Tchad Republic (Part I). *African Wild Life* **20** (2): 103–115.

Gillet, H. (1971): l'oryx algazelle et l'addax. Distribution géographique. Chances de survie. *Compte Rendu des Séances de la Société de Biogéographie* **405**: 177–189.

Glew, L., Hudson, M. & Osborne, P. (2010): *Evaluating the effectiveness of community-based conservation in northern Kenya.* A report to The Nature Conservancy. Centre for Environmental Sciences, University of Southampton, Southampton.

Gordon, I. (1989): A case of intense inter-specific aggression between scimitar-horned oryx (*Oryx dammah*) and addax (*Addax nasomaculatus*). *Journal of Zoology* **218**: 335–337.

Gordon, I. J. (1991): Ungulate re-introductions: the case of the scimitar-horned oryx. *Symposia of the Zoological Society of London* **62**: 217–240.

Gordon, I. J. & Gill, J. P. (1993): Reintroduction of scimitar-horned oryx *Oryx dammah* to Bou-Hedma National Park, Tunisia. *International Zoo Yearbook* **32**: 69–73.

Gordon, I., Hester, A. & Festa-Bianchet, M. (2004): The management of wild large herbivores to meet economic, conservation and environmental objectives. *Journal of Applied Ecology* **41**: 1021–1031.

Haltenorth, T. (1963): Klassifikation der Säugetiere: Artiodactyla. In *Handbuch der Zoologie.* **8**. *Teil* **1** (18): 1–167. Helmcke, J. & von Lengerken, H. & Starck, D. (Eds). Berlin: W. de Gruyter.

Hatchwell, M. (2014): Public–private partnerships as a management option for protected areas. *Animal Conservation* **17**: 3–4.

Heringa, A. (1989): Scimitar-horned oryx. *Gnusletter* **8**(2): 4–5.

Hibbard, C., Hogg, C. J., Ford, C. & Embury, A. (2011): Maintaining the status of species management in a changing operating environment: outcomes over outputs. *WAZA Magazine* **12:** 6–10.

Hu, H. & Jiang, Z. (2002): Trial release of Père David's deer *Elaphurus davidianus* in the Dafeng Reserve, China. *Oryx* **36**: 196–199.

Islam, M. Z., Knutson, C. & Boug, A. (2010): Strategy and action plan to reduce the risk of mass mortalities of reintroduced ungulates in the Mahazat as-Sayd Protected Area in Saudi Arabia. *Gnusletter* **28** (2): 9–15.

IUCN. (1987): *Scimitar-horned oryx (Oryx dammah)* (Cretzschmar, 1826). IUCN Database.

IUCN SSC Antelope Specialist Group. (2008): *Oryx dammah*. The IUCN Red List of Threatened Species. Version 2014.2. www.iucnredlist.org. (as at 10 October 2014).

IUCN SSC. (2013): *Guidelines for reintroductions and other conservation translocations. Version 1.0*. Gland, Switzerland: IUCN Species Survival Commission.

Iyengar, A., Gilbert, T., Woodfine, T., Knowles, M., Diniz, F. M., Brenneman, R. A., Louis Jr, E. E. & Maclean, N. (2007): Remnants of ancient genetic diversity preserved within captive groups of scimitar-horned oryx (*Oryx dammah*). *Molecular Ecology* **16**: 2436–2449.

Junhold, J. & Oberwemmer, F. (2011): How are animal keeping and conservation philosophy of zoos affected by climate change? *International Zoo Yearbook* **45:** 99–107.

Kacem, S., Muller, H. P. & Wiesner, H. (1994): *Gestion de la faune sauvage et des parcs nationaux en Tunisie*. Tunis: GTZ/DGF.

Karem, A. Ksantini, M. Schoenenberger, A. & Waibel T. (1993): *Contribution à la régénération de la végétation dans le parcs nationaux en Tunisie aride*. Tunis: GTZ/DGF.

Kelly, B. T., Beyer, A. & Phillips, M. K. (2014): *Canis rufus*. The IUCN Red List of Threatened Species. Version 2014.2. www.iucnredlist.org (as at 8 August 2014).

Le Houerou, H. (1997): Climate, flora and fauna changes in the Sahara over the past 500 million years. *Journal of Arid Environments* **37**: 619–647.

Leus, K., Bingaman Lackey, L., van Lint, W., de Man, D., Riewald, S., Veldkam, A. & Wijmans, J. (2011): Sustainability of European association of zoos and aquaria bird and mammal populations. *WAZA Magazine* **12**: 11–14.

Lewis, J. (2014): *Still wild, but without a wilderness*. http://www.pbs.org/wgbh/nova/next/nature/conservation-in-limbo/(as at 28 May 2014).

Loarie, S. R., Duffy, P. B., Hamilton, H., Asner, G. P., Field, C. B. & Ackerly, D. D. (2009): The velocity of climate change. *Nature* **462**, 1052–1055.

Maydon, H. C. (1923): (North Kordofan to South Dongola): *Geographical Journal* **61**: 34–41.

Millenium Ecosystem Assessment. (2005): *Ecosystems and human well-being: biodiversity synthesis*. Washington, DC: World Resources Institute.

Mills, S. & Johnson, H. (2013): Wildlife population dynamics. In *Wildlife Management and Conservation: Contemporary Principles and Practices*: 84–111. Krausman, P., & Cain, W. (Eds). Baltimore: Johns Hopkins University Press.

Molcanova, R. & Wacher, T. (2007): Scimitar-horned oryx social dynamics and the influence of management in a fenced protected area (Sidi-Toui National Park, Tunisia). In *Proceedings of the Seventh Annual SSIG Meeting 2007*: 10–15. Woodfine, T. (Ed.). Hannover, Germany: Sahara Conservation Fund.

Muller, H. P. & Engel, H. (2004): The reintroduction of herbivores to Souss Massa National Park, Morocco. In *The biology, husbandry and conservation of scimitar-horned oryx (Oryx dammah): 77–81*. Gilbert, T. & Woodfine, T. (Eds). Winchester, UK: Marwell Preservation Trust.

Newby, J. E. (1975): *Addax and scimitar-horned oryx in the Sahara Desert and Sahelian Zone: the addax and scimitar-horned oryx in Niger*. Report to the International Union for the Conservation of Nature and Natural Resources IUCN: 1–8.

Newby, J. E. (1984): Large mammals. In *Key environments – Sahara desert*: 277–290. Cloudsley-Thompson, J. L. (Ed.). Oxford and New York: Pergamon Press.

Newby, J. (1988): Aridland wildlife in decline: the case of the scimitar-horned oryx. In *Conservation and biology of desert antelopes*: 146–166. Dixon, A. & Jones, D. (Eds). London: Christopher Helm.

Olech, W. (IUCN SSC Bison Specialist Group) (2014): *Bison bonasus*. The IUCN red list of threatened species. Version 2014.2. www.iucnredlist.org (as at 8 August 2014).

Olsen, D. M., Dinerstein, E., Wikramanayake, E. D., Burgess, N. D., Powell, G. V. N., Underwood, E. C., D'Amico, J. A., Itoua, I., Strand, H. E., Morrison, J. C., Loucks, C. J., Allnutt, T. F., Ricketts, T. H., Kura, Y., Lamoreux, J. F., Wettengel, W. W., Hedao, P. & Kassem, K. R. (2001): Terrestrial ecoregions of the world: a new map of life on earth. *BioScience* **51**(11): 933–938.

Polak, T., Gutterman, Y., Hoffman, I. & Saltz, D. (2014): Redundancy in seed dispersal by three sympatric ungulates: a reintroduction perspective. *Animal Conservation* **17**(6): 565–572.

Rode, P. (1943): *Faune de l'empire français. ii. mammifères ongulés de l'Afrique noire*. Paris: Librairie Larose.

Rotshuizen, S. & Smith. M. (2013): Of warriors, poachers and peacekeepers: protecting wildlife after conflict. *Cooperation and Conflict* **48**(4): 502–521.

Schomber, H. W. (1963): Wildlife in the Sudan. 4. Desert and semi-desert game animals. *African Wild Life* **17**: 117–124.

Sclater, P. L. & Thomas, O. (1899/1900): *The book of antelopes 4*. London: R.H. Porter.

Seddon, P. (1999): Persistence without intervention: assessing success in wildlife reintroductions. *TREE* **14**(12): 503.

Stanley-Price, M. R. (1989): *Animal reintroductions: the Arabian oryx in Oman*. Cambridge: Cambridge University Press.

SCF. (2012): *Scimitar-horned oryx site evaluation matrix*. http://www.saharaconservation.org/IMG/pdf/SHO_Matrix_Chad_English.pdf (as at 29 August 2014).

Tarhouni, M., Ben Salem, F., Belgacem, A. & Neffati, M. (2014): Impact of livestock exclusion on Sidi Toui National Park vegetation communities, Tunisia. *International Journal of Biodiversity* **14**: 1–7.

Van Dierendonck, M. C. & Devries, M. F. W. (1996): Ungulate reintroductions: experiences with the Takhi or Przewalski horse (*Equus ferus przewalskii*) in Mongolia. *Conservation Biology* **10**: 728–740.

Wakefield, S., Engel, J. & Gilbert, T. (2004): Conservation. In *The biology, husbandry and conservation of scimitar-horned oryx (Oryx dammah)*: 14–20 Gilbert, T. & Woodfine, T. (Eds). Winchester, UK: Marwell Preservation Trust.

Wilson, R. T. (1980): Wildlife in Northern Darfur, Sudan: a review of its distribution and status in the recent past and at present. *Biological Conservation* **17**: 85–101.

Woodfine, T. & Wacher. T. (Eds.) (2008): *Proceedings of the Ninth Annual SSIG Meeting 2008, Al Ain, United Arab Emirates. L'Isle*, Switzerland: Sahara Conservation Fund.

Woodfine, T., Zahzah, K., Chetoui, A., Gilbert, T. & D'Alterio, G. L. (2009): *Reintroduction of scimitar-horned oryx to Dghoumes National Park, Tunisia.* Report prepared for Direction Générale des Forêts, Tunis.

Woodfine, T., Chetoui, A., Zahzah, K. & Gilbert, T. (2011): Re-introduction of scimitar-horned oryx to Dghoumes National Park, Tunisia. In *Global re-introduction perspectives: 2011. More case studies from around the globe*: 190–193. Soorae, P. S. (Ed.). Gland, Switzerland: IUCN/SSC Re-introduction Specialist Group; Abu Dhabi, UAE: Environment Agency-Abu Dhabi.

15

Two Decades of Saiga Antelope Research: What have we Learnt?

E. J. Milner-Gulland[1] and Navinder J. Singh[2]

[1]*Department of Zoology, University of Oxford, United Kingdom*

[2]*Department of Wildlife, Fish and Environmental Studies, Faculty of Forest Sciences, Swedish University of Agricultural Sciences, Sweden*

Background

In the face of a biodiversity crisis, conservationists must often act before knowing all the facts, taking decisions based more on experience and opportunity than on strategic planning (Soulé, 1985). In the face of urgent problems, conservationists must decide how much and what kind of science is needed in order to proceed with management decisions, and how best to allocate funding to baseline studies, conservation implementation, monitoring, and evaluation (McCarthy & Possingham, 2007). Basic understanding of the biology, habitat requirements, and threats to a species or assemblage may be required in order to diagnose the likely impacts of our actions. Long-term continuous monitoring of the system can give the best results, but potentially at substantial cost (Yoccoz *et al.*, 2001). Hence, Field *et al.* (2004), for example, demonstrated that when species have very high value and monitoring will not change actions, it is better to invest directly in the action rather than in establishing the status of the population. Scientific studies are also set within a social and political context, such that effective science needs to be conducted in such a way as to influence policy effectively; this may not necessarily equate to the most academically interesting or even the most management-relevant focus. Lawton (2007) suggests that scientists can, and do, influence government policy on the environment, but often in a complex and slow way, and that this may require fundamental

Antelope Conservation: From Diagnosis to Action, First Edition. Edited by Jakob Bro-Jørgensen and David P. Mallon.

changes in politicians' belief systems, values and norms. A transdisciplinary approach to conservation science, that involves both social scientists and also local stakeholders, is most likely to be effective (Reyers *et al.*, 2010).

Conservation interventions need to be evaluated in order to learn and improve. One important, but often neglected, approach is to take a historical perspective, reflecting on the lessons learnt during a long-term engagement with an area or species. Few published studies provide this longer-term perspective, particularly with respect to species conservation, with notable exceptions including Durant *et al.* (2007) on the Serengeti cheetah (*Acinonyx jubatus*) and Pusey *et al.* (2007) on chimpanzees (*Pan troglodytes*). Studies such as these enable conservation scientists to look beyond the immediate threats and actions required, and instead to reflect on factors promoting success or failure over a decadal timescale. In this chapter, we provide another such historical and personal perspective, on the relationship between research and conservation action for the saiga antelope (*Saiga tatarica*) based on the work done by E. J. Milner-Gulland (EJMG) and her collaborators over the period 1990–2012. This is very much a personal perspective, and focussed on our own research programme and attempts to engage with and inform policy and management. This approach to the chapter is not meant to downplay the major research and conservation contributions made by others, or to ignore the different perspectives that others would bring. However this long-term unbroken engagement with a single species and region is relatively rare, and we hope that by looking back over this period and reflecting on our own journey, we can draw out some themes that would not be apparent otherwise.

The story of the saiga antelope (Plate 3iv) is a particularly interesting case study, because it demonstrates how research and conservation have interacted and adapted to dramatic changes in political, social, and ecological circumstances, and responded to the new tools and perspectives which have come to the fore in the wider ecological and conservation science literature over these two decades. Here, we first describe the changes in the region's socio-political and ecological landscape and how they have affected saigas. Following this, we describe the research carried out by us and our collaborators on the ecology of the saiga and on the effects of profound social, political, and economic changes on the species. Next, we outline the conservation approaches which have been adopted over the last two decades. Finally we reflect on the lessons we have learnt over the last two decades of saiga research and conservation, and how they might translate to other antelopes, and other conservation targets more widely.

Saigas as victims of a turbulent history

The saiga antelope is a migratory goat-sized antelope of the semi-arid steppe regions of Eurasia. It is currently found in the pre-Caspian region of Russia, Kazakhstan, Uzbekistan, and Mongolia (Figure 1). The species undertakes

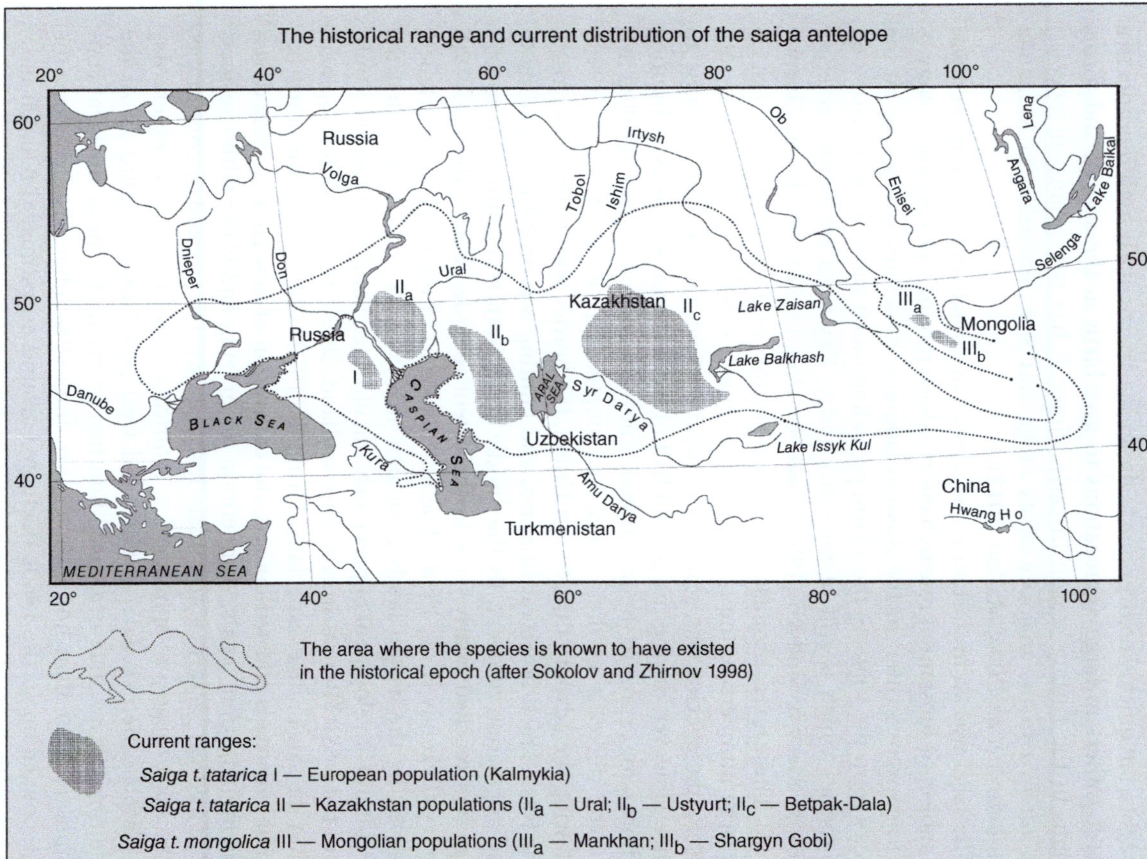

Figure 1 **Map of saiga distribution in the past and present. Adapted from Lushchekina & Struchkov (2001).**

seasonal migrations between summer pastures in the steppe zone and winter refuges in the desert zone; these migrations can reach up to 1000 km north-south in some populations. Populations undergo dramatic fluctuations as a result of the harsh continental climate, with crashes due to severe winters and droughts, but quickly recover due to the species' particularly high reproductive rate. It gathers in large aggregations to give birth in the spring, and forms harems in the rutting season in early winter. It has two extant sub-species; *S. t. tatarica* in the majority of its range and *S. t. mongolica* in Mongolia (Kholodova *et al.*, 2006; Mallon, 2012).

The saiga was an important game species in Soviet times, and as such was the subject of substantial research effort. Major monographs on the biology of the saiga were written in the Russian language by scientists working on the species from the 1950s to the 1990s, providing much valuable information on the species which is still of relevance today. Key texts are Bannikov *et al.* (1961); Fadeev & Sludskii (1982); Bekenov *et al.* (1998); Sokolov & Zhirnov (1998). These texts chronicle the dramatic history of the saiga: after the last ice age large numbers of the species roamed much of Eurasia, but by the 16th century the species was confined to open steppe habitat from Ukraine to China. However, during the second half of the 19th century massive hunting for horns, for export to the traditional Chinese medicine market, resulted in a dramatic decrease in their range and population (Bekenov *et al.*, 1998). Following the advent of the Soviet Union, the border with China was closed, there was a crack-down on firearms, and hunting was prohibited, leading to rapid population recovery. By the middle of the twentieth century the species range had reached nearly back to the extent recorded before the mid–19th century (Bekenov *et al.*, 1998). The northern part of the range was curtailed, however, by the 'virgin lands campaign' in the 1950s, which promoted the agricultural development of previously uncultivated land in the eastern republics of the USSR (Robinson & Milner-Gulland, 2003). Increased development within the saiga range, especially during the 1970–1990s, led to the construction of new roads and settlements, the cultivation of new land, mining development, irrigation canals, competition for grazing and water with livestock, and fenced-off pastures and military installations. This created discontinuities in the saiga range and split the *S. t. tatarica* subspecies into the present-day pre-Caspian, Betpak dala, Ustiurt, and Ural populations (Figure 1). At the same time, heavy state-organised hunting, primarily for meat for local consumption rather than horns for export, limited numbers. Nonetheless, the saiga population in the 1980s fluctuated around an estimated 820,000 individuals, with 82% of the animals in Kazakhstan (Figure 2; Milner-Gulland *et al.*, 2001).

The collapse of the Soviet Union in 1991 brought about massive economic and social hardship in rural areas, including the dissolution of the state-owned

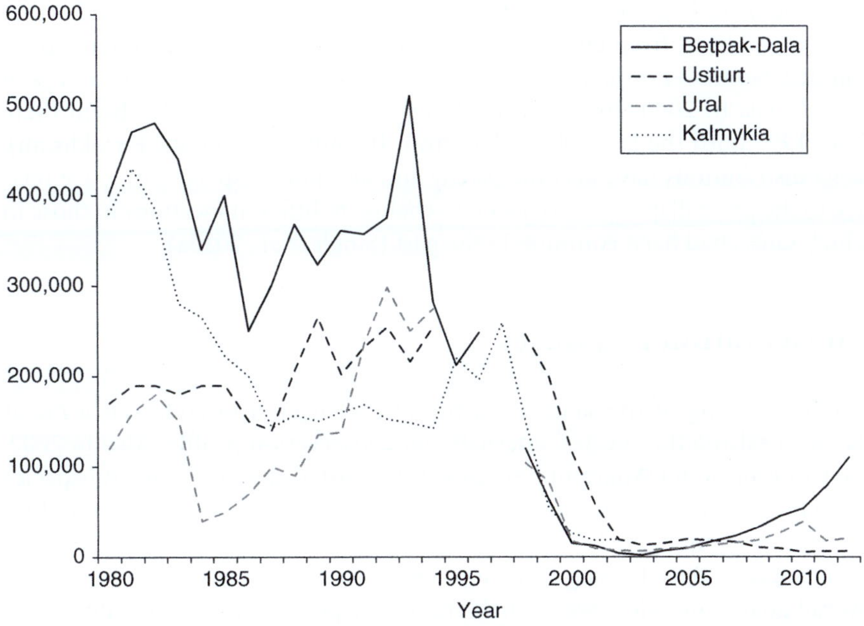

Figure 2 **Saiga antelope population changes since the 1980s. Data from the Institute of Zoology, Kazakhstan and Sokolov & Zhirnov (1998).**

collective farms in Kazakhstan, leaving local people without any defined land tenure system or means of support (Robinson & Milner-Gulland, 2003; Robinson *et al.*, 2012). Livestock numbers collapsed in Russia and Kazakhstan, leaving people without livelihoods. Saigas were a relatively easy and widely available commodity to hunt for meat and to trade, with willing buyers for the horn appearing as the borders with China were opened after 70 years. At the same time, funding for both science and wildlife management was virtually non-existent, so the saigas were left unmonitored and unprotected. The consequence was an unprecedentedly rapid decline in the global saiga population, by 95% in the ten years to 2003, and leading IUCN to uplist the species from Near Threatened to Critically Endangered, its highest category of threat in 2002 (Figure 2; Milner-Gulland *et al.*, 2001; IUCN, 2007). As well as direct mortality, the fact that poachers targeted males meant the sex ratio became severely distorted, leading to a reproductive collapse as females failed to conceive (Milner-Gulland *et al.*, 2003).

This intensive hunting also left other effects on the grouping behaviour, calving aggregations, and distribution of saigas in the early to mid 2000s, including substantial herd fragmentation and herd size reductions (Kuhl, 2008; McConville *et al.*, 2009), failure to form large, dense calving aggregations

(Fry, 2004), and age-sex structure in the herds strongly biased towards adult females and away from either males or juveniles (Fry, 2004; Iu. Grachev, pers. comm.). No males or juveniles were seen in the ground and aerial surveys of Betpak dala in 2000–2002, and birth aggregations were largely absent there from 1997 until the late 2000s (data from Institute of Zoology, Kazakhstan). Saiga distributions have become patchy and are more dispersed in space relative to the pre-collapse period, as well as being in different locations to those in which saigas had been common in the past (Singh *et al.*, 2010a).

Conservation responses

The Red Listing of the saiga as Critically Endangered in 2002 was a major factor in raising the species' international conservation profile. Also in 2002, the Convention for Migratory Species (CMS) listed the migratory subspecies *Saiga tatarica tatarica* on Appendix II and in September 2006 a Memorandum of Understanding (MoU) concerning the "Conservation, Restoration and Sustainable Use of the Saiga Antelope" entered into effect (CMS, 2006). The overall goal of the MoU was to halt and where possible reverse the decline in saiga populations in the next five years, with a long-term vision to restore the saiga populations to a point that sustainable use can be envisioned. The Medium Term International Work Programme for the MoU, which is a prioritised action plan, was revised and re-affirmed at the second Conference of the Parties in 2010, and the CMS MOU was extended to the whole *Saiga* genus, hence including the Mongolian population. The saiga was first included in the Convention on International Trade in Endangered Species of Wild Fauna and Flora (CITES) Appendix 2 well before the population collapse, in 1994. However, since the species became the object of conservation attention it has been subject to a Significant Trade review and to several resolutions at the Conference of the Parties, the most recent being in 2013. CITES has worked closely with CMS over the years to support international efforts for conservation. Currently, both international and bilateral cooperation for saiga conservation is well established, with mechanisms in place for coordination of conservation actions; this does not necessarily, however, mean that national-level action follows.

Still, a huge range of conservation activities have taken place in the last decade, ranging in scale from the Altyn Dala Conservation Initiative, a 5.2 million hectare protected landscape with saigas as the flagship, to small-scale, people-focussed initiatives such as holding Saiga Day festivals in schools within the saiga range. The government of Kazakhstan has been particularly active in investing in saiga conservation, through new protected areas and anti-poaching patrols. In 2010, the CMS commissioned a report on progress towards the implementation of the saiga MOU,[1] which lists and assesses the many initiatives that have been undertaken by governments, international and local NGOs, and scientific institutions.

One initiative which aims to link international, national, and local action is the Saiga Conservation Alliance (SCA), set up in 2006 by participants in the scientific research networks coordinated by EJMG. The SCA obtained UK charitable status in 2010, and works as an international network to support saiga conservation at all scales through a six-language newsletter (*Saiga News*), a small grants programme targeted at grassroots organisations working in the field, and support of in-country conservationists through obtaining grants for their work from international foundations and private donors, and collaborating with them to implement, monitor, and evaluate conservation projects. A new initiative is developing an online Saiga Resource Centre, to act both as a resource and repository for anyone interested in saigas, and also as a coordination hub for the CMS MOU.[2] The charity is funded from a range of sources, with long-term support from the Wildlife Conservation Network.

Our research, phase one: science for science's sake

EJMG's research on saigas began in 1990, when she was working for her PhD on elephant and rhino management. With a keen personal interest in visiting the Soviet Union, she chose to work on saigas because they were then seen by some as a potentially sustainable substitute for rhino horns in the Traditional Chinese Medicine trade.[3] She was also interested in the population dynamics of a species that lived in such a variable environment (Figure 3). Her early research therefore focussed on population dynamics in stochastic systems and the potential for sustainable use of the species, and particularly on the effects of male-biased harvesting on the population (e.g., Milner-Gulland, 1994), helped by the wealth of knowledge accumulated by Soviet scientists, but as yet not used to answer questions such as these. A move to another university department which had a strong wildlife epidemiology focus prompted a new research theme, on the parasites and diseases of the species, interesting because of the interactions between saigas and livestock, and the long history of fundamental research on this theme by colleagues in Kazakhstan (e.g., Morgan *et al.*,

[1] http://www.cms.int/species/saiga/2ndMtg_Mongolia/Mtg_docs/Doc_05_Rev3_MTW_Progress_Report_Oct2006_Sept2010_E.pdf (as at 19 February 2016).

[2] www.saigaresourcecentre.com (as at 19 February 2016).

[3] http://www.rhinoresourcecenter.com/pdf_files/121/1216650686.pdf (as at 19 February 2016). ; http://articles.latimes.com/1986-01-05/news/mn-24555_1_new-rhino-horn (as at 19 February 2016).

2005). A further research project, on rangeland degradation and the effect on saigas, was prompted by discussions with colleagues in Kazakhstan and the UK, and reflected issues we felt were scientifically interesting, areas of research strength in our respective departments, and of general policy relevance (e.g., Robinson & Milner-Gulland, 2003). She also carried out research with in-country colleagues on saiga genetics (e.g., Kholodova *et al.*, 2006) and life histories (e.g., Milner-Gulland, 2001, Figure 3). All these research programmes were funded by the EU through their International Association for the promotion of cooperation with scientists from the independent states of the former Soviet Union (INTAS) programme building the science base of the former Soviet countries, and were motivated by scientific curiosity and our mutual research interests.

During the fieldwork for these projects, alarming reports began to emerge of dramatic saiga population declines, lack of calving aggregations, heavy poaching, and no males. We published the first papers highlighting these dramatic events to the international scientific community, which prompted the policy reactions described above (e.g., Milner-Gulland *et al.*, 2001; Milner-Gulland *et al.*, 2003). The fact that we were already carrying out research programmes on this species was fortuitous, allowing rapid dissemination of the status of this species, and providing quantitative data, which may well have ensured that the international conservation response was relatively timely and appropriate in scale and target. The substantial international interest also highlighted to national governments the importance and urgency of action. Many other species in the region and elsewhere do not have the contingency of an established research programme being there on the spot to record and disseminate their conservation crises.

Phase two: research for conservation

The experience of helplessly charting a population decline to near-extinction was a shocking one for an ecologist, and led EJMG, from 2003 onwards, to move away from pure research questions towards applied research on saiga conservation. Additionally, the end of the EU's INTAS programme in 2006, after a decade of almost continuous funding for our research, meant there was no longer such easily obtainable funding for collaborative pure scientific research in the region, and the parlous state of the saiga population meant that it was difficult to carry out basic research on its ecology and life history, because it was much harder to find substantial herds for data collection (Figure 3).

Our first foray into conservation research, in 2003–2006, was part-funded by INTAS and by the UK government's Darwin Initiative, a conservation funder.

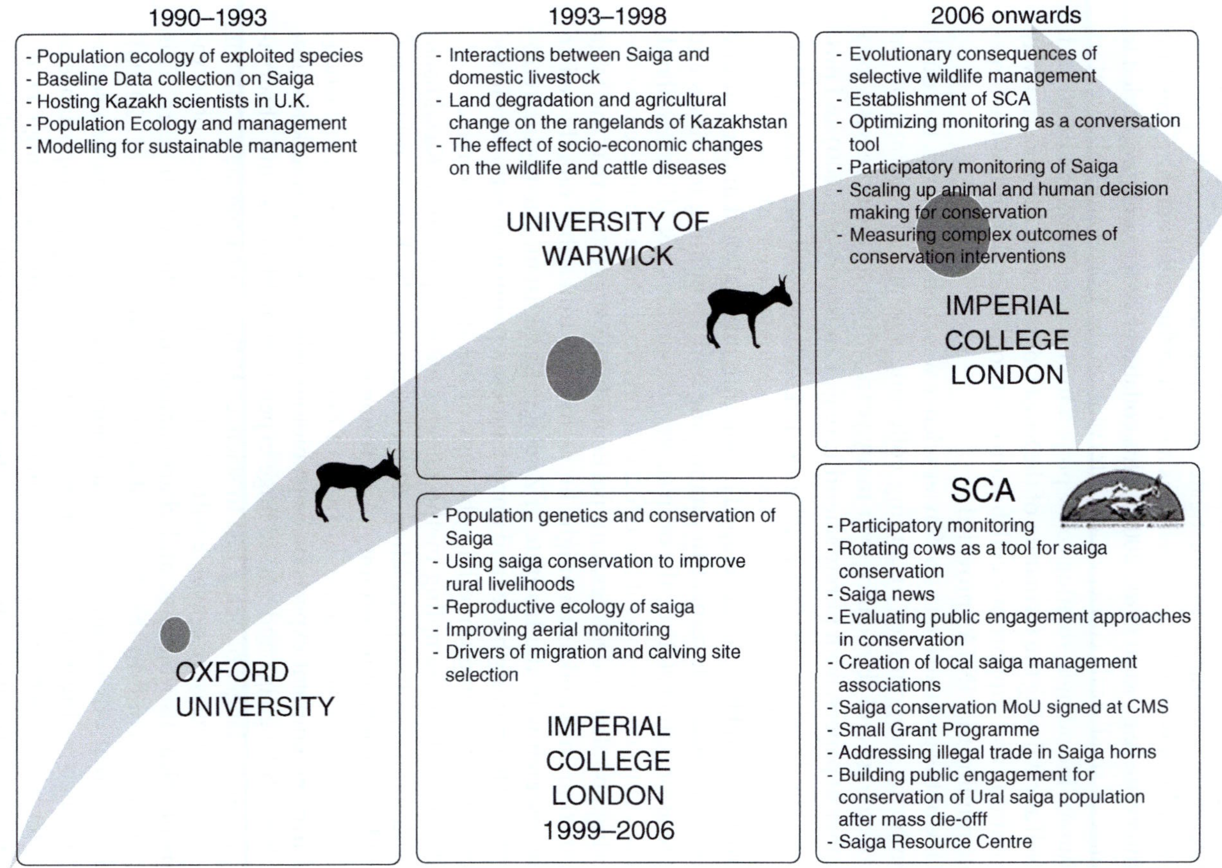

Figure 3 **The changing themes of EJMG's saiga research and conservation over her research career.**

This included work on the reproductive ecology of the saiga justified by a need to understand the effects of sex-biased poaching on saiga reproduction, but including some fundamental life history research (e.g., Kühl *et al.*, 2007). However we also started to investigate the human dimension of the problem, asking who was poaching and why, as a baseline for understanding how best to intervene (e.g., Kühl *et al.*, 2009). This collaborative research project led to a range of small-scale conservation interventions by EJMG and collaborators in Russia, including a 'rotating cows' project to give poor families a livelihood, as well as catalysing the formation of the Saiga Conservation Alliance, as a way to continue and support our collaborative network after the end of the large-scale funding that INTAS provided.

One of the great strengths of Soviet science was their long-term, thorough monitoring programmes, including collecting substantial information on saiga ecology and abundance over many years, which formed the basis for a number of papers, both purely scientific (e.g., Coulson *et al.*, 2000) and providing the baseline for our papers on population declines (e.g., Milner-Gulland *et al.*, 2003). However, the methodologies were unchanged over many years, and had not developed in line with international understanding of best practice with respect to quantifying and accounting for biases and imprecision, either statistically or in survey design (Singh & Milner-Gulland, 2011a). Our concerns about the effect that this might have on inference concerning population declines, and particularly the suggestion that the saiga population in Kazakhstan might be recovering unfeasibly rapidly, led to joint research with the Institute of Zoology, Kazakhstan, attempting retrospectively to quantify biases and uncertainties in population estimates in the light of the changing underlying biology of the species (McConville *et al.*, 2009). Since then, sampling methodologies have improved in Kazakhstan and Mongolia, so that population estimates are now both more robust and include estimates of uncertainty (e.g., Young *et al.*, 2010).

Our collaborations with the Institute of Zoology in Kazakhstan, in particular, involved eminent experts on saiga biology whose work was unknown in the English language and whose huge experience, field notes, and observations were in danger of being lost to science. At the same time, new tools for spatial modelling were available with which we could interpret their observations of saiga distributions, and chart changes over the last 40 years. This enabled us to confirm the drivers of saiga migrations (Singh *et al.*, 2010b), and show that calving sites had shifted north and away from human settlements since the break-up of the Soviet Union, confirming the importance of long-term data sets in understanding present-day changes (Singh *et al.*, 2010a). A particular concern was that the government of Kazakhstan's substantial investment in protected area planning was based on current saiga distributions, which may be very different from either past or potential future distributions; our work was able to highlight these discrepancies and make recommendations for future-proofing protected area networks (Singh & Milner-Gulland, 2011b).

Phase three: evaluating conservation outcomes

As conservation projects gained ground, and several years of experience built up, it was possible to step back and reflect upon the impact that our efforts were having. Starting with a Darwin Initiative follow up project in Russia in 2007–2009, our team began not just to chart saiga declines and lay out the issues to be faced in a generally passive way, but also to evaluate the impact of our conservation projects, and to build monitoring and evaluation research into the conservation proposals which we wrote (Figure 3). Generally the project funding came to the SCA, but the research was carried out, on a relatively small scale, by UK students working alongside in-country project implementers. This shift from pure research, through applied research to action-based research perhaps reflects wider trends as conservation science develops as a discipline separate from conservation biology.

Within our Darwin project, we analysed the effects of the public awareness, social intervention, and anti-poaching interventions that we had implemented over the previous three years, demonstrating that people were widely aware of the plight of the saiga and held positive attitudes towards the species (Howe & Medzhidov, 2011; Howe *et al.*, 2012). We also investigated the cultural associations that local people had with saigas, and the degree to which our support for a Buddhist centre's environmental teachings had changed their perceptions. We found the main effect to be a stronger sense that they themselves had the power to change their environment, not just the authorities or international organisations (Figure 3, Waylen *et al.*, 2012).

With the shift to small-scale evaluation of individual projects, came also a shift towards publishing in internal reports, Masters theses, and in the SCA's newsletter *Saiga News*. Although this research does not have the stamp of the international peer-reviewed journal, it is perhaps more widely read by those who are in a position to use it, particularly as *Saiga News* articles are translated into local languages and distributed personally to key decision makers from Ministers to village heads. Examples of reports used for planning interventions include a review of the potential for participatory monitoring in Uzbekistan (Offord, 2011) and a description of poaching levels and trade routes on the Ustyurt plateau (Phillipson & Milner-Gulland, unpublished). Evaluations include the effectiveness of monitoring by rangers in Russian protected areas (O'Neill, 2008), Saiga Day in the Ural population, Kazakhstan (Samuel, 2011), and the SCA's embroidery programme in Uzbekistan (Figure 3, Damerell *et al.*, 2013).

New stories, old stories

In May 2010, 12,000 saiga females dropped dead within a few days, just after giving birth, in the Ural population. This represented one third of the estimated population size, and was a huge shock for donors, the government of Kazakhstan, and scientists. Despite the long history of research into saiga diseases, which our team had itself contributed to (e.g., Morgan *et al.*, 2006), and the many episodes of mass mortality in saigas in the past (Bekenov *et al.*, 1998), the lack of recent outbreaks and the overwhelming influence of poaching as a threat to saigas meant that the international saiga conservation community had taken its eye off the ball. Disease had not been included in the first Medium Term International Work Programme under the CMS MOU on saiga conservation in 2006, despite that document aiming to address all key threats to the population, and despite our research team being instrumental in drafting it. The issue is certainly back on the agenda now. Although the mortality events are likely to be pasture-related rather than caused by an infectious agent (Kock *et al.*, 2012), many interesting and topical research questions emerge, and the government of Kazakhstan has launched a major research programme to address them. This represents a cautionary reminder that the presence of a primary threat does not negate the need to conserve species based on a broad understanding of its biology and limiting factors, developed over years of basic research.

A growing threat, but also opportunity, is the substantial increase in infrastructure as the region restarts its industrial development and exploitation of its natural resources. New and imminent infrastructural developments include major investments in railways by Kazakhstan, which will cut across prime saiga habitat; a new border fence erected by Kazakhstan across the Ustiurt population's migratory route; and oil and gas developments in Kazakhstan and Uzbekistan. Mitigation of the impacts of these developments is imperative, and the trend towards private companies being required to take responsibility for environmental damage needs to extend to the saiga range states and to be mirrored by similar commitments from governments. Lessons can be learnt from the effects of the previous wave of developments in the 1970s (described in Sokolov & Zhirnov (1998)). Lessons can also be learnt from experiences of mitigating infrastructural impacts on other ungulates, such as Gates et al.'s (2012) review on the effects of land use change and fences on pronghorn (*Antilocapra americana*) movements. Biodiversity offsetting is a new approach that aims to ensure no net loss of biodiversity once all possible mitigation has been implemented; we are investigating how this could be implemented in Uzbekistan with saigas as the focal species (Bull *et al.*, 2013). Work that we carried out under very different circumstances, investigating the relationships between rangeland vegetation and ungulate population dynamics as the Soviet Union collapsed (e.g., Robinson & Milner-Gulland, 2003) is now becoming policy-relevant again as a basis for current work on devising offsets. In a similar vein, the Government of Kazakhstan has recently announced their intention to resume legal saiga hunting in the next few years, assuming the population continues to recover. This brings back to relevance our very earliest work on the preconditions for sustainable hunting of saigas (e.g., Milner-Gulland, 1994).

The major research issue which we see as unresolved is around promoting and demonstrating behavioural change. For saigas, the main issue is still poaching. We have documented who poachers are (poor, unemployed young men with access to motorbikes; Kühl *et al.*, 2009). We have shown that our interventions change attitudes, behavioural intentions, and social norms, all of which should lead to changes in behaviour (Howe & Medzhidov, 2011; Howe & Milner-Gulland, 2012; Damerell *et al.*, 2012). However, we have not shown that our interventions have actually reduced poaching rates over what they would have been in the absence of the interventions. In the two places where we have carried out most of these social interventions (the pre-Caspian and the Uzbekistan part of the Ustiurt population), poaching is still, in fact, worryingly high, and the saiga populations are faring far worse than elsewhere in their range. That is not to say that the situation would not have been worse still in the absence of the interventions, but currently our return on conservation investment is negative. Because saiga poaching is illegal, it is very hard to obtain reliable information on the factors promoting behavioural change at the individual level, or on changes in the overall prevalence of poaching in the population (Phillipson & Milner-Gulland, unpublished report). This is our next major challenge.

What have we learnt in twenty years?

In the case of this long-term saiga research programme, our scientific progress has been driven as much by contingency, external factors, opportunity, and personal interest as by strategic decisions to tackle the most academically interesting questions of the day. There is a lot to be said for identifying a scientifically sexy research question and sticking to it, producing excellent and well-integrated research papers as you go. However, we have opted for the bumpy ride of taking advantage of funding opportunities, collaborators' interests, and gut feelings as to what is important. This is surely a common experience for conservation researchers.

The danger in this approach is producing science that is reactive rather than proactive, that is always a step behind the big issues rather than identifying them in time to contribute when advice is needed. Following fashions in conservation science can militate against planning conservation science research strategically and for the longer term. We have initiated participatory monitoring programmes in several areas now, but although this is a very fashionable area of conservation science at the moment (Danielsen *et al.*, 2010), we are in this for the longer term (five years already, starting with Whitebread, 2008) and hope that this approach is a sustainable basis for future monitoring and engagement of local people. Impact evaluation is also very high on the conservation agenda (e.g., Ferraro & Pattanayak, 2006), but is also something we are only now able to do for our saiga projects, after five or more years of conservation interventions, and which can now meaningfully inform our future strategies.

There is a substantial move towards adaptive approaches, in which conservation science research is driven by management-relevant questions, and promotes both scientific and practitioner learning. Our work has gone more in this direction in recent years. However it is worth reflecting that long-term monitoring that is consistent over time, such as characterised Soviet science, can produce enormously valuable insights that might not have been recognised as the research was done. Perhaps in the current climate of short-term funding and short-term outcome-driven research, this type of work is a luxury we can no longer afford, but it is worth reflecting upon what we have lost; and the story of the saiga gives many examples of this. For example a number of our most high profile and influential papers (e.g., Milner-Gulland *et al.*, 2003, demonstrating reproductive collapse) were founded on years of apparently pointless annual dissection of pregnant female saigas.

Our research programme also shows that it is possible to carry out high-impact research, from both an academic and a practitioner perspective, on a shoestring budget. Our early years were characterised by relatively large and well funded projects, but even then the presence of highly motivated PhD students, funded from scholarships, provided a lynchpin for the research programme. In recent years, our grants have been small and conservation-focussed, precluding us embarking on a long-term scientific programme, but by piecing together short-term projects within a broader strategic framework, we are able still to carry out novel and publishable science. Nonetheless this is not a satisfactory situation, and the foundation for this work was laid in the large and well-funded projects of the 1990s, without which it would have been impossible to build our collaborative networks. Although funding agencies must promote reactive science so that, in urgent situations, research can be rapidly carried out and used for formulating conservation actions (cf. the experience with the saiga mass mortality in 2010), long-term funding for more proactive, blue-skies research is also needed. Critically, the same collaborative teams need to be able to access both types of funding.

Conservation research is set within an ever-changing ecological, social and political landscape. Personal relationships of trust and mutual respect are key both for effective research and in order to ensure that research genuinely informs conservation practice, and vice versa. These relationships take years to develop and require researchers not to walk away to easier problems, however bleak the situation may appear.

Take-home messages

Key threats to saigas:
Saigas face a substantive ongoing threat from poaching.
Infrastructural development including railways, fences and roads threaten the integrity of the habitat.
Climate change and disease are potentially serious threats on the horizon.

Conservation priorities:
Promote behavioural change amongst poachers through a combination of incentives, engagement, and law enforcement.
Landscape-level planning to ensure development is sustainable and maintains ecosystem connectivity for saiga migrations.
Promote robust monitoring using a range of approaches (participatory and scientific).

Current research priorities:
Understanding the factors underlying saiga movement and distribution in a changing environment.
Evaluating the effectiveness of conservation actions.
Supporting the Government of Kazakhstan in developing a plan for sustainable hunting that benefits local people.

Insights for conservation science:
Contingency, opportunity, funding limitations and personal interest are key drivers of short-term research priorities, but do not preclude a strategic vision.
Long-term consistent monitoring of populations and vital rates can be invaluable (with caveats), so long as there is periodic evaluation.
Focussing on a primary current threat or issue should not eclipse others that may re-emerge unexpectedly, so maintaining research on less proximate issues is worthwhile.
Conservation science cannot be expected immediately to yield results in terms of species recovery or policy uptake; there is a need to stay engaged for the long term.

References

Bannikov, A. G., Zhirnov, L. V., Lebedeva, L. C. & Fandeev, A. A. (1961): *Biologiia Saigaka. [The Biology of the Saiga.]* Moscow: Publisher of Agricultural Literature, Journals and Posters.

Bekenov, A. B., Grachev, I. A. & Gulland, E. M. (1998): The ecology and management of the saiga antelope in Kazakhstan. *Mammal Review* **28**: 1–52.

Bull, J. W., Suttle, K. B., Singh, N. J. & Milner-Gulland, E. J. (2013): Conservation when nothing stands still: moving targets and biodiversity offsets. *Frontiers in Ecology and the Environment* **11**: 203–210.

CMS. (2006): Revised overview report of the first meeting of the signatories to the memorandum of understanding concerning conservation, restoration and sustainable use of the saiga antelope (*Saiga tatarica tatarica*) (CMS/SA-1/Report Annex 5): Convention of Migratory Species.

Coulson, T., Milner-Gulland, E. J. & Clutton-Brock, T. (2000): The relative roles of density and climatic variation on population dynamics and fecundity rates in three contrasting ungulate species. *Proceedings of the Royal Society B: Biological Sciences* **267**: 1771–1779.

Danielsen, F., Burgess, N. D. & Jensen, P. M. (2010): Environmental monitoring: the scale and speed of implementation varies according to the degree of peoples involvement. *Journal of Applied Ecology* **47**: 1166–1168.

Damerell, P., Bykova, E. & Milner-Gulland, E. J. (2012) Analysing environmental education on the Ustyurt Plateau. *Saiga News* **15**, 17–18.

Damerell, P., Bykova, E. & Milner-Gulland, E. J. (2013): Using embroidery to address saiga poaching on the Ustyurt Plateau. *Saiga News* **16**: 1–4.

Durant, S. M., Bashir, S. & Maddox, T. (2007): Relating long-term studies to conservation practice: the case of the Serengeti Cheetah Project. *Conservation Biology*, **21**, 602–611.

Fadeev, V. A. & Sludskii, A. A. (1982): *Saigak v Kazakhstane [The Saiga in Kazakhstan]*. Academy of Sciences, USSR, *Alma-Ata* [in Russian].

Ferraro, P. J. & Pattanayak, S. K. (2006): Money for nothing? a call for empirical evaluation of biodiversity conservation investments. *PLoS Biology* **4**: e105.

Field, S. A., Tyre, A. J., Jonzén, N. & Rhodes, J. R. (2004): Minimizing the cost of environmental management decisions by optimizing statistical thresholds. *Ecology Letters* **7**: 669–675.

Fry, M. (2004): *The status of the Saiga antelope in the Ustiurt region of western Kazakhstan*. Masters Thesis, Imperial College London, UK.

Gates, C. C., Jones, P., Suitor, M., Jakes, A., Boyce, M. S., Kunkel, K. & Wilson, K. (2012): The influence of land use and fences on habitat effectiveness, movements and distribution of pronghorn in the grasslands of North America. In *Fencing for Conservation*: 277–294. Somers, M. J. & Hayward, M. (Eds). New York: Springer.

Howe, C. & Gulland, E. M. (2012): Evaluating indices of conservation success: a comparative analysis of outcome- and output-based indices. *Animal Conservation* **16**: 139–140.

Howe, C. & Medzhidov, R. (2011): Evaluating the relative effectiveness of alternative conservation interventions in influencing stated behavioural intentions: the saiga antelope in Kalmykia (Russia). *Environmental Conservation* **38**: 37–44.

IUCN Species Survival Commission (IUCN). (2007): 2006 ICUN red list of threatened species. International Union for Conservation of Nature and Natural Resources. http://iucn.org/about/work/programmes/species/who_we_are/about_the_species_survival_commission_/(as at 7 February 2007).

Kholodova, M. V., Milner-Gulland, E. J. & Easton, A. J. (2006): Mitochondrial DNA variation and population structure of the Critically Endangered saiga antelope *Saiga tatarica*. *Oryx* **40**: 103–107.

Kock, R., Grachev, Y., Zhakypbayev, A., Usenbayev, A., Zuther, S., Klimanova, O., Dieterich, T., Sapanov, M. & Izimbergenova, G. (2012): A retrospective assessment of saiga antelope (*Saiga tatarica*) die-off in Western Kazakhstan 2010–2011. *Saiga News* **14**: 1–4.

Kühl, S. A. (2008): *The conservation ecology of the saiga antelope*. Ph.D. Thesis, Imperial College London, UK.

Kühl, A., Balinova, N., Bykova, E., Arylov, Y. N., Esipov, A., Lushchekina, A. A. & Milner-Gulland, E. J. (2009): The role of saiga poaching in rural communities: linkages between attitudes, socio-economic circumstances and behaviour. *Biological Conservation* **142**: 1442–1449.

Kühl, A., Mysterud, A., Erdnenov, G. I., Lushchekina, A. A., Grachev, I. A., Bekenov, A. B. & Milner-Gulland, E. J. (2007): The 'big spenders' of the steppe: sex-specific maternal allocation and twinning in the saiga antelope. *Proceedings: Biological Sciences* **274**: 1293–1299.

Lawton, J. H. (2007): Ecology, politics and policy. *Journal of Applied Ecology* **44**: 465–474.

Lushchekina, A. & Struchkov, A. (2001): The saiga antelope in Europe: once again on the brink. *Open Country* **3** 11–24.

Mallon, D. (2012): The taxonomy of the saiga antelope. *Saiga News* **15**: 1–2.

McCarthy, M. A. & Possingham, H. P. (2007): Active adaptive management for conservation. *Conservation Biology* **21**: 956–963.

McConville, A. J., Grachev, I. A. & Keane, A. (2009): Reconstructing the observation process to correct for changing detection probability of a critically endangered species. *Endangered Species Research* **6**: 231–237.

Milner-Gulland, E. J. (1994): A population model for the management of the saiga antelope. *Journal of Applied Ecology* **31**: 25–39.

Milner-Gulland, E. J. (2001): A dynamic game model for the decision to join an aggregation. *Ecological Modelling* **145**: 85–99.

Milner-Gulland, E. J., Bukreeva, O. M., Coulson, T., Lushchekina, A. A., Kholodova, M. V., Bekenov, A. B. & Grachev, I. A. (2003): Conservation: reproductive collapse in saiga antelope harems. *Nature* **422**: 135–135.

Milner-Gulland, Kholodova, M. V., Bekenov, A., Bukreeva, O. M., Grachev, A., Amgalan, L. & Lushchekina, A. A. (2001): Dramatic declines in saiga antelope populations. *Oryx* **35**: 340–345.

Morgan, E. R., Lundervold, M. & Medley, G. F. (2006): Assessing risks of disease transmission between wildlife and livestock: the saiga antelope as a case study. *Biological Conservation* **131**: 244–254.

Morgan, E. R., Shaikenov, B., Torgerson, P. R., Medley, G. F. & MIlner-Gulland, E. J. (2005): Helminths of saiga antelope in Kazakhstan: implications for conservation and livestock production. *Journal of Wildlife Diseases* **41**: 149–162.

Offord, S. (2011): *An evaluation of potential monitoring strategies for saiga antelopes on the Ustyurt Plateau.* Master Thesis, Imperial College London, UK.

O'Neil, H. (2008): *Designing robust ranger based monitoring strategies for the saiga antelope Saiga tatarica tatarica.* Master Thesis, Imperial College London, UK.

Pusey, A. E., Pintea, L., Wilson, M. L. & Kamenya, S. (2007): The Contribution of long-term research at Gombe National Park to chimpanzee conservation. *Conservation Biology* **21**: 623–634.

Reyers, B., Roux, D. J., Cowling, R. M., Ginsburg, A. E., Nel, J. L. & O' Farrell, P. (2010): Conservation planning as a transdisciplinary process. *Conservation Biology* **24**: 957–965.

Robinson, S. & Milner-Gulland, E. J. (2003): Political change and factors limiting numbers of wild and domestic ungulates in Kazakhstan. *Human Ecology* **31**: 87–110.

Robinson, S., Wiedemann, C., Michel, S., Zhumabayev, Y. & Singh, N. (2012): Pastoral tenure in Central Asia: theme and variation in the five former Soviet republics. In *Rangeland stewardship in Central Asia*: 239–274. Squires, V. (Ed.). Dordrecht: Springer Netherlands.

Samuel, C. (2011): Evaluating the success of a public engagement project for the conservation of the Ural saiga population in Kazakhstan. Master thesis, Imperial College London, UK.

Singh, N. J. & Milner-Gulland, E. J. (2011a): Monitoring ungulates in Central Asia: current constraints and future potential. *Oryx*, **45**, 38–49.

Singh, N. J. & Milner-Gulland, E. J. (2011b): Conserving a moving target: planning protection for a migratory species as its distribution changes. *Journal of Applied Ecology*, **48**, 35–46.

Singh, N. J., Grachev, I. A., Bekenov, A. B. & MIlner-Gulland, E. J. (2010a): Saiga antelope calving site selection is increasingly driven by human disturbance. *Biological Conservation* **143**: 1770–1779.

Singh, N. J., Grachev, I. A., Bekenov, A. B. & MIlner-Gulland, E. J. (2010b): Tracking greenery across a latitudinal gradient in Central Asia – the migration of the saiga antelope. *Diversity and Distributions* **16**: 663–675.

Sokolov, V. E. & Zhirnov, L. V. (1998) Saigak: filogeniya, sistematika, ekologiya, okhrana i ispol'zovanie [The saiga antelope: phylogeny, systematics, ecology, conservation, and use]. Moscow: Russian Academy of Sciences.

Soulé, M. E. (1985): What is conservation biology? *Bioscience* **35**: 727–734.

Waylen, K. A., Fischer, A., McGowan, P. J. K. & Milner-Gulland, E. J. (2012): Interactions Between a collectivist culture and Buddhist teachings influence environmental concerns and behaviors in the Republic of Kalmykia, Russia. *Society & Natural Resources* **25**: 1118–1133.

Whitebread, E. (2008): *Evaluating the potential for participatory monitoring of saiga antelope by local villagers in Kalmykia, Russia.* Master Thesis, Imperial College London, UK.

Yoccoz, N. G., Nichols, J. D. & Boulinier, T. (2001): Monitoring of biological diversity in space and time. *Trends in Ecology & Evolution* **16**: 446–453.

Young, J. K., Murray, K. M., Strindberg, S., Buuveibaatar, B. & Berger, J. (2010): Population estimates of endangered Mongolian saiga *Saiga tatarica mongolica*: implications for effective monitoring and population recovery. *Oryx* **44**: 285–292.

16

Synthesis: Antelope Conservation – Realising the Potential

Jakob Bro-Jørgensen

Mammalian Behaviour and Evolution Group, Department of Evolution, Ecology and Behaviour, Institute of Integrative Biology, University of Liverpool, United Kingdom

The drastic wildlife declines over the past century have had a severe impact on many ecosystems, with antelope extinctions at both local and global level a reality. However, natural habitat teeming with antelopes still covers significant parts of Africa and Asia, and iconic species on the verge of extinction are still hanging on in remote corners of the continents. So to what extent is our antelope heritage already doomed? Within the next 50 years, there is hope that the growth of the world's human population may slow down drastically or even fall (UN, 2015) and thereby provide a new robust foundation for sustainable conservation management. Actions over the next few decades are therefore likely to be critical, as they will decide our success in bringing representative populations through what may be an exceptionally challenging time. The chapters in this book provide a basis for reflection on the more specific aims and the most effective approaches.

What do we want to preserve?

Exactly which aspects of antelope biodiversity is it we want to safeguard for generations to come? A vision of the vast expanses of uninterrupted wilderness that existed 200 years ago may be alluring but is clearly not a realistic short-term conservation goal given societal changes and present-day human population pressures (Ford *et al.*, in Chapter 2). It is therefore crucial to identify the key components of antelope biodiversity on which we should focus our attention.

Antelope Conservation: From Diagnosis to Action, First Edition. Edited by Jakob Bro-Jørgensen and David P. Mallon.

Running through this book is a consensus that preventing further species loss is mandatory for biodiversity conservation (although it is worth noting that the delineation of antelope species is associated with a certain degree of arbitrariness and controversy; see Groves & Grubb, 2011; Heller *et al.*, 2013). The phylogeographic review by Lorenzen (Chapter 9) implies that species without close relatives are of particular concern due to their unique evolutionary history. Lorenzen moreover clearly demonstrates that significant biodiversity also exists *within* species, with considerable differences between populations in their amount of genetic variation and the degree to which they have differentiated genetically from each other. In some cases, genetic differentiation reflects adaptation to local conditions and incipient speciation. Since the ultimate goal of conservation is to preserve ecosystems and their evolutionary potential, the information becoming available about the genetic profile of antelope species is highly valuable. The landscape genetics approach advocated by Atickem *et al.* (Chapter 10) provides a practical solution to how genetic diversity can be preserved by managing species as meta-populations in a networked landscape, by combining population genetic analysis with telemetry, remote-sensing, and habitat modelling.

Biodiversity is, however, not to be equated with genetic diversity. Both between and within species significant natural diversity in ecology and behaviour is not necessarily well captured by molecular analysis, and certain antelope populations exhibit exceptional local distinctions in their behavioural ecology worth conservation attention. Ford *et al.* (Chapter 2) highlight how antelope migrations can have profound effects on the structure of entire ecosystems, the blue wildebeest (*Connochaetes taurinus*) and Thomson gazelles (*Eudorcas thomsonii*) of the Serengeti being a classic example (Plate 8), with other notable antelope migrations including the saiga (*Saiga tatarica*) and Mongolian gazelle (*Procapra gutturosa*) in central Asia, the tiang (*Damaliscus lunatus tiang*) and white-eared kob (*Kobus kob leucotis*) in Sudan, and springbok (*Antidorcas marsupialis*) and blue wildebeest in South Africa (although most of the latter have witnessed drastic declines in recent years; Bolger *et al.*, 2008). Such migrations depend on healthy population sizes and possibly cultural transmission. Moreover, some antelope populations have developed alternative mating systems (Isvaran, 2005), probably the most extreme example being the rare and enigmatic 'lekking' behaviour (also termed 'arena-breeding') found in selected populations of blackbuck (*Antilope cervicapra*), kob (*Kobus kob*), topi (*Damaliscus lunatus*) and lechwe (*Kobus leche*) (Plate 6). Again, lekking behaviour is associated with high population density (Bro-Jørgensen, 2003) as well as cultural transmission (J. Bro-Jørgensen, unpublished). Because of the extreme male mating skew characteristic of leks, lekking is expected to intensify sexual selection, and the behaviour may well have evolutionary consequences.

For the greater picture, antelope conservation cannot, and should not, be considered in isolation and must incorporate any collateral benefits and costs antelopes inflict on other species, be they plants, predators, or other herbivores. Prins (Chapter 4) throws the gauntlet down by stressing our limited understanding of trophic interactions between herbivore species – to what extent they are resource limited and compete with, or facilitate, each other often remains highly speculative. Such interspecific interactions are likely to vary between populations and can potentially affect their evolutionary trajectories. Also the presence or absence of certain predators in the ecological community can radically change selective pressures on antelopes (Gittleman & Gompper, 2001; Hopcraft in Chapter 3), and little is known as to whether evolution in predator-free environments is reversible. Moreover, Kock *et al.* (Chapter 7) point out that our understanding of the natural role of disease in antelope ecology in 'healthy' ecosystems is limited.

Where ecological conditions are altered due to human activities, a key question becomes how to define 'wilderness' and when a population ceases to be 'wild' (Mallon & Stanley Price, 2013; Stanley Price in Chapter 12). This in turn has implications for how we assess the conservation value of specific areas. A case can be made for defining certain populations as truly 'wild' (for all intents and purposes); however, rather than labelling the remainder simply as 'not wild', it is probably most constructive to regard them in a continuum depending on the degree of anthropogenic interference and the level of management they require (Redford *et al.*, 2011). Knight *et al.* (Chapter 8) present a comprehensive range of management interventions useful in preserving or restoring viable populations under various circumstances. However, for each intervention, the cost of its application in terms of reduced 'wildness' of the system needs to be carefully weighed against the benefits since tinkering with natural eco-evolutionary processes is to be avoided unless necessary. Undoubtedly, in some scenarios concern about reduced 'wildness' is a luxury that cannot be afforded, in the extreme a local population may have gone extinct with reintroduction by translocation the only option for its restoration (Chapters 2, 12 & 14). But when success in antelope conservation is evaluated solely by examining trends in population sizes, an inaccurate picture is obtained if heavily managed populations founded by translocation are given the same weight as pristine populations in their original habitat.

Having defined what we want to preserve, information on the threat level of different antelope populations is crucial to identify the appropriate conservation action. The IUCN Red List is the key indicator of extinction risk at the species and subspecies level, and it is based on rigorous assessment criteria which require reliable population monitoring. For many threatened taxa, sufficient information is currently available so that further conservation resources will be spent more wisely on other conservation action than simply documenting their decline in detail. However, certain species are particularly difficult to monitor due to dense habitat, nocturnal activity patterns, or war within their ranges; and lack of vital population information is severely hampering their assessment. Amin *et al.* (Chapter 11) present camera-trapping as a powerful tool to survey nocturnal and dense-habitat species, as illustrated by the recent discovery of vital populations of the critically endangered Aders' duiker (*Cephalophus adersi*) in the northern coastal forests of Kenya.

Using ecosystem services provided by antelopes to support their conservation

Why are so many taxa threatened? Throughout this book, over-exploitation and wildlife-incompatible land use changes are repeatedly confirmed as the main underlying causes of threat to antelopes. Of central importance to tackle these threats is the realization that antelopes, distinct from their biodiversity value, also have a socioeconomic value arising from the ecosystem services they provide. However, while ecosystem services delivered can be a major incentive for sustainable management of antelope populations, it is important to recognize that many key aspects of biodiversity are not represented by any value in the market economy. Therefore, from a conservation perspective, the financial value of antelopes may be more aptly seen as a means to be manipulated rather than as an objective measure of worth.

As indicated by Ford *et al.* (Chapter 2), a key motivation for biodiversity preservation comes from the profit that can be gained from ecotourism, an industry which especially in Africa is heavily dependent on healthy antelope populations. Not only do antelopes provide important wildlife spectacles in their own right, they are the *sine qua non* for the persistence of the large predators which are currently even more powerful tourist magnets). Where economically viable, ecotourism is likely to be the preferred land-use option from a conservation perspective as its impact on the structure and function of natural ecosystems is minimal when properly managed. In order to reverse the significant declines of wildlife populations within many African protected areas (PAs) (Craigie *et al.*, 2010), examples from eastern and southern Africa show how ecotourism can contribute to the funding of entire protected area networks, where the most profitable parks support those with lower tourist potential. As pointed out in Chapters 2 & 8, PA networks remain a cornerstone in antelope conservation, and a high priority for their sustainable management is that local communities receive a fairer share of the revenue. A positive development in eastern and southern Africa is the proliferation of community and privately owned conservancies, which make an increasingly important contribution to the protected area network; for instance, since the late 1990s, Kenya has witnessed an increase of about 50% in the area under protection for this reason.

Although there is still scope for further development of ecotourism in antelope range states, in many areas of vital importance for antelope conservation, such initiatives are challenged by low visibility of wildlife and poor infrastructure. In the absence of other sources of funding, preserving wildlife-compatible habitat under these circumstances may sometimes only be feasible if wildlife is associated with a value from exploitation, either by locals or foreign hunters. Bunnefeld & Milner-Gulland (Chapter 6) point out that adaptive management of any exploitation, whether for meat or trophies, depends on robust models that can be used to set quotas and determine law enforcement effort. Like Milner-Gulland & Singh (Chapter 15), they observe that such models should incorporate multiple sources and types of uncertainty if they are to produce results that can confidently be used to guide policy.

Fa (Chapter 5) shows that antelopes are the most important source of bushmeat in forests and as such, they are heavily relied upon as both a source of income and nutrition (Plate 10). Forest antelopes are very diverse in terms of their population dynamics, and sustainable quotas differ drastically between species. Non-selective and wasteful hunting methods such as snaring can therefore be highly detrimental, and at present unselective harvesting results in larger endangered species with slow reproductive rates being severely overhunted in many places. Fa emphasizes that more information on species-specific population sizes, reproductive rates and offtake rates is needed to assess the sustainability of current hunting pressures properly; this again brings attention to camera-trapping as a useful technique for monitoring population trends in dense habitats (Chapter 11). Fa argues further in support of spatiotemporal zoning of hunting activities, but concludes that so far sustainable management systems for bushmeat hunting have proven difficult to implement. To identify illegal hunting of threatened species, Lorenzen (Chapter 9) emphasizes the ability of genetic analysis to pinpoint the sources of bushmeat, especially as this forensic technique has recently become much more affordable. Unregulated bushmeat hunting is also a major issue in savanna systems (Lindsey *et al.*, 2013): although the ungulate biomass is usually higher (Robinson & Bennett, 2004), hunting is relatively easy in open habitats with good visibility, and reproductive rates are relatively low as antelopes are generally larger. A better understanding of the socioeconomic context of unsustainable subsistence hunting in both environments is therefore needed, and here a stronger link between the biological and social sciences can provide insights into decision making at the local level and its ecological consequences (e.g., Nuno *et al.*, 2013).

As an alternative land use option, Bunnefeld & Milner-Gulland (Chapter 6) show that trophy-hunting has the potential to generate substantial income to support conservation and preserve wide areas of wilderness where ecotourism is not financially viable. However, in their review of current trophy hunting practices, Bunnefeld & Milner-Gulland call for implementation of stronger benefit-sharing mechanisms as well as more transparency and regulation, pointing out the usefulness of models and approaches from fisheries. As illustrated by the saiga case study reviewed by Milner-Gulland & Singh (Chapter 15), long-term research projects can in this context provide crucial information to set appropriate quotas. A further concern in relation to current trophy hunting practices is the intentional genetic manipulation involved in the selective breeding for favoured traits in the target animals (IUCN SSC Antelope Specialist Group, 2015), such as abnormal horn length and coloration (e.g., black impala and golden wildebeest). Ample evidence also reveals unintentional selection from trophy hunting targeting long-horned individuals can lead to unnatural selection for shorter horn length in hunted populations of bovids (Mysterud, 2011; see also Allendorf & Hard, 2009). An additional concern is that hunters are known to prefer rarer species which can potentially lead to vicious cycles of population declines (Palazy *et al.*, 2012; Prescott *et al.*, 2012; Bro-Jørgensen, 2014).

At present, antelope populations in several parts of Africa and Asia have been depleted to such a severe degree that neither ecotourism nor any kind of offtake is sustainable financially, or even biologically. Political instability further limits the conservation options in several vital areas for antelope conservation. The Sahelo-Saharan belt and its extension into the Horn of Africa are of extreme concern, and at present, Newby *et al.* (Chapter 13) argue strongly against any kind of offtake of the most threatened desert antelopes, whose plight is the legacy of uncontrolled hunting. In the short term, they rather stress the importance of providing support for local communities to engage directly in non-consumptive conservation activities, and harvesting is considered relevant only following evidence of a marked population recovery.

Ironically, the physiological and behavioural adaptations of the desert antelopes illustrate particularly well the superior ecological potential of antelopes compared to traditional livestock species in many natural habitats: their productivity is higher, and their ecological impact is lower (Chapter 13). This ability to subsist on marginal lands can be translated into financial benefits through game-ranching and farming enterprises. Again, a concern that needs addressing in any antelope farming enterprise is the interference with natural genetic variation. Also here, the interference may be more or less intentional, with adverse consequences including selection for tameness as well as interbreeding between distinct genetic populations or even hybridization (e.g., in Chapter 14), sometimes following accidental escape from enclosures (IUCN SSC Antelope Specialist Group, 2015). Hence Knight *et al.* (Chapter 8) question whether game ranching should be considered true conservation at all.

Substantial profits can thus be made through both non-consumptive and consumptive exploitation of antelopes. However, as mentioned above, many ecosystem services provided by antelopes are not currently captured by the formal economy. Being an integral part of many ecosystems, antelope populations are often vital in maintaining habitat productivity and nutrient-cycling, which in turn is important for carbon sequestration and water retention. This function is not quantified as an economic value at present, stressing that the free market economy cannot be relied upon to accurately reflect the socioeconomic benefits provided by antelopes, let alone their biodiversity value. A key conservation challenge is to bring such ecosystem services firmly into economic decision making by introducing innovative mechanisms at the governmental level, such as the environmental offsetting imposed on private companies mentioned by Milner-Gulland & Singh (Chapter 15).

Integrating local and global action to preserve our antelope heritage

Can we arrive at an action plan that will work in the real world where the threats to antelope biodiversity seem intrinsically linked to short-term profit maximization by actors at various levels? A master plan for antelope conservation involves action locally, nationally as well as globally.

At the local level, Ford *et al.* (Chapter 2) and Bunnefeld & Milner-Gulland (Chapter 6) stress the need to build the capacity of communities and private landowners to engage in wildlife-friendly land-use. Crucial in this context is support to benefit-sharing mechanisms in favour of the local communities that experience human-wildlife conflicts and suffer opportunity costs from conservation. Bunnefeld & Milner-Gulland stress how co-management rather than top-down management enables locals to link economic benefits to compliant behaviour more easily. Community engagement is further stimulated if capacity-building initiatives are accompanied by strategic campaigns to raise conservation awareness. Evidence shows that outreach that is motivational and presents alternative conservation-friendly livelihood options can indeed change people's behaviour (Steinmetz *et al.*, 2014). Antelopes can have a significant cultural value, and outreach activities aimed at boosting pride in antelope heritage have considerable potential as an effective means of achieving conservation objectives, as shown by the Saiga Day run by the Saiga Conservation Alliance in Uzbekistan, Russia, and Kazakhstan (Chapter 15; Saiga Conservation Alliance 2015) and education programmes run by the

Bongo Surveillance Project in Kenya (Bongo Surveillance Project 2015). However, Milner-Gulland & Singh (Chapter 15) call for increased monitoring of the impact of such conservation initiatives to determine their effectiveness in leading to behavioural change.

At the national level, a priority of paramount importance is integration of conservation and development planning: a key issue in many antelope range states is the lack of coordinated national land-use plans (Chapter 2). Design and management of networks of protected areas and hunting concessions have to take into account, and wherever possible support, social development objectives, just as development plans have to accommodate priorities in biodiversity conservation. Lack of such coordination is responsible for loss of important antelope habitat due to perverse incentives supporting wildlife-incompatible land uses that would otherwise not be financially viable. Systematic zoning as suggested by Fa (Chapter 5) would allow optimization of land use by dividing the land surface – according to the underlying potential – into national parks, hunting concessions (with set quotas), livestock pastures (with defined stocking rates), and the range of other land uses required by society.

In this context, decisions have to be made about when to integrate or separate areas of wilderness and areas with human activities such as livestock grazing (Chapters 2, 7 & 8). Currently, the issue of fencing is highly controversial (Chapter 12) (Creel *et al.*, 2013; Packer *et al.*, 2013), and Knight *et al.* (Chapter 8) hint at how eastern and southern African countries could learn more from each other's experiences. In landscapes with low human density, any benefits from fencing will often be outweighed by detrimental ecological effects from preventing dispersal and seasonal movements of wildlife; for many antelope populations, a special concern is severing of migratory routes (Harris *et al.*, 2009). By contrast, where human-dominated landscapes adjoin wilderness at a hard boundary, fencing can be an effective way of protecting antelope populations from poaching and overgrazing by livestock and at the same time preventing crop damage and disease transmission between livestock and wildlife. That fencing can be beneficial is indicated by Prins (Chapter 4) who presents evidence that livestock can indeed outcompete antelope populations under poorly managed co-existence. Stanley Price (Chapter 12) moreover mentions that fences can allow natural vegetation to recover, and Kock *et al.* (Chapter 7) conclude that livestock diseases can also sometimes have severe impacts on antelopes, perhaps especially on the most closely related species and naïve populations. However, on the other hand Woodfine & Gilbert (Chapter 14) point out that confinement of wildlife within too small ranges by fencing can result in excessive intra- and inter-specific competition, increased disease transmission, and abnormal social behaviour. When it comes to decisions about whether to integrate or separate, the landscape genetics approach presented by Atickem *et al.* (Chapter 10) is also valuable in allowing managers to quantify the benefit of maintaining connections between populations from a genetic perspective. Knight *et al.* (Chapter 8) bring attention to the importance of establishing empirically whether antelopes indeed utilize the corridors identified as priorities theoretically. At present, however, the complex ecological and socioeconomic impact of fencing is insufficiently understood to make clear recommendations for its wise use (Durant *et al.*, 2015).

Natural movements and densities of many antelope populations are severely affected also by infrastructural developments such as roads, railways, and pipelines. Again, migratory populations are at particular risk, as illustrated by the recent plans to construct a main road across the migratory routes of the Serengeti wildebeest (Hopcraft *et al.*, 2015). The detrimental effects of roads are exacerbated by the fact that they encourage adjacent human settlements and facilitate access to wildlife populations, which often promotes unsustainable hunting.

Integrated land use planning depends on close cooperation between management at the central and local level, and a key priority is to provide legislative conditions conducive for communities and private landholders to engage in conservation-friendly land use. Knight *et al.* (Chapter 8) advocate for providing ownership of wildlife resources to communal and private land owners, with the right to manage these resources accordingly. However, they also allude to the fact that the expertise required for sustainable wildlife management often will not be available locally and the contribution of the private sector would need to be under central regulation and oversight to assure that conservation – rather than purely profit-maximizing – objectives are met. Concerns include the complex tasks of setting and enforcing sustainable quotas (Chapters 5 & 6) and genetic management (Chapters 9 & 10), but also challenges such as fire management, eradication of invasive aliens and ecologically sound reintroduction and restoration initiatives will benefit from expert supervision (Chapters 2, 8 & 12). An adaptive management approach will allow the best use of knowledge as it emerges, and at a wider scale Prins (Chapter 4) draws attention to the benefit of evidence-based conservation as a basis for informed decisions. Milner-Gulland & Singh (Chapter 15) in this context stress the importance of reflecting historically on the lessons learned from long-term conservation projects.

At the global level, increased recognition is moreover needed that conservation often is a global rather than a local priority. Local communities in the antelope range states are often poor and have to bear costs of conservation in terms of human-wildlife conflict, including crop damage and disease transmission, as well as opportunity costs. Therefore more investment in biodiversity conservation needs to come from international donors.

At the global level, action is also needed to counter climate change, the perturbing effects of which are discussed by Hopcraft (Chapter 3). Compounding the adverse impact of climate change is the increasingly fragmented landscapes that prevent many species from coping by shifting their ranges, and the most affected species may be those already categorized as at risk due to other threats (Payne & Bro-Jørgensen, 2016). The recent northwards shift of the saiga range illustrates that, in the absence of obstacles, antelope populations may well be able to adjust their ranges (Singh & Milner-Gulland, 2011), which again speaks against fencing as a general conservation solution (Woodroffe *et al.*, 2015).

From an antelope perspective, a specific concern in terms of global priorities is furthermore a bias in the conservation community towards prioritizing forests rather than savannas and deserts, as reflected by Newby *et al.* (Chapter 13; see also Durant *et al.*, 2012). Hotspot-based priority scoring exercises repeatedly score forests above savannas and especially deserts based on species richness. However, savannas and deserts are qualitatively different ecosystems, and the merit of using quantitative measures based on species richness to assess their conservation value is questionable. Hence, it may well be argued that such approaches give a misleading sheen of objectivity that has influenced the conservation agenda unduly. Also initiatives such as REDD offer financial rewards to projects that turn open habitats into forests, whether by natural regeneration or plantation. The scheme is aimed at reducing climate change by CO_2-sequestration; however, it can undermine the dynamics of savanna systems, many of which grow into forests if intermittent disturbance is prevented (Parr *et al.*, 2014).

Finally, increased international collaboration is essential to tackle several practical conservation issues. In terms of integrated land-use planning, Hopcraft (Chapter 3) points out that coordination between nation states is critical to establish and manage transboundary parks and implement the most effective design of protected area networks. Where conservation efforts in situ are failing, international collaboration on reintroduction schemes is critical in order to avoid extinctions. Stanley Price (Chapter 12) notes that reintroduction initiatives aimed at antelopes inhabiting more fertile savannas may face most obstacles because the habitats required are associated with higher opportunity costs of biodiversity conservation, a fact that may explain the observed lack of reintroduction initiatives for these species. Specific international collaborations that in particular have benefitted antelope conservation include CMS (Convention on the Conservation of Migratory Species of Wild Animals) and CITES (Convention on International Trade in Endangered Species of Wild Fauna and Flora).

Priorities for research

Although the stark evidence of antelope population declines and documentation of their underlying causes already provide an imperative for immediate action, the coverage in this book also draws attention to particular areas in which we need to know more. Most information is available on easily observed species in open habitats. By contrast, the basic biology and population dynamics of many forest antelopes remains poorly understood, and this hampers efforts to assess sustainability of any offtake and set quotas. In Asia, the neglect of forest species is illustrated by the fact that the spectacular saola (*Pseudoryx nghetinhensis*) evaded discovery right up until 1992 (Dung *et al.*, 1993). Although more studies have focused on antelopes in open habitats, for many of these species we still have insufficient knowledge about their spatial ecology, dietary requirements, habitat specificity, and competitive relations to assess the impact of human landscape modifications such as fragmentation due to infrastructural development, fencing, and changes in water management. Also the effect of livestock on antelope populations due to resource competition, disease transmission, and general disturbance is often highly controversial owing to lack of rigorous data collection. Furthermore, from vast expanses of wilderness, critical information on hunting intensity on different antelope species is lacking, and a better socioeconomic understanding of the motivational drivers underlying unsustainable hunting is needed. Finally, our understanding of the impact of climate change on antelopes remains fairly theoretical and needs firmer linking to environmental monitoring on the ground. In filling these gaps, technological advances are opening up for new frontiers in antelope research: satellite monitoring of animal movements and habitats, camera-trapping, breakthroughs in molecular techniques, and dietary analysis of faecal pellets are among the technologies that are generating crucial new insights into the ecology of many species.

Conclusion

The 21st century is likely to prove critical in terms of securing our antelope heritage for the long-term future. Main action points include promotion of conservation-friendly land-use through capacity-building and awareness-raising at the local level, improved integration of social development and conservation planning at the national level, and a stronger recognition of funding responsibility and the need for collaboration at the global level. The key decision makers have to balance a complex set of interests on which they are more or less well informed; and as part of this process, it is vital to express clearly the priorities from an antelope perspective: failing to do so would undoubtedly bring about further extinctions as slippery baselines lead to ever lower ambitions and increased marginalization of antelopes on the conservation agenda. Scientists and conservationists with a passion for antelopes therefore need to shout louder about how amazing and important antelopes are. Unfortunately, for many of the most threatened antelope species, civil conflicts are currently a severe obstacle to effective conservation management; however, action in practice will depend on individual conservationists making use of available opportunities. A high priority is to encourage and support in-country conservationists with an interest in antelope preservation. If the global community of antelope specialists keep paying special attention to the significant threatened populations that are not already benefitting from conservation initiatives, there is hope that lives of future generations will also be enriched by sharing the planet with the charming and ecologically important aspect of biodiversity that is antelopes.

References

Allendorf, F. W. & Hard, J. J. (2009): Human-induced evolution caused by unnatural selection through harvest of wild animals. *PNAS* **106**: 9987–9994.

Bolger, D. T., Newmark, W. D., Morrison, T. A. & Doak, D. F. (2008): The need for integrative approaches to understand and conserve migratory ungulates. *Ecology Letters* **11**: 63–77.

Bongo Surveillance project (2015): http://www.mountainbongo.org (as at 27 May 2015).

Bro-Jørgensen, J. (2003): The significance of hotspots to lekking topi antelopes (*Damaliscus lunatus*). *Behavioral Ecology and Sociobiology* **53**: 324–331.

Bro-Jørgensen, J. (2014): Will their armaments be their downfall? Large horns increases extinction risk in bovids. *Animal Conservation* **17**: 80–87.

Craigie, I. D., Baillie, J. E. M., Balmford, A., Carbone, C., Collen, B., Green, R. E. & Hutton, J. M. (2010): Large mammal population declines in Africa's protected areas. *Biological Conservation* **143**: 2221–2228.

Creel, S., Becker, M. S., Durant, S. M., M'Soka, J., Matandiko, W., Dickman, A. J., Christianson, D., Dröge, E., Mweetwa, T., Pettorelli, N., Rosenblatt, E., Schuette, P., Woodroffe, R., Bashir, S., Beudels-Jamar, R. C., Blake, S., Borner, M., Breitenmoser, C., Broekhuis, F., Cozzi, G., Davenport, T. R., Deutsch, J., Dollar, L., Dolrenry, S., Douglas-Hamilton, I., Fitzherbert, E., Foley, C., Hazzah, L., Henschel, P., Hilborn, R., Hopcraft, J. G., Ikanda, D., Jacobson, A., Joubert, B., Joubert, D., Kelly, M. S., Lichtenfeld, L., Mace, G. M., Milanzi, J., Mitchell, N., Msuha, M., Muir, R., Nyahongo, J., Pimm, S., Purchase, G., Schenck, C., Sillero-Zubiri, C., Sinclair, A. R. E., Songorwa, A. N., Stanley-Price, M., Tehou, C. A., Trout, C., Wall, J., Wittemyer, G. & Zimmermann, A. (2013): Conserving large populations of lions – the argument for fences has holes. *Ecology Letters* **16**: e1–3.

Dung, V. V., Giao, Pm., Chinh, N. N., Tuoc, D., Arctander, P. & Mackinnon, J. (1993): A new species of living bovid from Vietnam. *Nature* **363**: 443–445.

Durant, S. M., Becker, M. S., Creel, S., Bashir, S., Dickman, A. J., Beudels-Jamar, R. C., Lichtenfeld, L., Hilborn, R., Wall, J., Wittemyer, G., Badamjav, L., Blake, S., Boitani, L., Breitenmoser, C., Broekhuis, F., Christianson, D., Cozzi, G., Davenport, T. R. B., Deutsch, J., Dollar, L., Dolrenry, S., Douglas-Hamilton, I., Droge, E., FitzHerbert, E., Foley, C., Hazzah, L., Hopcraft, J. G. C., Ikanda, D., Jacobson, A., Joubert, D., Kelly, M. J., Milanzi, J., Mitchell, N., M'Soka, J., Msuha, M., Mweetwa, T., Nyahongo, J., Rosenblatt, E., Schuette, P., Sillero-Zubiri, C., Sinclair, A. R. E.,

Price, M. R., Zimmermann, A. & Pettorelli, N. (2015): Developing fencing policies for dryland ecosystems. *Journal of Applied Ecology* **52**: 544–551.

Durant, S. M., Pettorelli, N., Bashir, S., Woodroffe, R., Wacher, T., De Ornellas, P., Ransom, C., Abáigar, T., Abdelgadir, M., El Alqamy, H., Beddiaf, M., Belbachir, F., Belbachir-Bazi, A., Berbash, A. A., Beudels-Jamar, R., Boitani, L., Breitenmoser, C., Cano, M., Chardonnet, P., Collen, B., Cornforth, W. A., Cuzin, F., Gerngross, P., Haddane, B., Hadjeloum, M., Jacobson, A., Jebali, A., Lamarque, F., Mallon, D., Minkowski, K., Monfort, S., Ndoassal, B., Newby, J., Ngakoutou, B. E., Niagate, B., Purchase, G., Samaïla, S., Samna, A. K., Sillero-Zubiri, C., Soultan, A. E., Stanley Price, M. R. & Baillie, J. E. M. (2012): Forgotten biodiversity in desert ecosystems. *Science* **336**: 1379.

Gittleman, J. L. & Gompper, M. E. (2001): The risk of extinction – what you don't know will hurt you. *Science* **291**: 997–999.

Groves, C. & Grubb, P. (2011): *Ungulate taxonomy*. Baltimore: Johns Hopkins University Press.

Heller, R., Frandsen, P., Lorenzen, E. D. & Siegismund, H. R. (2013): Are there really twice as many bovid species as we thought? *Systematic Biology* **62**: 490–493.

Harris, G., Thirgood, S., Hopcraft, J. G. C., Cromsight, J. P. M. & Berger, J. (2009): Global decline in aggregated migrations of large terrestrial mammals. *Endangered Species Research* **7**: 55–76.

Hopcraft, J. G. C., Mduma, S. A. R., Borner, M., Bigurube, G., Kijazi, A., Haydon, D. T., Wakilema, W., Rentsch, D., Sinclair, A. R. E., Dobson, A. & Lembeli, J. D. (2015): Conservation and economic benefits of a road around the Serengeti. *Conservation Biology* **29**: 932–936.

Isvaran, K. (2005): Variation in male mating behaviour within ungulate populations: patterns and processes. *Current Science* **89**: 1192–1199.

IUCN SSC Antelope Specialist Group (2015): *IUCN SSC Antelope Specialist Group position statement on the intentional genetic manipulation of antelopes*. Ver. 1.0.

Lindsey, P. A., Balme, G., Becker, M., Begg, C., Bento, C., Bocchino, C., Dickman, A., Diggle, R. W., Eves, H., Henschel, P., Lewis, D., Marnewick, K., Mattheus, J., McNutt, W., McRobb, R., Midlane, N., Milanzi, J., Morley, R., Murphree, M., Opyene, V., Phadima, J., Purchase, G., Rentsch, S., Roche, C., Shaw, J., van der Westhuizen, H., Van Vliet, N. & Zisadza-Gandiwa, P. (2013): The bushmeat trade in African savannas: impacts, drivers, and possible solutions. *Biological Conservation* **160**: 80–96.

Mallon, D. P. & Stanley Price, M. R. (2013): The fall of the wild. *Oryx* **47**: 467–468.

Mysterud, A. (2011): Selective harvesting of large mammals: how often does it result in directional selection? *J. Appl. Ecol.* **48**: 827–834.

Nuno, A., Bunnefeld, N., Naiman, L. C. & Milner-Gulland, E. J. (2013): A novel approach to assessing the prevalence and drivers of illegal bushmeat hunting in the Serengeti. *Conservation Biology* **27**: 1355–1365.

Packer, C., Loveridge, A., Canney, S., Caro, T., Garnett, S. T., Pfeifer, M., Zander, K. K., Swanson, A., Macnulty, D., Balme, G., Bauer, H., Begg, C. M., Begg, K. S., Bhalla, S., Bissett, C., Bodasing, T., Brink, H., Burger, A., Burton, A. C., Clegg, B., Dell, S., Delsink, A., Dickerson, T., Dloniak, S. M., Druce, D., Frank, L., Funston, P., Gichohi, N., Groom, R., Hanekom, C., Heath, B., Hunter, L., Deiongh, H. H.,

Joubert, C. J., Kasiki, S. M., Kissui, B., Knocker, W., Leathem, B., Lindsey, P. A., Maclennan, S. D., McNutt, J. W., Miller, S. M., Naylor, S., Nel, P., Ng'weno, C., Nicholls, K., Ogutu, J. O., Okot-Omoya, E., Patterson, B. D., Plumptre, A., Salerno, J., Skinner, K., Slotow, R., Sogbohossou, E. A., Stratford, K. J., Winterbach, C., Winterbach, H. & Polasky S. (2013): Conserving large carnivores: dollars and fence. *Ecology Letters* **16**: 635–641.

Palazy, L., Bonenfant, C., Gaillard, J. M. & Courchamp, F. (2012):. Rarity, trophy hunting and ungulates. *Animal Conservation* **15**: 4–11.

Parr, C. L., Lehmann, C. E. R., Bond, W. J., Hoffmann, W. A. & Andersen, A. N. (2014): Tropical grassy biomes: misunderstood, neglected, and under threat. *Trends in Ecology & Evolution* **29**: 205–213.

Payne, B. L. & Bro-Jørgensen, J. (2016): Disproportionate climate-induced range loss forecast for the most threatened African antelopes. *Current Biology* **26**: 1200–1205.

Prescott, G. W., Johnson, P. J., Loveridge, A. J. & MacDonald, D. W. (2012): Does change in IUCN status affect demand for African bovid trophies? *Animal Conservation* **15**: 248–252.

Redford, K. H., Amato, G., Baillie, J., Beldomenico, P., Bennett, E. L., Clum, N., Cook, R., Fonseca, G., Hedges, S., Launay, F., Lieberman, S., Mace, G. M., Murayama, A., Putnam, A., Robinson, J. G., Rosenbaum, H., Sanderson, E. W., Stuart, S. N., Thomas, P. & Thorbjarnarson, J. (2011): What does it mean to successfully conserve a (vertebrate) species? *BioScience* **61**: 39–48.

Robinson, J. G. & Bennett, E. L. (2004): Having your wildlife and eating it too: an analysis of hunting sustainability across tropical ecosystems. *Animal Conservation* **7**: 397–408.

Saiga Conservation Alliance (2015): http://www.saiga-conservation.com (as at 1 June 2015).

Singh, N. J. & Milner-Gulland, E. J. (2011): Conserving a moving target: planning protection for a migratory species as its distribution changes. *Journal of Applied Ecology* **48**: 35–46.

Steinmetz, R., Srirattanaporn, S., Mor-Tip, J. & Seuaturien, N. (2014): Can community outreach alleviate poaching pressure and recover wildlife in South-East Asian protected areas? *Journal of Applied Ecology* **51**: 1469–1478.

UN (2015): World Population Prospects: The 2012 Revision. http://esa.un.org/wpp/unpp/panel_population.htm (as at 1 June 2015).

Woodroffe, R., Hedges, S. & Durant, S. M. (2014): To fence or not to fence. *Science* **344**: 46–48.

Appendix

IUCN Red List Status of Antelope Species April 2016

Scientific name	Common name	IUCN category
Addax nasomaculatus	Addax	CR
Aepyceros melampus	Impala	LC
Alcelaphus buselaphus	Hartebeest	LC
Ammodorcas clarkei	Dibatag	VU
Antidorcas marsupialis	Springbok	LC
Antilope cervicapra	Blackbuck	NT
Beatragus hunteri	Hirola, Hunter's antelope	CR
Boselaphus tragocamelus	Nilgai	LC
Cephalophus adersi	Aders' duiker	CR
Cephalophus callipygus	Peters' duiker	LC
Cephalophus dorsalis	Bay duiker	LC
Cephalophus harveyi	Harvey's duiker	LC
Cephalophus jentinki	Jentink's duiker	EN
Cephalophus leucogaster	White-bellied duiker	LC
Cephalophus natalensis	Natal red duiker	LC
Cephalophus niger	Black duiker	LC
Cephalophus nigrifrons	Black-fronted duiker	LC
Cephalophus ogilbyi	Ogilby's duiker	LC
Cephalophus rufilatus	Red-flanked duiker	LC
Cephalophus silvicultor	Yellow-backed duiker	LC
Cephalophus spadix	Abbott's duiker	EN
Cephalophus weynsi	Weyns' duiker	LC
Cephalophus zebra	Zebra duiker	VU
Connochaetes gnou	Black wildebeest	LC
Connochaetes taurinus	Common wildebeest	LC
Damaliscus lunatus	Topi/Tsessebe/Tiang/Korrigum	LC
Damaliscus pygargus	Blesbok/Bontebok	LC

(*continued overleaf*)

Antelope Conservation: From Diagnosis to Action, First Edition. Edited by Jakob Bro-Jørgensen and David P. Mallon.

(*Continued*)

Scientific name	Common name	IUCN category
Dorcatragus megalotis	Beira	VU
Eudorcas albonotata	Mongalla gazelle	LC
Eudorcas rufifrons	Red-fronted gazelle	VU
Eudorcas rufina	Red gazelle	DD
Eudorcas thomsonii	Thomson's gazelle	NT
Gazella arabica	Arabian gazelle	DD
Gazella bennettii	Chinkara	LC
Gazella bilkis	Queen of Sheba's gazelle	EX
Gazella cuvieri	Cuvier's gazelle	EN
Gazella dorcas	Dorcas gazelle	VU
Gazella gazella	Mountain gazelle	VU
Gazella leptoceros	Slender-horned gazelle	EN
Gazella saudiya	Saudi gazelle	EX
Gazella spekei	Speke's gazelle	EN
Gazella subgutturosa	Goitered gazelle, including sand gazelle	VU
Hippotragus equinus	Roan antelope	LC
Hippotragus leucophaeus	Bluebuck	EX
Hippotragus niger	Sable antelope	LC
Kobus ellipsiprymnus	Waterbuck	LC
Kobus kob	Kob	LC
Kobus leche	Southern lechwe	LC
Kobus megaceros	Nile lechwe	EN
Kobus vardonii	Puku	NT
Litocranius walleri	Gerenuk	NT
Madoqua guentheri	Guenther's dikdik	LC
Madoqua kirkii	Kirk's dikdik	LC
Madoqua piacentinii	Silver dikdik	DD
Madoqua saltiana	Salt's dikdik	LC
Nanger dama	Dama gazelle	CR
Nanger granti	Grant's gazelle	LC
Nanger soemmerringii	Soemmerring's gazelle	VU
Neotragus batesi	Bates' pygmy antelope	LC
Neotragus moschatus	Suni	LC
Neotragus pygmaeus	Royal antelope	LC
Oreotragus oreotragus	Klipspringer	LC
Oryx beisa	Beisa oryx	NT
Oryx dammah	Scimitar-horned oryx	EW
Oryx gazella	Gemsbok	LC

(*Continued*)

Scientific name	Common name	IUCN category
Oryx leucoryx	Arabian oryx	EN
Ourebia ourebi	Oribi	LC
Pantholops hodgsonii	Tibetan antelope, chiru	EN
Pelea capreolus	Grey rhebok	LC
Philantomba maxwellii	Maxwell's duiker	LC
Philantomba monticola	Blue duiker	LC
Procapra gutturosa	Mongolian gazelle	LC
Procapra picticaudata	Tibetan gazelle	NT
Procapra przewalskii	Przewalski's gazelle	EN
Pseudoryx nghetinhensis	Saola	CR
Raphicerus campestris	Steenbok	LC
Raphicerus melanotis	Cape grysbok	LC
Raphicerus sharpei	Sharpe's grysbok	LC
Redunca arundinum	Southern reedbuck	LC
Redunca fulvorufula	Mountain reedbuck	LC
Redunca redunca	Bohor reedbuck	LC
Saiga tatarica	Saiga	CR
Sylvicapra grimmia	Bush duiker	LC
Tetracerus quadricornis	Four-horned antelope, chousingha	VU
Tragelaphus angasii	Lowland nyala	LC
Tragelaphus buxtoni	Mountain nyala	EN
Tragelaphus derbianus	Giant eland	LC
Tragelaphus eurycerus	Bongo	NT
Tragelaphus imberbis	Lesser kudu	NT
Tragelaphus oryx	Common eland	LC
Tragelaphus scriptus	Buchbuck	LC
Tragelaphus spekii	Sitatunga	LC
Tragelaphus strepsiceros	Greater kudu	LC

EX: Extinct; EW: Extinct in the Wild; CR: Critically Endangered; EN: Endangered; VU: Vulnerable; NT: Near-Threatened; LC: Least Concern; DD: Data Deficient.

Index

Page numbers in *italics* refer to figures, those in **bold** refer to tables.

Antelope Conservation: From Diagnosis to Action, First Edition. Edited by Jakob Bro-Jørgensen and David P. Mallon.

Plate 1 **Antelope tribes. Boselaphini: (i) Four-horned antelope, Kanha National Park (NP), India (©Kalyan Varma). Pseudorygini: (ii) Saola, central Laos (©William Robichaud). Tragelaphini: (iii) Common eland, Masai Mara National Reserve (NR), Kenya (©Jakob Bro-Jørgensen). (iv) Mountain nyala, Bale Mountains NP, Ethiopia (©Anagaw Attickem). (v) Greater kudu, Kruger NP, South Africa (©Jakob Bro-Jørgensen). Peleini: (vi) Rhebok from South Africa (©Brent Huffman). Reduncini: (vii) Waterbuck, Queen Elizabeth NP, Uganda (©Jakob Bro-Jørgensen).**

Antelope Conservation: From Diagnosis to Action, First Edition. Edited by Jakob Bro-Jørgensen and David P. Mallon.

Plate 2 **Antelope tribes (*continued*). Cephalophini: (i) Aders' duiker, coastal forests of Kenya (©Rajan Amin). Alcelaphini: (ii) Hirola, coastal savannahs of Kenya (©Abdullahi Hussein Ali). (iii) Topi, Masai Mara NR (©Jakob Bro-Jørgensen). Hippotragini: (iv) Addax from the Sahelo-Saharan region (©Olivier Born). (v) Scimitar-horned oryx from the Sahelo-Saharan region (©Olivier Born). (vi) Giant sable antelope, woodlands of Angola (©Pedro Vaz Pinto). (vii) Gemsbok, Kgalagadi Transfontier Park, South Africa (©Gernot Lorber). Pantholopini: (viii) Tibetan antelope, Tibet (©George B. Schaller).**

Plate 3 **Antelope tribes (*continued*). Aepycerotini: (i) Impala, Masai Mara NR (©Jakob Bro-Jørgensen). Neotragini: (ii) Royal antelope from the West African rainforests (©Brent Huffman). (iii) Kirk's dikdik, Masai Mara NR (©Jakob Bro-Jørgensen). Antilopini: (iv) Saiga antelope, Kalmykia, Russia (©Navinder Singh). (v) Thomson's gazelle, Serengeti NP (©Daniel Rosengren). (vi) Cuvier's gazelle from the Atlas region (©Bert Buxbaum). (vii) Slender-horned gazelle, Sidi Toui NP, Tunisia (©Tim Wacher). (viii) Dama gazelle, R'Mila Royal Reserve, Morocco (©Tim Wacher).**

Plate 4 **Antelope habitats. (i) Bushbuck, southern Zambia. (ii) Roan antelope, Waterberg NP, Namibia. (iii) Bush duiker, Aberdares NP, Kenya. (iv) Wildebeest, Serengeti NP. (v) Klipspringer, Serengeti NP. (vi) Bohor reedbuck, Bale Mountains NP. (vii) Blue duiker, Mahale NP, Tanzania. ([i,ii,iv,v,vi,vii] ©Daniel Rosengren; [iii] ©Jakob Bro-Jørgensen).**

Plate 5 **Antelope habitats (*continued*). (i) Addax in typical Sahara desert habitat, Termit/Tin Toumma, Niger (©Thomas Rabeil). (ii) Dorcas gazelles in typical Sahel grassland habitat, Ouadi Rimé-Ouadi Achim, Chad (©John Newby).**

Plate 6 **Arena-breeding ('lekking') in antelopes. (i) Blackbuck, Tal Chhapar wildlife sanctuary, Rajasthan, India. (ii) Kob, Queen Elizabeth NP. (iii) Topi, Masai Mara NR. Sexual displays (iv) and intense fighting (v) characterize leks. ([i,iv] ©Jayabharathy Ranganathan; [ii,iii,v] ©Jakob Bro-Jørgensen).**

Plate 7 **Mother/offspring relations. (i) Wildebeest calves quickly follow their mothers after birth. Ngorongoro Conservation Area, Tanzania. (ii) Crèche of eland calves, Masai Mara NR. (iii) Newborn Thomson fawn, Serengeti NP. (iv) Thomson fawns mainly hide on their own during their first weeks. Masai Mara NR. (v) Thomson mother defending her fawn against spotted hyenas, Masai Mara NR. (vi) Hartebeest giving birth, Serengeti NP. (vii) Hartebeest consuming her afterbirth, Serengeti NP. ([i,iii,vi,vii] ©Daniel Rosengren; [ii,iv,v] ©Jakob Bro-Jørgensen).**

Plate 8 **Antelope ecology. (i),(ii) Wildebeest migration, Serengeti NP. (iii) Predation on topi by lions and spotted hyenas, Masai Mara NR. (iv) Impala displaced by domestic cattle, Masai Mara NR. (v) Hartebeest on a burning savannah, Serengeti NP. ([i,ii,v] ©Daniel Rosengren; [iii,iv] ©Jakob Bro-Jørgensen).**

Plate 9 **Antelopes in changing climates. (i) Wildebeest carcass, Serengeti NP. (ii) Grant's gazelle by Lake Turkana, Northern Kenya. (iii) Wildebeest, Etosha NP, Namibia. (iv) Impala, Serengeti NP. (v) Impala stuck in mud, Serengeti NP. (vi) Wildebeest, Serengeti NP. ([i,iii,vi,v,vi] ©Daniel Rosengren; [ii] ©Gernot Lorber).**

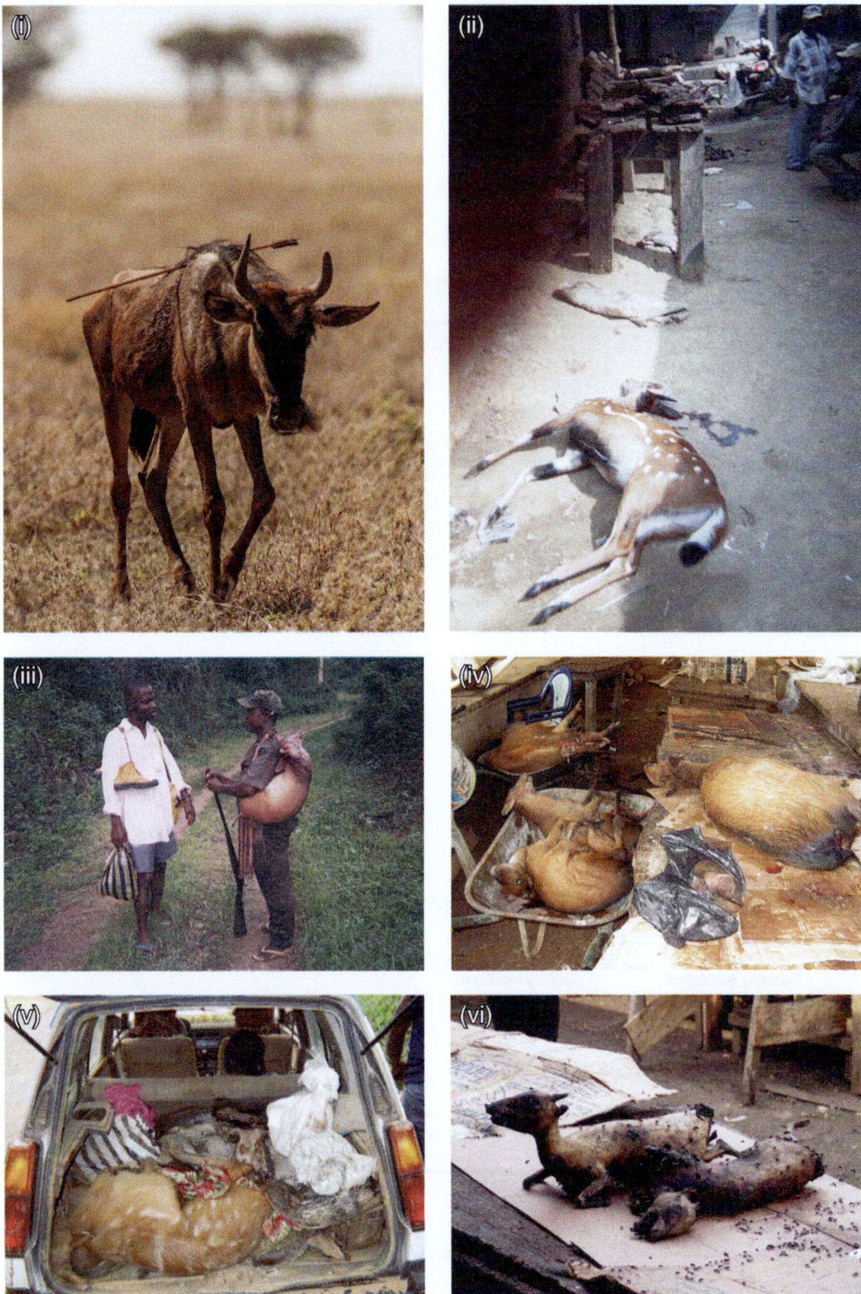

Plate 10 **Antelopes as bushmeat. (i) Wildebeest pierced by arrow, Serengeti NP (©Daniel Rosengren). (ii) Bushbuck carcass, market in Uyanga, Nigeria (©John E. Fa). (iii) Hunter with duiker carcass, Ekuri forest, Nigeria (©John E. Fa). (iv) Peters' duikers, market in the regional capital Makokou, Gabon (©Nathalie van Vliet). (v) Sitatunga and other carcasses in Niefang on their way to a bushmeat market in Bata, the largest town in Equatorial Guinea (©Lise Albrechtsen). (vi) Duikers for sale, market in Malabo, the capital of Equatorial Guinea on Bioko Island (©Lise Albrechtsen).**

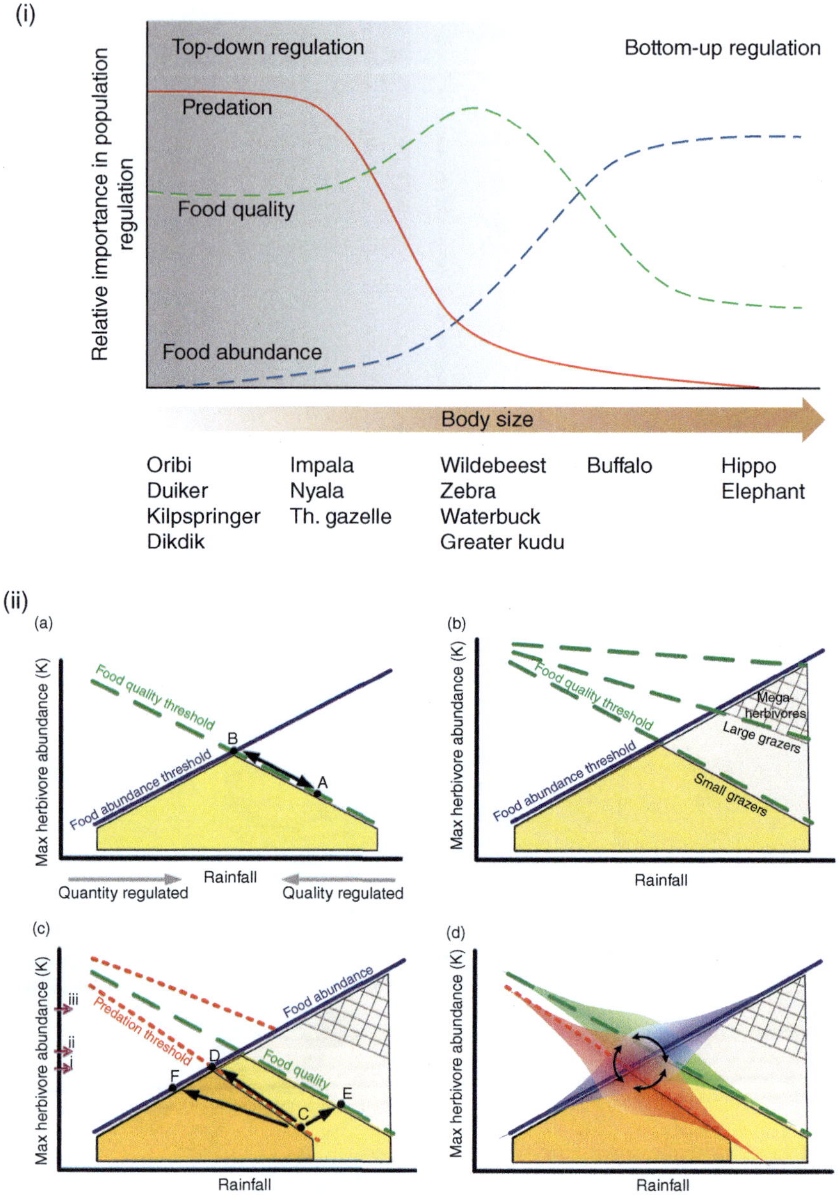

Plate 11 **Population regulation in herbivores (chapter 3). (*Legend continues overleaf*). (i) Population regulation of different sized herbivores. Large herbivores tend to be regulated by food abundance (dashed blue line), whereas small herbivores tend to be regulated by food quality (dotted green line). The smallest herbivores tend to be regulated by predation (solid red line). (Reproduced from (Hopcraft *et al.*, 2010)).**

(iii)

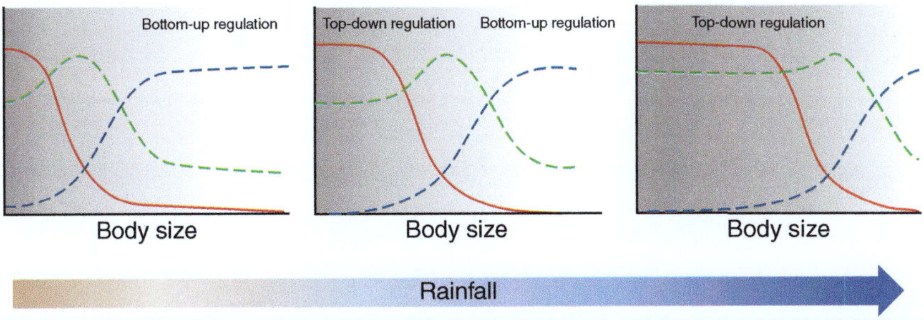

Plate 11 ***(continued)*** **(ii) A phase-plane diagram illustrating the thresholds that limit the overall size of a population of ungulates. (a) The abundance and quality of grass responds oppositely to rainfall, which limits a population of grazers to the shaded area. Migration between point A and B enables a population to seasonally maximize access to high-quality grass and have larger populations than otherwise expected. (b) Large grazers can occupy the light and dark shaded areas because they have evolutionary adaptations that enable them to extract more nutrients from poor-quality food. Mega-herbivores (such as elephant) can occupy the hatched area in addition, because they are not regulated by the quality of food, but regulated by its abundance instead. (c) Moist savannas have greater rates of predation because there is more vegetation that conceals stalking predators; however, the evolution of large body-size allows these grazers to escape most predators (i.e., the largest can occupy both the light and dark shaded areas; the size-dependent predation threshold is indicated by multiple broken red lines). A population at point C could become more abundant than otherwise expected by either migrating up and down a rainfall-induced predation gradient as zebra do (C to D), or adapting to moister environments with low-quality food as oribi do (C to E), or adapting to dry low-biomass environments as oryx do (C to F). Note that the slope and intercept of the lines are dependent on the body-size of the animal with each species having its own thresholds, therefore the relative abundance of the population at points A–E may not be the same across all taxa. By congregating in herds, antelope reduce their exposure to predators and can expand from the dark shaded area in panel c into the moderate and light shaded areas. Regardless, the maximum size a population can attain occurs at the intersection of these thresholds and is regulated by either by [i] predation, [ii] food quality, or [iii] food abundance. (d) Disturbances such as fire remove vegetation and are most frequent in areas with moderate rainfall. Sudden changes in vegetative quality and biomass lead to frequent switches between modes of regulation from predation to food quality to food abundance, which can lead to multiple alternative states.**

(iii) Alternative modes of population regulation of different-sized herbivores under climate change scenarios (solid red line is regulation by predation, dotted green is food quality, and dashed blue is food abundance). If climate change leads to less rainfall in Africa, this would push an ecosystem down the x-axis and lead to less vegetative biomass, greater visibility, and better-quality forage, which would favour small antelope over large (i.e., fewer constraints on the populations of small antelope but more constraints on populations of large antelope). Conversely, if climate change leads to more rainfall, this would push an ecosystem up the x-axis and lead to more vegetative biomass, more cover for stalking predators, and poorer-quality forage, which would favour large antelope over small ones.

Printed and bound by CPI Group (UK) Ltd, Croydon, CR0 4YY
27/09/2026
14970447-0001